# DNA Repair: Advanced Concepts

# DNA Repair: Advanced Concepts

Edited by **Nas Wilson**

New York

Published by Callisto Reference,
106 Park Avenue, Suite 200,
New York, NY 10016, USA
www.callistoreference.com

**DNA Repair: Advanced Concepts**
Edited by Nas Wilson

International Standard Book Number: 978-1-63239-150-6 (Hardback)

Printed in the United States of America.

# Contents

# Preface

Every book is initially just a concept; it takes months of research and hard work to give it the final shape in which the readers receive it. In its early stages, this book also went through rigorous reviewing. The notable contributions made by experts from across the globe were first molded into patterned chapters and then arranged in a sensibly sequential manner to bring out the best results.

The procedure of regeneration of damaged DNA molecule by the identification of its genome by the cell is referred to as DNA repair. This book is a collection of several chapters divided into various distinct categories. Each chapter has been researched by leading experts in their respective fields. The range of the book varies from the DNA damage response and DNA repair methods to transformational characteristics of DNA repair, providing a picture of present perceptions of the DNA repair procedure and the function of microRNA in DNA damage response. This book consists of several case studies and also examines the evolution of DNA repair in plants and meiosis as an evolutionary adaptation for DNA Repair.

It has been my immense pleasure to be a part of this project and to contribute my years of learning in such a meaningful form. I would like to take this opportunity to thank all the people who have been associated with the completion of this book at any step.

Editor

# Part 1

# Mechanisms of DNA Repair

# The Gratuitous Repair on Undamaged DNA Misfold

Xuefeng Pan[1,2], Peng Xiao[1], Hongqun Li[2], Dongxu Zhao[1] and Fei Duan[2]
*[1]School of Life Science, Beijing Institute of Technology, Beijing*
*[2]Health Science Center, Hebei University, Baoding*
*China*

*A Paradox of Life*
*B-DNA is needed for maintenance of genetic stability, while it will convert*
*into non-B DNA in replication, repair, transcription or recombination, leading to*
*exposure of bases, single strands, and even introduction of distortions. All these*
*could intrigue gratuitous repair on undamaged DNA using the*
*conventional repair, recombination mechanisms.*
*Repair or not Repair, turns to be a question?*

## 1. Introduction

In natural genomes, tens of DNA structure analogous to B-DNA conformation have been found to be formed through compiling weak interacting forces, including hydrophobic, Van der Waals and hydrogen-bond accepters and donors and inductions of certain agents (Rao et al., 2010). Of which, hairpins, cruciform junctions, Z-DNA, G-tetrads/quadruplexes, helices, loops and bulges are most studied so far.

Since the late 1950s, the roles of the non-B DNA structures in biological functions have begun to be enlightened (Watson & Crick, 1953; Wilkins et al., 1953a, 1953b; Svozil et al., 2008). Piling up results suggest that non-B conformations, such as cruciforms, triplexes, tetraplexes, can interact with proteins involving DNA metabolism, including replication, gene expression and recombination, or influence nucleosomes and other supramolecular structures formation (Wang & Griffith, 1996; Shimizu et al. 2000). However, non-B DNA secondary structures may also be treated as DNA mis-folds by DNA repair systems. Because of which the non-B DNA secondary structures can serve as end points for several types of genome rearrangements seen in some diseases (Wang & Vasquez, 2006; Wells, 2007; Bacolla & Wells, 2009; Chen et al., 2010).

## 2. DNA sequences which are susceptible to abnormal folding

The non-B DNA structure forming sequences are found to be rich in genomes from divergent organisms (Table 1) (Cox & Mirkin, 1997; Svozil et al., 2008; Cerz et al., 2011). For example, nearly half of the human genome consists of repetitive sequences, which can be arranged as inverted, direct tandem, and homopurine–homopyrimidine mirror repeats.

These repeat sequences are major contributors to forming non-B DNA structures, although the unusual structures can also be formed by various other sequences that are not repeating tracts (Svozil et al., 2008; Cerz et al., 2011). Repeat DNA sequences may adopt either orthodox right-handed B-DNA or non-B DNA conformations at specific sequence motifs as a function of negative supercoil density, created by transcription, protein binding, and other reasons. For example, inverted repeats can form B conformation in cells, while also forming hairpin structures, slipped structures with looped-out bases, four-stranded G-quartet structures, left-handed Z-DNA and intramolecular triplex DNA structures (H-DNA) depending on the base compositions and the arrangements.

| Structural Feature | human | Chimpanzee | Macaque | Dog | Mouse |
|---|---|---|---|---|---|
| Cruciform | 197910 | 190736 | 128334 | 172032 | 188532 |
| Slipped Motif | 347969 | 314516 | 305285 | 404750 | 695150 |
| Triplex Motif | 179623 | 105640 | 140580 | 303385 | 565479 |
| Z-DNA Motif | 294320 | 278928 | 280982 | 261012 | 690276 |
| G-Tetraduplex | 374545 | 314171 | 298142 | 492535 | 559280 |
| Direct repeats | 871045 | 787335 | 765798 | 968955 | 1593107 |
| Inverted repeats | 1044533 | 998249 | 843889 | 814080 | 801242 |
| Mirror Repeats | 1651723 | 1485135 | 1455025 | 1849897 | 1651723 |

Table 1. Non-B DNA motifs in different mammalian genomes (Cer et al., 2011)

## 2.1 Cruciform motif

DNA sequence that reads the same from 5' to 3' in either strand of a duplex is called as inverted repeat or palindrome DNA sequence. This subset of inverted repeat sequences may fold-back and form intramolecular, antiparallel, double helices stabilized by Watson–Crick hydrogen bonds (van Holde & Zlatanova, 1994; Courey, 1999; Smith, 2008).

As a whole, the interstrand hydrogen bonds in the inverted repeats must be broken, and intrastrand hydrogen bonds form between the complementary bases in each single strand, forming two hairpin-like arms with small (3-4 unpaired bases) loop at their tips. The structure looks similar to a four-way junction, of which the nucleobases in and around the junction are fully involved in base pairing.

## 2.2 Potential quadruplex sequences

Potential quadruplex sequences are usually G-rich, such as the DNA sequences in eukaryotic telomeres, and in non-telomeric genomic DNA, like the nuclease-hypersensitive promoter regions (Burge et al., 2006; Rawal et al., 2006; Qin & Hurley, 2008; Sannohe & Sugiyama, 2010). To form a quadruplex, the DNA sequences have to form overlapping four G-blocks. Each contains the same number (n) of G bases (n vary from 3 to 7), on each strand, and/ or separated by 1–7 nt (Burge et al., 2006). The potential unimolecular G-quadruplex forming sequences (i.e. intramolecular) can be expressed as follows (Burge et al., 2006):

$$G_aX_bG_aX_cG_aX_dG_a$$

Where "a" is the number of G residues in each short G-tract, which are usually directly involved in G-tetrad. Xb, Xc and Xd can be any combination of residues, including G, forming the loops.
The potential quadruplex sequences were therefore restricted to:

$$G_{3-5}NLoop1G_{3-5}NLoop2G_{3-5}NLoop3G_{3-5}$$

Where NLoop1-3 are loops of unknown length, within the limits 1<NLoop1-3 <7 nt.

## 2.3 Z-DNA motif

In 1979, DNA sequence of d (CpGpCpGpCpG) was crystallized and found to adopt a left-handed conformation (the Z-DNA conformation) with altered helical parameters relative to right-handed B-form (Rich et al., 1983; Mirkin, 2008). Later, it was realized that DNA sequences with alternating pyrimidines and purines, such as $(CA:TG)_n$ and $(CG:CG)_n$, may wind a double helix into a left-handed zigzag form (Z-DNA). Z-DNA is thinner (18 Å) than B-DNA (20 Å), due to its bases shifting to the outskirts of a double helix. It has only one deep, narrow groove equivalent to the minor groove in B-DNA.
In general, five or more tandem repeats, each comprising an alternating pyrimidine–purine dinucleotide motif, in which the pattern YG is preserved on at least one of the DNA strands can adopt Z-DNA.

## 2.4 Triplex motif

A subset of mirror repeat sequences comprise only purines (A and G, R) or pyrimidines (C and T, Y) on the same strand of a double stranded DNA, separated by few (0~8) nucleotides. These DNA motifs can adopt various intramolecular three-stranded analogous (triplex, H-DNA) stabilized by Hoogsteen hydrogen bonds (Casey & Glazer, 2001; Mukherjee & Vasquez 2011).
For a sequence requirement in forming triplex DNA is thought to be that only R· Y-containing mirror repeats can yield A: A*T and G: G * C triads. When the hydrogen bonds in the A · T and G· C base pairs are formed in canonical B-form DNA, several hydrogen bond forming groups in the bases can still be free unpaired. Each purine base has two hydrogen bond forming groups on the edges that are posed in the major groove. These unpaired bases can be used to form base triads that are unit blocks of triple-stranded DNA (see the following explanation for detail).
In theory, a homopurine-homopyrimidine duplex can form triplexes of either purine (Pu) motif (purine, antiparallel motif) or pyrimidine (Py) motif (pyrimidine, parallel motif). However, under physiological conditions, cytosine protonation is not favored, and C·G*G becomes therefore the most stable triad in a Pu motif. To form an intermolecular or intramolecular triplex, adjoining homopurine-homopyrimidine tracts of at least 10 base pairs are normally required for a duplex acceptor, since shorter than that the triplexes formed can be unstable under physiological conditions (Fox & Brown, 2011).
A triplex may be mutagenic *in vivo*, as double-strand breaks may occur in or near the triplex site, which if with DNA replication, recombinational repair may produce triplex mediated mutagenesis (Chan et al., 1999; Faruqi et al., 2000).
Triplex can also be formed in RNA transcription, although it is a kinetically unfavored compared to duplex annealing. However triplex RNA and DNA are stable, showing half-lives on the order of days, which may involve the molecular mechanism of Friedreich's ataxia (FRDA) (Pan et al., 2009).

## 3. The non-B DNA structures and non-B DNA structure-induced genetic instability

### 3.1 DNA loops/ bulges and slipped DNA

DNA loops and bulges are similar non-B DNA structures sharing common features of unpaired bases of different number (Fig. 1). They can be formed in anywhere by any DNA sequence in natural genome, therefore they may be the most frequent non-B DNA conformations in genomes. For example, $(CA \cdot TG)_n$ DNA sequences are found to exist everywhere in eukaryotic genomes as of 60 base pairs tracts. $(CA \cdot TG)_n$ forms both classical right-handed DNA double helix, and diverse alternative conformations including small DNA loops or bulges (Kladde et al., 1994; Ho, 1994).

Genomic instabilities can also be caused by DNA loops and bulges, which are often seen as slippage instabilities or insertion/deletion ( I/ D) instabilities (Pan, 2004). Proteins that bind DNA loops and bulges are also found and mainly known to be mismatch repair proteins (Parker & Marinus, 1992; Carraway & Marinus, 1993; Fang et al., 2003; Kaliyaperumal et al., 2011).

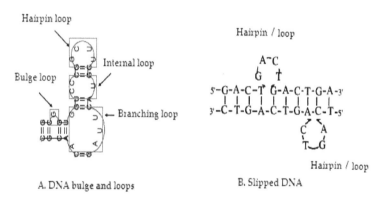

A. DNA bulge and loops                    B. Slipped DNA

Fig. 1. DNA Loops and bulges

### 3.2 Branched structures

A branched DNA structure refers to a non-B DNA secondary structure with structured or unstructured "branch". For example, DNA intermediates appeared in homologous recombination as 3- and 4- way junctures are such branched DNA structures with differently oriented double helix arms. Similarly, flapped DNA structures appeared in processing Okazaki fragment in the lagging strand DNA replication also belong to branched DNA. Branched DNA migrates more slowly than their B-DNA conformation having same molecular weight and base composition. Importantly, branched DNA structures can also make genomic instability when in processing.

### 3.3 Hairpin/ cruciform and genetic instability

A hairpin can be formed at one strand of an inverted repeat, whereas a cruciform consists of two hairpin structures, both in each strand at the same position of the DNA (Fig.2)(Courey, 1999). Similarly some tandem arranged trinucleotide repeats such as CAG, CTG, CCG, CGG,

AAT, ATT etc. can also adopt hairpin structures with mismatched base pairs in the stem (McMurray, 1999; Trotta, et al., 2000).

To form a hairpin/cruciform, DNA duplex needs to be unwound in replication, transcription, and/or DNA repair processing; affording single-stranded repeat sequences the opportunity to base pair with itself in an intramolecular fashion. The term of "cruciform" originates from forming two duplex arms, which adopts either an "open" form, allowing strand migration or a "stacked" (locked) form, where the helices stack on each other (Courey, 1999; Khuu et al., 2006; Lilley, 2010). In both cases, the overall conformation and the intraduplex angles behave like the Holliday junction recombination intermediates (Fig.2A) (Courey, 1999; Khuu et al., 2006;; Lilley, 2010).

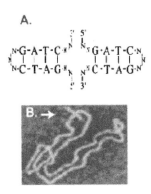

Fig. 2. Hairpin/cruciform of DNA

Both inverted repeats and tandem arranged trinucleotide repeats were found to be mutagenic, causing genomic instability. Inverted repeats were initially found to cause deletions in E. coli (Sinden et al., 1991), and then were seen in humans as (8; 22) (q24.13; q11.21), and many types of t (11; 22) translocations. The breakpoints of these translocation mutations were localized at the center of AT-rich palindromic sequences on 11q23 and 22q11, respectively. So far, t (11; 22) is the only known recurrent, non-Robertsonian translocation in humans, in some cases leads to male infertility and recurrent abortion (Kurahashi et al., 2000, 2006, 2010; Kurahashi & Emanuel, 2001). Furthermore, deletions stimulated by a poly (R.Y) sequence from intron 21 of the polycystic kidney disease 1 gene (PKD1) have also been characterized ( Bacolla et al., 2001; Patel et al., 2004). And a long (CCTG-CAGG)$_n$ repeat in E.coli was also found to form cruciform (Pluciennik et al., 2002; Dere & Wells, 2006). Interestingly, cruciform-forming inverted repeats have mediated many of the microinversions in evolution that distinguish the human and chimpanzee genomes (Kolb et al., 2009).

In cells, DNA double strand breaks can be derived from cruciform, because hairpin/ cruciform are substrates for several structure-specific nucleases and/ or repair enzymes, such as SbcCD in E.coli and Mre11-Rad50 in eukaryotes. The actions of such enzymes make strand breaks, which may result in rearrangements or translocation of chromosomes (Smith, 2008).

In addition, proteins working in nucleotide excisional repair (NER) can also recognize the helical distortions in hairpin, therefore NER may recognize DNA hairpin to resolve the hairpin in the DNA.

Besides, some other proteins were also found to bind the structural elements in cruciforms. For example, HMG proteins, replication initiation protein RepC, cruciform binding protein CBP, and four-way junction resolvases have all been indentified to bind cruciforms (Pearson et al., 1996; Jin et al., 1997; Novac et al., 2002; Lange et al., 2009; lilley, 2010).

## 3.4 Z-DNA and genetic instability

Z-DNA can be seen as the high-energy conformers of B-DNA that forms *in vivo* during transcription as a result of torsion strain generated by a moving polymerase (Wang, 1984; Casasnovas & Azorin, 1987; Johnston, 1988; Hebert & Rich, 1996). It has been thoroughly studied since 1957, how a right-handed B-DNA adopting a Z-DNA *in vitro* through "flipping" the base pairs upside down, and rotating every other purine from *anti* to *syn* conformation (Johnston, 1988; Hebert & Rich, 1996). Compared to B-DNA, Z-DNA does not have a major groove, therefore could potentially impact transcription by physically blocking RNA polymerase, or by relaxing negative supercoiling turns, or by acting as an enhancer through recruiting transacting factors.

In Z-DNA, the guanosine nucleotides are in *syn* position where the bases are found over the sugar without protection, thus more accessible to DNA damaging factors, more resistant to processing by DNA repair enzymes. For example, alkylating damage such as $N^7$-methylguanine, which is typically removed by a DNA glycosylase in B-DNA is not efficiently repaired when present in Z-DNA (Pfohl-Leszkowicz et al., 1983; Boiteux et al., 1985).

Further, DNA sequences with the potential to adopt Z-DNA are associated with recombination hot spots in eukaryotic cells (Wang et al., 2006). A hot spot of 1000 bp in the major histocompatibility complex (MHC) in mice, containing several copies of long GT repeats, may account for up to 2% of the recombination events occurring on the chromosome (Crouau-Roy, 1999). In *E.coli*, the RecA molecules show a much higher binding affinity for Z-DNA than for normal B-DNA and single-stranded DNA, and show a Z-DNA structure-stimulated ATPase activity, implicating a recombination hot spot of Z-DNA in prokaryotes as well. Genetic recombination in Z-DNA can potentially induce deletion instability and/ or produce DNA double-strand breaks. For example, a CG (12) sequence forming Z-DNA induces high levels of genetic instability in both bacterial and mammalian cells (Casasnovas & Azorin, 1987).

Recently, proteins binding Z-DNA are found, including specific proteins, such as Zα domain-containing proteins ADAR1 and ESL, and fairly low specific proteins, such as HMG proteins (Suda et al., 1996; Lange et al., 2009).

## 3.5 H-DNA and H-DNA induced DSBs and genetic instability

H-DNA, alternatively known as triplex DNA can be classified into either pyrimidine motif or purine motif according to the orientation and composition of the third strand in a triple stranded DNA structure (Fig. 3). The third strand can form either Hoogsteen or reverse-Hoogsteen hydrogen bonds with the purine-rich strand of the duplex DNA. Therefore, the third strand can be both pyrimidine-rich and parallel to the complementary strand (Y* R: Y) or purine-rich and antiparallel to the complementary strand (R* R: Y), producing either pyrimidine motif or purine motif triple stranded DNA (as described previously).

Whereas (R* R: Y) triplexes form under conditions of physiological pH, triplex of the (Y* R: Y) composition form most readily under conditions of acidic pH. At physiological pH, triplex may be stabilized by negative supercoiling, modified with phosphorothioate groups, or polyvalent cations such as spermine and spermidine. For the R* R: Y intramolecular triplexes and T: A* T and C⁺: G* C triplets for the Y* R: Y intramolecular triplexes are included since these are considered the most stable triplet combinations.

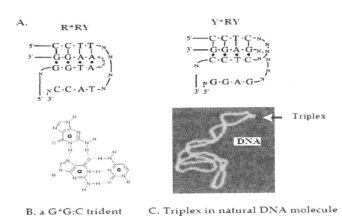

B. a G*G:C trident          C. Triplex in natural DNA molecule

Fig. 3. H-DNA (Star/ Dot marks Hoogsteen hydrogen bonded bases; colon/ line shows Watson–Crick hydrogen bonded bases)

In general, formation of a triplex DNA was a role of sequence, topology (supercoil density), ionic conditions, protein binding, methylation, carcinogen binding, and other factors. Global negative supercoil density acts in concert with local transient waves of topological changes produced by replication or transcription, and both have a critical influence on forming and stabilizing triplex DNA *in vivo*. It has been reported that a higher negative supercoiling destabilized long CTG· CAG, CCG· CGG, and GAA· TTC repeats in *Escherichia coli*. Similarly a 2.5-kb poly (R· Y) tract from the human PKD1 gene lowered the viability of the host cells (Bacolla et al., 2001; Patel et al., 2004).

Several types of DNA damages induced by H-DNA have been reported, including single and/ or double strand breaks. For example, the endogenous H-DNA forming sequences from the human c-*myc* promoter was shown to be intrinsically mutagenic in mammalian cells because of the generation of either single or double strand breaks in the H-DNA, or near the H-DNA locus. Besides, the single-stranded area, or the triplex region is also a target of various nucleases, resulting in single or DSBs formation, and the increased mutagenesis or recombination (Wang & Vasquez, 2006).

Although triplex (H-DNA) DNA occurs mainly at poly (purine ·pyrimidine) ((R ·Y) n) tracts, it can also be induced to form with the sequence specific DNA recognition and binding of some synthetic triplex-forming oligonucleotides (TFOs) (Casey & Glazer, 2001; Mukherjee & Vasquez, 2011). TFOs bind to the major groove of homopurine-homopyrimidine stretches of double-stranded DNA to induce forming the triplex (Casey & Glazer, 2001; Mukherjee & Vasquez, 2011). During which the duplex DNA may have to undergo helical distortions on TFO binding and the distortions trigger endogenous recombination and repair mechanisms in the cell (Raghavan et al., 2004, 2005).

Indeed it has been reported that formation of TFO-induced triplex can induce sequence-specific DNA damages both in cells and in animals (Chan, et al., 1999; Kalish et al., 2005).

However, mismatch repair proteins are not involved in this TFO-induced mutagenesis. Several reports have now shown that cells that are deficient in the MutS and MutL homologues MSH2, MLH1, MSH3, or MSH6, do not show any change in TFO-induced mutagenesis. In contrasts, NER factors can recognize the intermolecular triplex at least in part. Therefore NER was involved in the triplex-induced mutagenesis and recombination in cells. For example, in *E.coli*, NER proteins, such as UvrB and UvrC, were necessary for H-DNA-induced cell growth retardation and cell lysis, similarly, recombination induced by TFOs depends also on the NER pathway (Faruqi et al., 2000).

### 3.6 G-tetraduplex and genetic instability

G-quadruplexes are higher-order DNA or RNA structures formed from G-rich DNA or RNA sequences that are built around tetrads of hydrogen-bonded guanine bases (Lipps & Rhodes, 2009; Sannohe & Sugiyama, 2010). Despite the wide prevalence of genomic sequences that have G-rich property and that can potentially fold into tetraplex / quadruplexes structures, a direct demonstration of their existence *in vivo* proved to be a difficult undertaking. Only recently has there evidence started to increase for their presence and role *in vivo* (Lipps & Rhodes, 2009), since most of the tetraplex/ quadruplexes forming sequences are fairly short and quadruplexes are likely to be transiently formed. G-quadruplexes (tetraduplex) may have several isomers which can be formed intramolecularly and intermolecularly (Fig. 4).

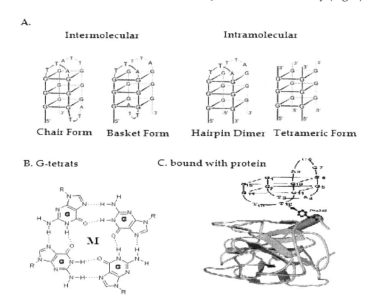

Fig. 4. Tetraplex DNA

Recent progress of the related studies revealed that G-quadruple could provide a nucleic acid based mechanism, such as regulating telomere maintenance, transcription, replication as well as translation. In the same time, various G-quadruplexes binding proteins, such as, a G4 quadruplex and purine Motif triplex nucleic acid-binding protein have also been characterized (Dyke et al.,2004), many others have been summarized in the reference (Fry, 2007).

The existence of cellular proteins that preferentially interact with tetraplex DNA provides a strong argument for the existence of quadruplex formations in genomic DNA.

### 3.7 Unwound DNA

Unwound DNA is known to be formed by A+T -rich sequences (Fig. 5). Since A · T base pairs contain two hydrogen bonds and C · G base pairs contain three, A · T-rich tracts are less thermally stable than C · G -rich tracts in DNA. In the presence of superhelical energy, A+T - rich regions can unwind and remain unwound under conditions normally found in the cell. Such sites often provide places for DNA replication proteins to enter DNA to begin the chromosome duplication. Unwound DNA can therefore be alternatively called as DNA unwinding elements (DUEs) that have been identified in both prokaryotic and eukaryotic DNA sequences. DUEs are AT-rich sequences about 30-100 bp long. They share little sequence similarity except for being AT-rich. Under torsion stress, unwinding of the double helix occurs first in AT-rich sequences; therefore, DUEs can be maintained as unpaired DNA regions in the presence of negative supercoiling. The single-stranded area of the unwound structure may be target of nuclease activity resulting in single or DSBs, leading to enhanced mutagenesis or recombination.

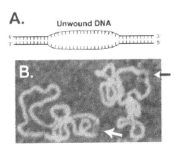

Fig. 5. Unwound DNA

### 3.8 Curved DNA

Normally, curved DNA is often seen in DNA segment containing runs of three or four bases of A in one strand and a similar run of T in the other and spaced at 10-base pair intervals. Interestingly we have recently found that trinucleotide repeats AAT can also adopt curved DNA in *E.coli*, which can be repressed by H-NS and its stimulated IS1E transposition (Pan et al., 2010)

## 4. Biological significance of DNA abnormal folding

Apart from the roles in DNA replication, transcription and gene regulation, non-B DNA may also lead to gene instability, including chromosomal translocation, deletion and amplification in cancer and other human diseases ( Bacolla & Wells, 2009; Chen et al., 2010). Since non-B DNA abnormal folds have been addressed with generating DNA breaks, including both single and double strand DNA breaks. Non-B DNA structures recruit DNA repair machinery to the breaking sites, which then make gene mutations and chromosomal rearrangements during repair.

### 4.1 Effects of non-B DNA structures on DNA replication / transcription

Some regions of DNA forming non-B DNA structures in replication or transcription, which may turn to affect the DNA transactions (Van Holde & Zlatanova, 1994; Samadashwily et al., 1997; Krasilnikova et al., 2004; Lin et al., 2006; Mirkin & Mirkin, 2007)

One of the well-studied effects of the non-B structures on replication is a block to polymerases because of template folding, which was shown for cruciforms/ hairpins and H-DNA (Samadashwily et al., 1997; Krasilnikova et al., 2004; Voineaqu et al., 2009). It has been found that triplex DNA can adversely affect DNA replication and potentially lead to replication fork collapse (Samadashwily et al., 1997; Krasilnikova et al., 2004; Voineaqu et al., 2009). The polypurine strand of a triplex forming duplex may not be a potential template, therefore giving increased chance of being single stranded, and forming intermolecular or intramolecular triplex (Hile & Eckert, 2004; Urban et al.,2010). Besides,a non-B DNA structure itself may also directly slow the progression of replication fork (Samadashwily et al., 1997; Mirkin & Mirkin, 2007; Trinh & Sinden, 1991). Such non-B DNA structures may be an obstacle to fork progression or a target for nucleolytic attack, thus allowing DNA breakage leading to deletion or recombination (Mirkin, 2006; Kim et al., 2006).

In contrast, the single-stranded parts in a cruciform or H-DNA may serve as the recognition elements for replication initiation proteins. For example, cruciform binding proteins (CBP), such as 14-3-3sigma in HeLa cells recruits replication proteins to a cruciform to start replication (Alvarez et al., 2002; Novac et al., 2002). Therefore, it is possible for a hairpin/cruciform DNA sequence behaves like a replication "origin", inducing an origin independent DNA replication. The similar way of DNA replication has been found in *E.coli* and named as stable DNA replication. More interestingly, the origin independent DNA replication has also been proposed as a mechanism for the production of expanded DNA repeats (Pan 2006).

In addition, certain non-B DNA structures can also interfere with RNA transcription and recombination (Van Holde & Zlatanova, 1994; Broxson et al., 2011). Similarly RNA transcription can also promote forming non-B DNA structures, including hairpin, triplexs and $G_4DNA$ (Van Holde & Zlatanova, 1994; Broxson et al., 2011).

## 4.2 Modulation of supercoiling and promoting transcription

The extent of supercoiling in a DNA segment is known to affect transcription, recombination, and replication such that an ideal DNA topology may be critical for them. It has been found that formation of cruciforms, Z-DNA and H-DNA caused partial relaxation of excessive superhelicity in a topological domain. Specific cases of DNA replication and gene expression have also been described as superhelicity dependent events induced by formation of cruciforms, Z-DNA and H-DNA.

## 4.3 Accumulation of DNA Damages causing increased mutability within non-B DNA structure forming sequences or their flanking sequences

DNA sequences that are prone to adopting non-B DNA secondary structures are associated with hot spots of genomic instability, where repeat expansions, chromosomal fragility, or gross chromosomal rearrangements can be often seen. For example, long repeating tracts of CTG · CAG, CCTG · CAGG, and GAA · TTC are associated with the etiology of myotonic dystrophy type 1 (DM1), type 2(DM2), and Friedreich's ataxia (FRDA) (Wells, 2007). The repeating sequences involved have potentials to adopt a variety of non-B DNA secondary structures (McMurray, 1999; Pan, 2004, 2006, 2009). Studies in various model systems, including *Escherichia coli* and mammalian cell lines, such as COS-7, CV-1, and HEK-293, have revealed that conditions promoting formation of non-B DNA structures enhanced the repeats instabilities. Such instabilities can occur both within the repeat sequences and in the flanking sequences of up to ~4 kbp (Wojciechowska et al., 2006).

Indeed, it has been found that DNA double-stranded breaks (DSBs) can sometimes be accumulated at or around the repeating sequences, and error-prone repair pathways were also proposed to be involved in forming gross DNA rearrangements (Kurahashi et al., 2006). Moreover, DNA breaks may also happen in the single-stranded area, or the structured region when they serve as targets of nuclease activity, leading to enhanced mutagenesis or recombination. The breakpoints of the disease-causing translocation cluster within a 150-bp genomic region of the *bcl*-2 gene were seen potentially form a triplex DNA structure (Adachi & Tsujimoto, 1990; Raghavan et al., 2004, 2005).

It has long been found that, the efficacies of DNA replication in the leading and lagging strand templates were differently performed in *E. coli* chromosome. Replication errors and SOS mutator effects occurred preferentially in the lagging strand, while intermolecular strand switch events during DNA replication occurred preferentially in the leading strand (Iwaki et al., 1995; Trinh & Sinden, 1995; Iwaki et al., 1996; Fijalkowska et al., 1998; Sinden et al., 1999; Maliszewska-Tkaczyk, 2002; Gawel et al., 2002; Hashem & Sinden, 2005). Similarly, unequal fidelities have also been found with deletions between direct repeats in the leading strand template (Hashem & Sinden, 2005). This may attribute to potential of non-B DNA structure formation in the leading and lagging strand template in DNA replication. Similarly, the replication fidelities of various inverted repeats, direct repeats, including trinucleotide repeats can also be compromised if they adopt non-B DNA conformations, such as hairpin, cruciform, triplex, tetra-duplex DNA, leading potentially to mutations or rearrangements (Pan & Leach, 2000; Sinden et al., 2002).

## 4.4 Nucleosome exclusion

In eukaryotes, chromosomal DNA wrapping around histones in nucleosomes interferes with the protein binding to promoters and origins of replication. Nucleosome formations, on one hand, and formation of cruciform, Z-DNA and triplex DNA, on the other hand, are mutually exclusive. Thus, the alternative structure-forming DNA sequences may expose nucleosome-free DNA, making them accessible to transcription, replication, recombination proteins as well as nucleases, producing fragile sites in chromosome (chwartz et al., 2006; Lukusa & Fryns, 2008).

Fragile sites are specific loci that appear as constrictions, gaps, or breaks on chromosomes from cells exposed to partial inhibition of DNA replication (Schwartz et al., 2006; Lukusa & Fryns, 2008). In chromosomal level, fragile sites always lack nucleosomes, and sometimes can be associated with trinucleotide repeats (TNRs) of CGG · CCG, CAG · CTG, GAA · TTC and GCN · NGC, with specific G-rich tetra- to dodecanucleotide repeats or with long AT-rich repeats, such as the 33 or 42 minisatellites in the FRA16B and FRA10B common fragile sites (Wang & Griffith, 1996). In the same time, fragile sites can be classified as rare or common, depending on their frequency within the population and their specific mode of induction. So far, there are more than 89 common fragile sites listed in GDB (Gene Databases), which are considered to be an intrinsic part of the chromosomal structure presented in all individuals. Six common fragile sites have been cloned and characterized, including FRA3B (Huebner & Croce, 2001; Lettessier et al., 2011), FRA7G, FRA7H, FRA16D (Shah et al., 2010), FRAXB , and FRA6F. Common fragile site instability was attributed to the fact that they contain sequences prone to form secondary structures that may impair replication fork movement, possibly leading to fork collapse and resulting in DNA breaks.

Most rare fragile sites are induced by folate shortage, and others are induced by DNA minor groove binders. So far, seven folate sensitive (FRA10A, FRA11B, FRA12A, FRA16A, FRAXA,

FRAXE and FRAXF) and two nonfolate sensitive (FRA10B and FRA16B) fragile sites have been molecularly characterized. Interestingly, almost all these fragile sites are found to have expanded DNA repeats resulting from mutation involving the normally occurring polymorphic CCG/CGG trinucleotide repeats and AT-rich minisatellite repeats (Balakumaran et al., 2000; Voineagu et al., 2009).

The expanded repeats were also demonstrated to have the potentials, at least under certain circumstances, to form stable secondary non-B DNA structures, including intrastrand hairpins, slipped strand DNA or tetrahelical structures, or to present flexible repeat sequences. Both of which are expected to affect the replication. In addition, these DNA sequences are also found to decrease the efficiency of nucleosome assembly, resulting in decondensation defects seen as fragile sites (Wang & Griffith, 1996; Freudenreich, 2007).

## 5. Genes and gene products that are involved in abnormal folding

A numerous proteins that interact with non-B DNA secondary structures have been characterized recently. These proteins may also be called as DNA structure-specific proteins, such as Rad1, Rad2, Rad10, Msh2, Msh3, BLM, WRN and Sgs1 (Bhattacharyya & Lahue, 2004; Nag & Cavallo, 2007; Kantelinen et al., 2010; Pichierri et al., 2011). These DNA structure-specific proteins can be further classified by function into several distinct groups, depending on their possible effects on the formation/ stability of non-B DNA structure. Some of the binding proteins may increase the stability of the bound non-B DNA secondary structures; and some may promote forming non-B DNA secondary structures; or destabilize non-B DNA secondary structures. Indeed, the available data implicate various proteins participating in mismatch repair, nucleotide excision repair, base excision repair, homologous recombination, recognize non-B DNA secondary structures in trying to avoid "so called" structure-directed mutagenesis.

As discussed previously, DNA structures can often induce DNA mutations. This DNA structure mediated mutagenesis may be because of the following reasons: the abnormal positioning of the bases and sugar in non-B DNA conformations, which impact the function of some DNA repair proteins on damaged DNA. For example, alkylating damage such as $N^7$-methylguanine or $O^6$-methylguanine is not repaired as efficiently in Z-DNA as it is in B-DNA. Alternatively, forming DNA secondary structures near DNA damage sites might influence the damage repair processing, depending on the types of damages, the environments, and the nature of the secondary structures (Pfohl-Leszkowicz et al., 1983; Boiteux et al., 1985).

### 5.1 MMR proteins

It has long been studied that MMR deficiency is associated with microsatellite sequence instability and human disease. For example, the instability of TNRs and AT-rich minisatellites is associated with their capacity of adopting unusual secondary structures, such as hairpins or DNA triplexes. This feature is common to different types of repeated DNA. Therefore, repeat instability is dependent on MMR in mice and yeast, consistent with the observation that sequences at repetitive DNA sites form short hairpins or small loops that are targets of the Msh2–Msh6 MMR (Modrich, 2006).

MMR proteins bind to non-B DNA secondary structures mainly through its capacity of recognizing mismatched base pairs. It has been found that MMR binds mismatches in a CNG triplet repeats hairpin stem. Although the MSH2–MSH3 complex of MMR also binds perfect hairpin formed by inverted repeats (lacking mismatched regions), affinity is low, suggesting that mismatches are important for the MMR protein binding (Kantelinen et al., 2008). In addition, MutS has also been reported to bind parallel G4 DNA in humans (Fry, 2007).

### 5.2 NER and HR proteins

NER proteins, such as the UvrB and UvrC in *E.coli*, and the XPA, XPG, XPC in eukaryotes and homologous recombination proteins, such as RecA, HsRad51, were found to be involved in H-DNA mediated repair and recombination (Bacolla et al., 2001). UvrB and UvrC may preferentially recognize the helical distortions, while RecA recognizing single stranded DNA region in an H-DNA.

### 5.3 Helicases and junction resolvases

Proteins that preferentially catalyze the unwinding of DNA non-B DNA secondary structures are DNA helicases in ATP-hydrolysis dependent manner. Helicases are DNA unwinding enzymes that preferentially melt some of the non-B DNA structures. The selectivity of helicases on non-B DNA secondary structures has been identified in simian virus 40 (SV40), yeast and human cells. The most studied helicases are members of RecQ family, whose roles are found in a broad range of organisms from *E. coli* RecQ to humans WRN, BLM and RecQL4 (Mohaghegh et al., 2001; Bachrati & Hickson, 2003; Cobb & Bjergbaek, 2006; Masai, 2011). All the non-B DNA secondary structure unwinding helicases act catalytically and all require for their hydrolysis of nucleotide triphosphate, normally ATP, and the presence of $Mg^{2+}$ ions. For example, G-quadruplex DNA substrates are unwound by RecQ helicase with a 3'→5' polarity and need the tetraplex to hold a short 3' single-stranded tail that serves as a "loading dock" for these enzymes (Jain et al., 2010). It should be emphasized, however, that none of the described helicases unwinds tetraplex DNA only and all the enzymes are also able to unfold, although at a lower efficiency, other DNA structures such as duplex DNA, Holliday junctions or triplex. Recently, DHX9 helicase from human cells was found to co-immunoprecipitate with triplex DNA, suggesting a role in maintaining genome stability (Jain et al., 2010). DHX9 displaced the third strand from a specific triplex DNA and catalyzed the unwinding with a 3' to 5' polarity for the displaced third strand ((Jain et al., 2010).

### 5.3.1 RecQ helicases BLM, WRN, RECQL4 and Sgs1

RecQ helicases are a group of DNA helicases that are conserved from bacteria to man (Bachrati & Hickson, 2003). *RecQ* helicase is named after the *recQ* gene of *Escherichia coli* and has the activity of unwinding DNA in the 3'–5' direction in relation to the DNA strand in which the enzyme is bound (Mohaghegh et al., 2001). There are at least five homologues in humans, three of which are associated with genetic diseases. The yeast homologue of RecQ is Sgs1, whose function was found to be similar to most of the members in the RecQ family (Bachrati & Hickson, 2003; Cejka & Kowalczykowski, 2010; Masai, 2011).

It has been reported that, without a functional RecQ helicase, DNA replication does not advance normally. In humans, lacking of WRN or BLM protein accumulates aberrant replication intermediates (Harrigan et al., 2003; Cheok et al., 2005), this may allow for certain non-B DNA structure forming (Mohaghegh et al., 2001; Bacolla et al., 2011). Therefore, it is not surprising to see that more and more reports are going to be published which specify the important roles of RecQ in resolving the non-B DNA structures, including those G4-DNA (Kamath-Loeb et al., 2001; Fry & loeb, 1999). Similarly the large T antigen and Dna2 helicase/ exonuclease have also been found to unwind the G-tetraduplex (Masuda-Sasa et al., 2008).

### 5.3.2 Junction resolvases

A cruciform is similar in appearance to a recombination intermediate, a four-way Holliday junction. Therefore, Holliday junction resolvases, RuvABC in prokaryotes, or Mus81, Sgs1 and Sgs2 in yeast might also have activity on cruciforms formed at inverted repeats (Cejka & Kowalczykowski, 2010; Lilley, 2010; Ashton et al., 2011; Mankouri et al., 2011),.

### 5.4 Topoisomerase

Non-B DNA structures can be substrates for DNA topoisomerase I and II (Howard et al., 1993; Froelich-Ammon et al., 1994). It has shown that DNA topoisomerase II binds and cleaves hairpins (e.g., hairpin formed at a negatively supercoiled 52-bp palindromic sequence in the human β-globin gene), but not cruciforms. DNA topoisomerase II cleavage sites near human immunodeficiency virus integration sites in the human genome consist of Z-DNA forming sequences and other repetitive sequence (Howard et al., 1993); in contrast, DNA topoisomerase I promotes forming parallel G4 DNA in humans. Similarly RAP1, Hop1 in yeast, and Thrombin in humans are also found to promote form of G4 DNA.

### 5.5 Single strand binding protein (SSB/RPA)

RPA–ssDNA serves as intermediate in many DNA repair processes. For example, ssDNA-RPA can be made through nuclease and helicase actions in repair of UV-induced thymine dimers by nucleotide excision repair, and in a replication fork where DNA polymerase is paused but without pausing DNA helicase accompanied. RPA may prevent or destabilize a non-B DNA structure formation. For example, RPA in humans has been found to destabilize a G'4 DNA (Fig. 1). As for a triplex, the polypyrimidine strands are preferred to bind with RPA, which will then form complex with XPA, XPC-hHR23B (Vasquez et al., 2002; Thomas et al., 2005). In mammalian cells, RPA binds 50-fold more strongly to pyrimidines than to purines, therefore, makes the polypyrimidine strand single-stranded in an intramolecular triplex structure at neutral pH. Moreover, persistent RPA binding may lead to RPA hyper-phosphorylation that triggers repair reactions (Thomas et al., 2005). In addition, RPA-ssDNA and an ssDNA–dsDNA junction can also act as initial signals for cells response to DNA damages, which activates the ATR pathway (Ball et al., 2004; Choi et al., 2010)   .

### 5.6 DNA structure-specific nucleases

Proteins consist of nucleases that specifically cleave DNA next to or within a non-B DNA secondary structures have been well studied. The earliest protein having such functions was identified in *Saccharomyces cerevisiae*, the gene *KEM1* (also called SEP1, DST2, XRN1 and RAR5) (Liu et al., 1994, 1995). *KEM1* was initially characterized as a telomere binding protein, and later, it was found to cleave DNA that includes a four-stranded G4 domain but show low or no nucleolytic activity toward single- or double-stranded DNA substrates. Other well-known DNA structure specific nucleases are SbcCD (Connelly & Leach, 1992,

1996, 2004; Connelly et al., 1998, 1999) and its eukaryotic homologue of Mre11-Rad50 (Paull & Gellert, 1998, 2000; Sonoda et al., 2006; Carter et al., 2007; Delmas et al., 2009).

### 5.6.1 SbcCD

It is now known that influences of repetitive DNA sequences on genomic instabilities were often attributable to forming non-B DNA secondary structures *in vivo*. Once a non-B DNA structure is stable, which will interfere with DNA replication, repair and/ or transcription *in vivo*, resulting in unstable genome. These deleterious non-B DNA secondary structures have already been found to form in *E.coli*, such as the large hairpin formed by the long palindrome DNA sequences (Leach, 1994). The stable hairpin can be cleaved by SbcCD, leading to forming DNA double strand breaks, and then be repaired by using homologous recombination (Connelly & Leach, 1996,; Connelly et al., 1992, 1998, 1999).

Long palindrome sequences are significantly more stable in nuclease-deficient (SbcCD) strains of *E. coli* than in wild-type strains. The SbcCD protein complex is a member of the structural maintenance of chromosomes (SMCs) family found in bacteriophage, bacteria, yeast, *Drosophila*, mouse, and human. SbcCD has both 3'–5' exonuclease activity on double-stranded DNA and endonuclease activity on single-stranded DNA (Connelly et al., 1999). *In vitro*, it can recognize and bind hairpin structures and cleave at the loop, 5' immediately next to the loop/ stem junction.

Further degradation of the hairpin cleavage products can occur by the ATP-dependent double-stranded DNA exonuclease activity of the SbcCD protein complex. This structure-specific endonuclease activity does not need a 3' or 5' terminus (Connelly & Leach, 1992, 1996; Connelly et al., 1998, 1999).

### 5.6.2 Mre11-Rad50-Nbs1 (MRN) / Mre11-Rad50-Xrs2 (MRX)

Rad50 and Mre11 are the eukaryotic homologues of SbcCD that have not been shown to bind hairpin/cruciform directly. Mre11 and Rad50, forming complex with Nbs1 (in human cells) or Xrs2 (in yeast), show a hairpin structure cleaving activity *in vitro*. And which participate in processing double strand breaks *in vivo* by homologous recombination or non-homologous end-joining (Paull & Gellert, 1998, 2000; Sonoda et al., 2006; Delmas et al., 2009). In hairpin cleavage, MRN/ MRX interacts with BRCA1 which preferentially binds four-way branched DNA, similar to cruciforms. Mre11 shows an incision activity at hairpin/ cruciform, and acts as a selective endonuclease in yeast to bind to G4 DNA or to G'2 quadruplex DNA and cleaves the G4 DNA.

### 5.6.3 other nucleases

Besides the DNA structure specific nucleases such as SbcCD and its eukaryotic homologue Mre11-Rad50-Nbs1 (Xrs1), many other DNA structure-specific DNA nucleases have also been determined. These nucleases recognize and cleave the non-B DNA structures or even the DNA sequences that have non-B DNA secondary structures adopted, playing important roles in various DNA transactions including DNA replication, repair and recombination. For example, Rad1-Rad10 (XPF or ERCC1) has shown to cleave branched intermediates/ Flapped DNA in repair (Li et al., 2008; Muñoz et al., 2009). And Rad2 family of nucleases, such as human XPG (Class I), FEN1 (Class II), and HEX1/ hEXO1 (Class III), have shown both substrate specific 5' to 3' exonuclease activity and endonuclease activity in repair, recombination, and/ or replication. Among them, Rad2 domain of human exonuclease 1

(HEX1-N2) has high activity on single- and double-stranded DNA substrates as well as a flap structure-specific endonuclease activity but does not have specific endonuclease activity at 10-base pair bubble-like structures, G:T mismatches, or uracil residues (Lee & Wilson, 1999). FEN-1, a structure-specific endonuclease is essential for DNA replication and repair, removes RNA and DNA 5' flaps (Tsutakawa et al., 2011). FEN-1 was thought to be involved in hairpin structure processing, and was found to be involved in CNG triplet repeat stability in the lagging strand template (Spiro et al., 1999; Singh et al., 2007). Similarly, Deletions in PCNA, RPA, and the Bloom protein (BLM), a 3'-5' helicase can also increase CNG repeat expansion or deletion, which reportedly interacts with FEN-1 in cleaving flaps. Recently NucS from *Pyrococcus abyssi* was found to be the equivalent of FEN-1 that cleaves the flapped DNA in Okazaki frangment processing in the lagging strand DNA replication (Ren et al., 2009; Creze et al., 2011).

SLX1 and SLX4 are other structure-specific endonucleases acting as heteromer that cleave branched DNA substrates, particularly simple-Y, 5'-flap, or replication fork structures. It also cleaves the strand bearing the 5' nonhomologous arm at the branch junction and generates ligatable nicked products from 5'-flap or replication fork substrates (Fricke & Brill, 2003).

RAGs is a complex consisting of RAG1, RAG2, and HMGB1 that cleaves 3' overhangs in multiple locations at the duplex/ single-stranded transitions (Fugmann, 2001). RAGs complex is able to cleave different non-B DNA structures such as symmetric bubbles, heterologous loops and proposed triplex DNA. For example, RAGs complex cleaves the *bcl-2* Mbr at 3' overhang and non-B DNA structures under physiological buffer conditions (Adachi & Tsujimoto, 1990; Fugmann, 2001; Raghavan et al., 2004, 2005).

In addition, many single-strand specific nucleases, like S1, P1, and mung bean nucleases, are also efficient at cleaving single stranded DNA in the non-B DNA structures, though at low pH. Since some non-B DNA structures, e.g. H-DNA and G4 DNA disclose an unstructured single-stranded DNA region, which therefore serve as substrates for those single-strand specific nucleases. Recently, a more specific nuclease that cuts single-stranded DNA 5' to a G4 domain was isolated from human cells. This enzyme, initially named G quartet nuclease 1 (GQN1) is thought to be involved in immunoglobulin heavy chain class switch recombination in B cells, does not digest single- or double-stranded DNA, Holliday junctions or tetraplex RNA. It specifically cuts single-stranded DNA located few nucleotides 5' to either G'2 or G4 domains (Sun et al., 2001). However, GQN1 cannot incise tetraplex RNA, showing a significant difference from a mouse cytoplasmic exoribonuclease (mXRN1p) which cleaves G4 RNA (Bashkirov et al., 1997).

## 6. Gratuitous repair on undamaged DNA misfolds by multiple proteins

DNA damage and repair are always active in living cells regardless of the proliferation status of the cells. And unpaired bases and the helix distortions/ junctions in most of the non-B DNA secondary structures can therefore be targets for the structure specific proteins working in DNA repair, e.g. mismatch repair, nucleotide excision repair etc., launching DNA repairs or activating checkpoints repair (Voineagu et al., 2009).

### 6.1 Repair by singular pathway of DNA repair

Small DNA loops/ bulges, triplex DNA may be readily corrected by an individual repair, such as a mismatch repair or a nucleotide excision repair. For example, helix distortion and/ or mismatched base pairs in a hairpin, which sometime also occurs with imperfect hairpin structures at CAG repeats, can be recognized by mismatch repair machinery (Yang, 2006). Msh2/ Msh3 complex in eukaryotic cells specifically binds CAG-hairpins, and the ATP-ase activity of the Msh2 / Msh3 complex can be altered by the binding. However, the repair is dependent on the number of loops/ bulges. A few of them may be repaired by MMR, but too many may not because of interfering MMR by multiple MutS binding, suggesting that repair on a particular non-B DNA conformation will be conditional, depending on locations and environments. Further, nucleotide excision repair (NER) proteins can bind intermolecular triplex, which are involved in the triplex mediated mutagenesis and recombination (Wang & Vasquez, 2006). In bacterial cells, NER proteins UvrB and UvrC were responsible for triplex-induced cell growth retardation. Given the likenesses of the intermolecular and intramolecular triplex, it is possible for NER contributing to the H-DNA-induced mutagenesis and recombination.

## 6.2 Competitions among multiple repair proteins

Apart from initiating an individual pathway of DNA repair, some non-B DNA structures can also be recognized by more than one repair proteins working in different repair pathways, resulting in competitions between proteins on same DNA structures.

Competition of repair proteins on a non-B DNA structure may be needed for a cooperative repair, setting up a cooperative new DNA repair to repair; in contrast, the competition may sometimes be internecine, failing in repair of either pathway. Under this circumstance, the repair on a non-B DNA structure by the compositing actions of the DNA structural recognition proteins would be compromised. For example, a stable hairpin may be needed for starting DNA replication, but such a stable hairpin would also be repaired by SbcCD or Mre11-Rad50, making a DNA break for homologous recombination to repair (Leach, 1994). Similarly, unwound DNA or small DNA loops may also be needed for DNA replication or for transcription. While they may also be recognized and bound by repair proteins, such as DNA mismatch and nucleotide-excision repair proteins, recombination proteins, instead of SSB/ RPA (Kirkpatrick & Petes, 1997).

A good demonstration for the internecine competition between multiple repair proteins was the foldings of TGG and AGG repeats in the lagging strand template in a replication fork (Pan & Leach, 2000; Pan et al., to be published results). TGG, AGG and CGG repeats are a group of NGG repeats which own significant potential of folding into non-B DNA secondary structures (Usdin, 1998; Pan & Leach, 2000). AGG repeats formed triplex (Suda et al., 1996; Mishima et al., 1996, 1997), homoduplex (Suda et al.,1995), tetra-duplex (Yang & Hurley, 2006), and a special G-quadruplex, known as tetrad:heptad:heptad:tetrad ((G:H:H:G) or (T:H:H:T)) (Matsugami et al., 2001a, 2001b, 2002, 2003), while CGG and TGG repeats formed pseudo-hairpin and tetra-duplex, respectively (Darlow & Leach, 1998; Usdin, 1998; Pan & Leach, 2000; Zemánek et al., 2005).

It was shown by Pan and Leach, that replication of TGG repeats in the lagging strand template experiences repeats misfolding, during which both MutS and SbcCD were found to affect the later processing by homologous recombination. Binding MutS to the non-B DNA structure formed by TGG repeats may stabilize the structure, while hindering SbcCD cleaving the structure. Interestingly, the roles of MutS and SbcCD in this case seemed complex, since TGG repeats can replicate either without MutS or SbcCD, suggesting that they also play same role in stabilizing the TGG repeat structure. In contrast, similar sized AGG repeats was found also to fold into non-B DNA structures in a similar lagging strand template of a replication fork.

However, the non-B DNA structure formed by AGG repeats was found to be incapable of binding with MutS protein, and being cleaved by SbcCD. This made consistence with the reports though AGG repeats belong to a same group of NGG trinucleotide repeats with TGG repeats, they form various G-rich DNA secondary structures, including quadruplex, triple helical, homoduplex and tetrad:heptad:heptad:tetrad ((G:H:H:G) or (T:H:H:T)). Obviously, some of these non-B DNA structures folded may not be recognized by MutS protein *in vivo*, making significant differences in DNA structure formation between AGG repeats and TGG repeats (Pan et al., unpublished results).

The examples of a coordinated repair by different repair proteins on the same non-B DNA structures are the repair of DNA loops by MMR and NER proteins (Kirkpatrick & Petes, 1997; Zhao et al., 2009, 2010). It has been found that both MSH2 and XPA proteins are involved in the instabilities of CAG repeats, possibly through some so far unidentified roles (Kirkpatrick & Petes, 1997; Lin & Wilson, 2009; Zhao etal., 2009, 2010). Knocking down both MSH2 and XPA proteins did not further reduce CAG repeat contraction, suggesting a new role for these proteins in the same pathway. Similarly, it has also been reported the MSH2 and XPA are also involved in H-DNA metabolism but once again the DNA structure may not be processed via canonical MMR or NER mechanisms (Zhao et al., 2009, 2010).

### 6.3 Repair proteins can be defeated by DNA secondary structure

It may be feasible by postulating that more non-B DNA structures might be formed by DNA sequences in the genomes. However the repair machinery in the cells may only be limited to a few types, such as those MMR, NER single / double strand breaks etc. It therefore raises a question as if all non-B DNA structures possibly form could be recognized and processed by those repair proteins? The answer to this question is presently unknown; however some of the known secondary structures cannot easily be repaired, including large DNA loops and the flapped DNA etc.

### 6.3.1 Large loops

Stable base pairing prevents recognition by repair enzymes of bases or junctions requiring repair. For example, in *E.coli*, small loops (or secondary structure) may allow mispairing of bases that are corrected by MMR enzymes, leading to loss of base interruption (Parker & Marinus, 1992; Carraway & Marinus, 1993). However, DNA loops made up of less than four unpaired bases are efficiently corrected by methyl-directed mismatch repair (MMR), but loops larger than that cannot be repaired effectively (Parker & Marinus, 1992; Carraway & Marinus, 1993; Fang et al., 2003). The reason for this inefficacy was found to be due to the failure in loop recognition using MutS proteins, leaving the large looped DNA unrepaired by MMR.

### 6.3.2 Flapped DNA

Flap endonuclease (RAD27 in *Saccharomyces cerevesiae*; FEN-1 in humans) can destabilize simple tandem repeat loci. The 5' to 3' flap endonuclease FEN-1/ RAD27 is a structure-specific nuclease required for Okazaki fragment processing in the lagging strand DNA replication. FEN-1, a structure-specific endonuclease is also thought to be involved in CNG triplet repeat stability. It has been reported that a stable hairpin formed by CTG or CAG repeats at the flap region can block the activity of FEN-1. Which then join the upstream Okazaki fragment, resulting in repeats expansion during the next cycle of replication, marking the activity of FEN-1 can be defeated by stable DNA structure (Spiro et al., 1999; Singh et al., 2007).

## 6.4 Cellular response to non-B DNA structures by activating checkpoints

The existence of cellular proteins that interact with non-B DNA structures provides both strong argument for the existence of non-B DNA structure formations in genomic DNA, and suggestion for cell having intrinsic response to the formation of non-B DNA structures. However, it seems that not all non-B DNA secondary structures, unless they make severe troublesome to DNA metabolism such as making DNA double strand breaks, or generating long single stranded region, were recognized as "DNA damage". Even if cruciforms / hairpins, triplexes, slipped conformations, quadruplexes, and left-handed Z-DNA have all been reported to be chromosomal targets for DNA repair, recombination, and aberrant DNA synthesis, leading to repeat expansion or genomic rearrangements associated with neurodegenerative and genomic disorders. Some of them may also raise more severe response by cells (Voineagu et al., 2009).

The situations for a non-B DNA secondary structure intriguing a cellular response may be addressed at the competing recognition and processing by multiple repair proteins, resulting in incomplete / partial / opposing processing of the non-B DNA structure. Such intermediates may be recognized by proteins capable of activating a cellular response. Alternatively non-B DNA structure bears components that can be recognized by proteins capable of activating a cellular response (Voineagu et al., 2009). In support of this idea, DNA structure-specific proteins Rad1, Msh2, Msh3, and Sgs1 were found to play opposite roles in yeast gene targeting, a triple stranded DNA mediated process. During which Rad1, Msh2, and Msh3 facilitated forming triplex DNA, while Sgs1 prevented forming triplex DNA (Langston & Symington, 2005), therefore should a cellular response be intrigued in gene targeting may have to wait for processing the structure-specific proteins.

The ssDNA region in a non-B DNA structure may likely be coated by single-stranded DNA-binding protein (RPA) directly, or RPA coats the ssDNA after the non-B DNA structure is processed. Either way makes a common intermediate of ssDNA-RPA that activates ATR signaling in response to all of the genotoxic lesions (Krejci et al., 2003; Hu et al., 2007). Indeed, the ssDNA-RPA complex has been found to be a common intermediate in the processing of many types of damaged DNA, including DSBs, UV-induced thymidine dimers, intrastrand cross-links, and mismatches in base-pairing (Ball et al., 2005; Choi et al., 2010 ). The RPA–ssDNA complex will promote the loading of the 9–1–1 and ATR–ATRIP complexes (Dore et al., 2009). The juxtaposition of these complexes allows ATR to phosphorylate Chk1, which then promotes cell cycle arrest, causing a cellular response to non-B DNA structure formation. Alternatively, ssDNA-RPA complex can recruit Cut5, by which ATR (ATR-ATRIP) (Mec1-Ddc2 in yeast), DNA polymerase α, Rad50-Mre11-Nbs1 (MRN) and clamp loader Rad24 (Rad17 in mammals) can all be recruited to the ssDNA-RPA (Cortez et al., 2001; Zou & Elledge, 2003; Robison et al., 2004).

The purpose of activating DNA damage checkpoint in response to the formation of non-B DNA secondary structure is to regulate cell cycle events, for mediating appropriate repair and fork restart processes. While non-B DNA structure forming sequences per se are probably an infrequent trigger of DNA damage checkpoint responses, and, thus, should not be regarded as a real DNA damage by cells. There has extensive evidence suggesting that non-B DNA structure forming sequences can only induce checkpoint-triggering events when stable non-B DNA structures are adopted. The stable DNA structures may affect normal DNA metabolism, making DSBs or causing more severe effects on DNA metabolism, such as replication fork stalling, formation of nucleosome free sites (Chromosomal Fragile Sites) etc.

Consisting with that, mutations in checkpoint genes, such as Mec1, Ddc2, Rad9, Rad17, Rad24, or Rad53, produce repeat instabilities by a $CAG_{\sim70}$, including both expansion and contraction instabilities. These suggested that DNA structure formed by long CAG repeats activated checkpoints in eukaryotes (Lahiri et al., 2004; Sundararajan & Freudenreich, 2011). Similarly, a $CAG_{175}$ repeat on plasmids can also be recognized as "DNA damage" in E. coli, as witnessed by inducing SOS response (Majchrzak et al., 2006).

Surprisingly, it was found that even those shorter CAG repeats (containing 13–20 triplets) can also intrigue DNA damage checkpoint. By which repeats expansion can be prevented when the repeats formed non-B structures, suggesting that cells have endowed the checkpoint mechanism of responding to non-B DNA structure formation (Razidlo & Lahue, 2008).

Another example as intriguing cellular response for non-B DNA structure formation by derived structure processing is also found with human PKD1 gene. The 2.5-kb polypurine-polypyrimidine tract in intron 21 in human PKD1 gene potentially forms H-DNA structure, contributing to the high mutation rate of the PKD1 gene (Bacolla et al., 2001; Patel et al., 2004). A plasmid carrying this polypurine–polypyrimidine tract induced a strong SOS response and severely delayed the host cell growth, resulting in a dramatic decrease in colony formation (Patel et al., 2004). However, the effect was largely reduced without UvrA (100-fold decrease in colony formation), and nearly vanished without UvrB or UvrC. These suggested the polypurine–polypyrimidine repeat sequence or the structure formed by the repeats per se was not involved in the effects, while the NER processing was essential (Bacolla et al., 2001).

## 6.5 Mre11-Rad50-Nbs1 (MRN)/ Mre11-Rad50-Xrs2 (MRX)

Apart from the nucleolytic activity, MRN / MRX can also play roles in activating the checkpoints as mentioned above (van den Bosch, et al., 2003; Sundararajan & Freudenreich, 2011). It was believed that a single stranded region in a non-B DNA structure forms ssDNA-RPA to the amount of triggering a checkpoint response (normally exceeds 300 bp). One way of Rad50-Mre11-Nbs1 (MRN) contributing to checkpoint response might be through Cut5 recruitment. Rad50-Mre11-Nbs1 (MRN) can be recruited to the single stranded region in the non-B DNA structure, and then participates in ATR checkpoint. Alternatively Rad50-Mre11-Nbs1 (MRN) can also secure DNA replication as implicated by its ortholog SbcCD in E.coli (Darmon et al., 2007; Zahra et al., 2007). Indeed, the MRN / MRX complex has been co localized in the replication machinery. In this context, the resection role of MRN / MRX on DSB initiated recombination repair may be no more necessary as long as the checkpoints mechanism prevented the DSB formation by checkpoint proteins (Mimitou & Symington, 2008; Zhu et al., 2008).

Non-B DNA structure forming sequences are potential triggers of DNA damage checkpoint responses mainly by inducing replication fork stalling and chromosomal breaks. Since the non-B DNA structures have specific DNA conformations at the damaged site, which may influence the checkpoint signaling, and the dynamics of checkpoint activation are likely to differ at different types of non-B DNA structure forming sequences.

## 7. Future perspectives

Many lines of evidence suggest that unusual DNA structures can form *in vivo* and play significant roles in DNA metabolism, while they may also serve as a source for the generation of genomic instability. Strikingly, unusual DNA structures were often found to trigger some kinds of repair actions or avoidance responses that promote their removal of the structures once formed. Under this later circumstance, it becomes obvious that formation of non-B DNA structures *in vivo* was somehow similar to the appearances of some real DNA damages as induced by environmental DNA damaging agents. Certain unusual DNA structures have unpaired bases and regions with helix distortions/junctions etc., which may experience unprovoked repair in cells. Therefore triggering cellular responses of a non-B DNA structure is subject to its morphological/ topological properties, which could attract recognizing repair proteins. In fact, a non-B DNA structure is often recognized by more than one repair proteins, such as the proteins working in MMR, NER and recombination. Questions rose therefore as if individual pathways of DNA repair accounts enough for the repair of the non-B DNA structures? Or does it need multiple proteins working in different repair pathways reconstitute synthesized pathway(s) to repair? Nevertheless, progress in this field seems support an idea that enzymes/ proteins that recognize and/ or process the possible non-B DNA structures may be different because of the non-B DNA structures formed. Proteins that have been found to associate with non-B DNA instability might take part in an unexpected way in processing the non-B DNA structures. Therefore studies in the coming future may have to focus on the identifications of the types of non-B DNA structures that elicit certain kinds of mutations and the enzyme systems involved. It could be expected that more diseases will be recognized as because of mutations at non-B DNA structures. Also, strategies will have to make toward developing therapeutics to appease the devastating effects of the syndromes.

## 8. References

Adachi, M. & Tsujimoto. Y. (1990). Potential Z-DNA elements surround the breakpoints of chromosome translocation within the 5' flanking region of *bcl*-2 gene. *Oncogene*, 5:1653–1657.

Alvarez, D., Novac, O., Callejo, M., Ruiz, M. T., Price, G. B. & Zannis-Hadjopoulos, M. (2002).14-3-3sigma is a cruciform DNA binding protein and associates in vivo with origins of DNA replication. *J Cell Biochem.* 87(2):194-207.

Antony, S., Arimondo, P. B., Sun, J-S. & Pommier, Y. (2004). Position- and orientation-specific enhancement of topoisomerase I cleavage complexes by triplex DNA structures. *Nucleic Acids Res.*, 32 (17): 5163-5173.

Arimondo, P. B., Riou, J-F., Mergny, J-L., Tazi, J., Sun, J-S., Garestier, T. & Hélène, C. (2000). Interaction of human DNA topoisomerase I with G–quartet structures. *Nucleic Acids Res.*, 28 (24): 4832-4838.

Ashton, T. M., Mankouri, H. W., Heidenblut, A., McHugh, P. J. & Hickson, I. D. (2011). Pathways for Holliday Junction Processing during Homologous Recombination in *Saccharomyces cerevisiae*. *Mol Cell Biol.*, 31(9):1921-1933.

Bachrati, C. Z. & Hickson, I. D. (2003). RecQ helicases: suppressors of tumorigenesis and premature aging. *Biochem J.*, 374(Pt 3):577-606.

Bacolla, A., Jaworski, A., Connors, T. D. & Wells, R. D. (2001). Pkd1 unusual DNA conformations are recognized by nucleotide excision repair. *J Biol Chem.*, 276:18597–18604.

Bacolla, A. & Wells, R. D. (2009). Non-B DNA conformations as determinants of mutagenesis and human disease. *Mol Carcinog.*, 48:273-285.

Bacolla, A., Wang, G., Jain, A., Chuzhanova, N. A., Cer, R. Z., Collins, J. 1 R., Cooper, D. N., Bohr, V. A. & Vasquez, K. M. (2011). Non-B DNA-forming sequences and WRN deficiency independently increase the frequency of base substitution in human cells. *J Biol Chem.*, 286(12):10017-10026.

Balakumaran, B. S., Freudenreich, C. H. & Zakian, V. A. (2000). CGG / CCG repeats exhibit orientation-dependent instability and orientation-independent fragility in Saccharomyces cerevisiae. *Hum Mol Genet.*, 9(1):93-100.

Ball, H. L, Myers, J. S. & Cortez, D. (2005). ATRIP binding to replication protein A-single-stranded DNA promotes ATR-ATRIP localization but is dispensable for Chk1 phosphorylation. *Mol Biol Cell.*, 16(5):2372-2381.

Bashkirov, V. I., Scherthan, H., Solinger, J. A., Buerstedde, J. M. & Heyer, W. D. (1997). A mouse cytoplasmic exoribonuclease (mXRN1p) with preference for G4 tetraplex substrates. *J Cell Biol.*, 136(4):761-773.

Bhattacharyya, S. & Lahue, R. S. (2004). *Saccharomyces cerevisiae* Srs2 DNA helicase selectively blocks expansions of trinucleotide repeats. *Mol Cell Biol.*, 24(17):7324-7330.

Boiteux, S., Costa de Oliveira, R. & Laval, J. (1985). The *Escherichia coli* O6-methylguanine-DNA methyltransferase does not repair promutagenic O6-methylguanine residues when present in Z-DNA. *J Biol Chem.*, 260(15):8711-8715.

Boiteux, S. & Laval, F. (1985). Repair of O6-methylguanine, by mammalian cell extracts, in alkylated DNA and poly (dG-m5dC)·(polydG-m5dC) in B and Z forms. *Carcinogenesis*, 6:805–807.

Broxson, C., Beckett, J. & Tornaletti, S. (2011). Transcription arrest by a G quadruplex forming-trinucleotide repeat sequence from the human *c-myb* gene. *Biochemistry.*, [Epub ahead of print]

Burge, S., Parkinson, G. N., Hazel, P., Todd, A. K. & Neidle, S. (2006). Quadruplex DNA: sequence, topology and structure. *Nucleic Acids Res.*, 34(19):5402-5415..

Carraway, M. & Marinus, M. G. (1993). Repair of heteroduplex DNA molecules with multibase loops in *Escherichia coli. J Bacteriol.*, 175(13):3972-3980.

Casasnovas, J. M. & Azorin, F. (1987). Supercoiled induced transition to the Z-DNA conformation affects the ability of a d(CG/GC)$_{12}$ sequence to be organized into nucleosome-cores. *Nucleic Acids Res.*, 15:8899–8918.

Casey, B. P. & Glazer, P. M. (2001). Gene targeting via triple-helix formation. *Prog Nucleic Acid Res Mol Biol.*, 67:163-192.

Cejka, P. & Kowalczykowski, S. C. (2010). The full-length *Saccharomyces cerevisiae* Sgs1 protein is a vigorous DNA helicase that preferentially unwinds holliday junctions. *J Biol Chem.*, 285(11):8290-82301.

Cer, R. Z., Bruce, K. H., Mudunuri, U. S., Yi, M., Volfovsky, N., Luke, B. T., Bacolla, A., Collins, J. R. & Stephens, R. M. (2011). Non-B DB: a database of predicted non-B DNA-forming motifs in mammalian genomes. *Nucleic Acids Res.*, 39 (Database issue):D383-391.

Chan, P. P., Lin, M., Faruqi, A. F., Powell, J., Seidman, M. M. & Glazer, P. M. (1999). Targeted correction of an episomal gene in mammalian cells by a short DNA fragment tethered to a triplex-forming oligonucleotide. *J Biol Chem.*, 274, 11541-11548.

Chen, J. M., Cooper, D. N., Ferec, C., Kehrer-Sawatzki, H. & Patrinos, G. P. (2010). Genomic rearrangements in inherited disease and cancer. *Semin Cancer Biol.*, 20(4):222-233.

Cheok, C. F., Bachrati, C. Z., Chan, K. L., Ralf, C., Wu, L. & Hickson, I. D. (2005). Roles of the Bloom's syndrome helicase in the maintenance of genome stability. *Biochem Soc Trans.*, 33(Pt 6):1456-1459.

Choi, J. H., Lindsey-Boltz, L. A., Kemp, M., Mason, A. C., Wold, M. S. & Sancar, A. (2010). Reconstitution of RPA-covered single-stranded DNA-activated ATR-Chk1 signaling. *Proc Natl Acad Sci USA.*, 107(31):13660-13665.

Connelly, J., de Leau, E. S. & Leach, D. (1992). DNA cleavage and degradation by the SbcCD complex from *Escherichia coli. Nucleic Acid Res.*, 27, 1039-1046

Connelly, J. & Leach, D. (1996). The *sbc*C and *sbc*D genes of *Escherichia coli* encode a nuclease involved in palindromic inviability and genetic recombination. *Genes to Cell.*, 1,285-291.

Connelly, J., Kirkham, L. A. & Leach, D. (1998). The SbcCD nuclease of *Escherichia coli* is a structural maintenance of chromosomes (SMC) family protein that cleaves hairpin DNA. *Proc Natl Acad Sci U S A.*, 95, 7969-7974.

Connelly, J., delau, E.S. & Leach, D. R. F. (1999). DNA cleavage and degradation by the SbcCD complex from *Escherichia coli, Nucleic Acid Res.*, 27:1039–1046.

Courey, A. J. (1999). Analysis of altered DNA structures: cruciform DNA. *Methods Mol Biol.* 94:29-40.

Cox, R. & Mirkin, S. M. (1997). Characteristic enrichment of DNA repeats in different genomes. *Proc Natl Acad Sci USA.*, 94: 5237–5242.

Creze, C., Lestini, R., Kühn, J., Ligabue, A., Becker, H. F., Czjzek, M., Flament, D. & Myllykallio H. (2011). Structure and function of a novel endonuclease acting on branched DNA substrates. *Biochem Soc Trans.*, 39(1):145-9.

Crouau-Roy, B. (1999). Trans-speciation maintenance in the MHC region of a polymorphism which includes a polymorphic dinucleotide locus, and the *de novo* arisal of a polymorphic tetranucleotide microsatellite. *Tissue Antigens.*, 54(6):560-4.

Darmon, E., Lopez-Vernaza, M. A., Helness, A. C., Borking, A., Wilson, E., Thacker, Z., Wardrope, L. & Leach, D. R. F., (2007). SbcCD regulation and localization in *Escherichia coli. J Bacteriol.*, 189(18), 6686-6694.

Dere, R. & Wells, R. D. (2006). DM2 CCTG*CAGG repeats are crossover hotspots that are more prone to expansions than the DM1 CTG*CAG repeats in *Escherichia coli. J Mol Biol.*, 360(1):21-36.

Dore′, A. S., Kilkenny, M. L., Rzechorzek, N. J. & Pearl, L. H. (2009). Crystal structure of the Rad9-Rad1-Hus1 DNA damage checkpoint complex— implications for clamp loading and regulation. *Mol Cell*, 34, 735–745.

Dyke, M. W. V., Nelson, L. D., Weilbaecher, R. G. & Mehta, D. V. (2004). Stm1p, a G4 Quadruplex and purine motif triplex nucleic acid-binding protein, interacts with ribosomes and subtelomeric Y' DNA in *Saccharomyces cerevisiae. J Biol Chem.*, 279, 24323-24333.

Fang, W. H., Wang, B. J., Wang, C. H., Lee, S. J., Chang, Y. T., Chuang, Y. K. & Lee, C. N. (2003). DNA loop repair by *Escherichia coli* cell extracts. *J Biol Chem.*, 278(25):22446-22452.

Faruqi, A. F., Datta, H. J., Carroll, D., Seidman, M. M. & Glazer, P. M. (2000). Triple-helix formation induces recombination in mammalian cells via a nucleotide excision repairdependent pathway. *Mol Cell Biol.*, 20, 990-1000.

Fox, K. R. & Brown, T. (2011). Formation of stable DNA triplexes. *Biochem Soc Trans.*, 39(2):629-34.

Freudenreich, C. H. (2007). Chromosome fragility: molecular mechanisms and cellular consequences. *Front. Biosci.*, 12:4911-24.

Froelich-Ammon, S. J., Gale, K. C. & Osheroff, N. (1994). Site-specific cleavage of a DNA hairpin by topoisomerase II. DNA secondary structure as a determinant of enzyme recognition/cleavage. *J Biol Chem.*, 269:7719-7725.

Fry, M. & Loeb, L. A. (1999). Human werner syndrome DNA helicase unwinds tetrahelical structures of the fragile X syndrome repeat sequence d(CGG)$_n$. *J Biol Chem.*, 274(18):12797-802.

Fry, M. (2007). Tetraplex DNA and its interacting proteins. *Frontiers in Bioscience*, 12, 4336-4351.

Fugmann, S. D. (2001). RAG1 and RAG2 in V (D) J recombination and transposition. *Immunol Res.*, 23(1):23-39.

Gur-Arie, R., Cohen, C. J., Eitan, Y., Shelef, L., Hallerman, E. M. & Kashi, Y. (2000). Simple Sequence Repeats in *Escherichia coli*: Abundance, Distribution, Composition, and Polymorphism. *Genome Res.*, 10: 62-71.

Harrigan, J. A., Opresko, P. L., von Kobbe, C., Kedar, P. S., Prasad, R., Wilson, S. H. & Bohr, V. A. (2003). The Werner syndrome protein stimulates DNA polymerase beta strand displacement synthesis via its helicase activity. *J Biol Chem.*, 278(25):22686-22695.

Hebert, A. & Rich, A. (1996). The Biology of Left-handed Z-DNA. *J Biol Chem.*, 271(20): 11595-11598.

Ho, P. S. (1994). The non-B DNA structure of d(CA/TG)$_n$ does not differ from that of Z-DNA. *Proc Natl Acad Sci U S A.*, 91(20):9549-9553.

Howard, M.T. & Griffith, J. D. (1993). A cluster of strong topoisomerase II cleavage sites is located near an integrated human immunodeficiency virus. *J Mol Biol.*, 232:1060-1068.

Inagaki, H., Ohye, T., Kogo, H., Kato, T., Bolor, H., Taniguchi, M., Shaikh, T. H., Emanuel, B. S. & Kurahashi, H. (2009). Chromosomal instability mediated by non-B DNA:Cruciform conformation and not DNA sequence is responsible for recurrent translocation in humans. *Genome Res.*, 19:191-198.

Jain, A., Bacolla, A., Chakraborty, P., Grosse, F., Vasquez, K. M. (2010). Human DHX9 helicase unwinds triple-helical DNA structures. *Biochemistry.*, 49(33):6992-6999.

Jin, R., Fernandez-Beros, M. E. & Novick, R. P. (1997). Why is the initiation nick site of an AT-rich rolling circle plasmid at the tip of a GC-rich cruciform? *EMBO J.*, 16(14):4456-66.

Johnston, B. H. (1988). Chemical probing of the B-Z transition in negatively supercoiled DNA. *J Biomol Struct Dyn.*, 6(1):153-166.

Kalish, J. M., Seidman, M. M., Weeks, D. L. & Glazer, P. M. (2005). Triplex-induced recombination and repair in the pyrimidine motif. *Nucleic Acids Res.*, 33(11):3492-502.

Kaliyaperumal, S., Patrick, S. M. & Williams, K. J. (2011). Phosphorylated hMSH6: DNA mismatch versus DNA damage recognition. *Mutat Res.*, 706(1-2):36-45.

Kamath-Loeb, A. S., Loeb, L. A., Johansson, E., Burgers, P. M. & Fry, M. (2001). Interactions between the Werner syndrome helicase and DNA polymerase delta specifically facilitate copying of tetraplex and hairpin structures of the $d(CGG)_n$ trinucleotide repeat sequence. *J Biol Chem.*, 276(19):16439-16446.

Kantelinen, J., Kansikas, M., Korhonen, M. K., Ollila, S., Heinimann, K., Kariola, R., Nystrm, M. (2010). MutSbeta exceeds MutSalpha in dinucleotide loop repair. *Br J Cancer.*, 102(6):1068-1073.

Khuu, P. A., Voth, A. R., Hays, F. A. & Ho, P. S. (2006). The stacked-X DNA Holliday junction and protein recognition. *J Mol Recognit.*, 19(3):234-242.

Kirkpatrick, D. T. & Petes T. D. (1997). Repair of DNA loops involves DNA mismatch and nucleotide-excision repair proteins. *Nature*, 387:929-931.

Kladde, M. P., Kohwi, Y., Kohwi-Shigematsu, T. & Gorski J. (1994). The non-B DNA structure of $d(CA/TG)_n$ differs from that of Z-DNA. *Proc Natl Acad Sci U S A.*, 91(5):1898-902.

Kolb. (2009). Cruciform-forming inverted repeats appear to have mediated many of the microinversions that distinguish the human and chimpanzee genomes. *Chromosome Res.*, 17:469-483

Kurahashi, H., Shaikh, T. H., Hu, P., Roe, B. A., Emanuel, B. S. & Budarf, M. L. (2000). Regions of genomic instability on 22q11 and 11q23 as the etiology for the recurrent constitutional t(11;22). *Hum Mol Genet*, 9:1665-1670.

Kurahashi, H. & Emanuel, B. S. (2001). Long AT-rich palindromes and the constitutional t(11;22) breakpoint. *Hum Mol Genet.*, 10(23):2605-17.

Kurahashi, H., Inagaki, H., Ohye, T., Kogo, H., Kato, T. & Emanuel, B. S. (2006). Palindrome-mediated chromosomal translocations in humans. *DNA Repair (Amst).*, 5(9-10):1136-45.

Kurahashi, H., Inagaki, H., Ohye, T., Kogo, H., Tsutsumi, M., Kato, T., Tong, M. & Emanuel, B. S. (2010). The constitutional t(11;22): implications for a novel mechanism responsible for gross chromosomal rearrangements. *Clin Genet.*, 78(4):299-309.

Lahiri, M., Gustafson, T. L., Majors, E. R. & Freudenreich, C. H. (2004). Expanded CAG repeats activate the DNA damage checkpoint pathway. *Mol Cell.*, 15(2):287-93.

Lange, S. S., Reddy, M. C. & Vasquez, K. M. (2009). Human HMGB1 directly facilitates interactions between nucleotide excision repair proteins on triplex-directed psoralen interstrand crosslinks. *DNA Repair (Amst).*, 8(7):865-72.

Langston, L. G. & Symington, L. S. (2005). Opposing roles for DNA structure-specific proteins Rad1, Msh2, Msh3, and Sgs1 in yeast gene targeting. *EMBO J.*, 24, 2214-2223

Lee, B. I. & Wilson, D. M. 3rd. (1999). The RAD2 domain of human exonuclease 1 exhibits 5' to 3' exonuclease and flap structure-specific endonuclease activities. *J Biol Chem.*, 274(53):37763-9.

Letessier, A., Millot, G. A., Koundrioukoff, S., Lachagès, A. M., Vogt, N., Hansen, R. S., Malfoy, B., Brison, O. & Debatisse, M. (2011). Cell-type-specific replication initiation programs set fragility of the FRA3B fragile site. *Nature.*, 470(7332):120-3.

Li, F., Dong, J., Pan, X., Oum, J. H., Boeke, J. D. & Lee, S. E. (2008). Microarray-based genetic screen defines SAW1, a gene required for Rad1 / Rad10-dependent processing of recombination intermediates. *Mol Cell.*, 30(3):325-335.

Lilley, D. M. (2010). The interaction of four-way DNA junctions with resolving enzymes. *Biochem Soc Trans.*, 38(2):399-403.

Lin, Y. & Wilson, J. H. (2009). Diverse effects of individual mismatch repair components on transcription-induced CAG repeat instability in human cells. *DNA Repair (Amst).*, 8(8):878-885.

Lipps, H. J. & Rhodes, D. (2009). G-quadruplex structures: *in vivo* evidence and function. Trends *Cell Biol.* 19(8):414-22.

Liu, Z. & Gilbert, W. (1994). The yeast KEM1 gene encodes a nuclease specific for G4 tetraplex DNA: implication of in vivo functions for this novel DNA structure. *Cell*, 77(7):1083-1092.

Liu, Z., Lee, A. & Gilbert, W. (1995). Gene disruption of a G4-DNA-dependent nuclease in yeast leads to cellular senescence and telomere shortening. *Proc Natl Acad Sci USA.*, 92(13):6002-6002.

Lukusa, T. & Fryns, J. P. (2008). Human chromosome fragility. *Biochim Biophys Acta.*, 1779(1):3-16.

Majchrzak, M., Bowater, R. P., Staczek, P. & Parniewski, P. (2006). SOS repair and DNA supercoiling influence the genetic stability of DNA triplet repeats in *Escherichia coli*. *J Mol Biol.*, 364(4):612-24.

Mankouri, H. W., Ashton, T. M., Hickson, I. D. (2011). Holliday junction-containing DNA structures persist in cells lacking Sgs1 or Top3 following exposure to DNA damage. *Proc Natl Acad Sci U S A.*, 108(12):4944-4949.

Masai, H. (2011). RecQL4: a helicase linking formation and maintenance of a replication fork. *J Biochem.*, 2011:149(6):629–631.

Masuda-Sasa, T., Polaczek, P., Peng, X. P., Chen, L. & Campbell, J. L. (2008). Processing of G4 DNA by Dna2 helicase / nuclease and replication protein A (RPA) provides insights into the mechanism of Dna2 / RPA substrate recognition. *J Biol Chem.*, 283(36):24359-24373.

Mäueler, W., Bassili, G., Epplen, C., Key, H. G. & Epplen, J. T. (1999). Protein binding to simple repetitive sequences depends on DNA secondary structure(s). *Chromosome Res.*, 7(3):163-166.

McMurray, C. T. (1999). DNA secondary structure: A common and causative factor for expansion in human disease. *Proc Natl Acad Sci U S A.*, 96: 1823–1825.

Mimitou, E. P. & Symington, L. S. (2008). Sae2, Exo1 and Sgs1 collaborate in DNA double-strand break processing. *Nature*, 455(7214):770-774.

Mirkin, S. M. (2008). Discovery of alternative DNA structures: a heroic decade (1979-1989). *Front Biosci.*, 13:1064-1071.

Mishima, Y., Kaizu, H. & Kominami, R. (1997). Pairing of DNA fragments containing (GGA: TCC)$_n$ repeats and promotion by high mobility group protein 1 and histone H1. *J Biol Chem*, 272, 26578–26584

Mukherjee, A. & Vasquez, K. M. (2011). Triplex technology in studies of DNA damage, DNA repair, and mutagenesis. *Biochimie*, 2011:93(8):1197-1208.

Muñoz, I. M., Hain, K., Déclais, A. C., Gardiner, M., Toh, G. W., Sanchez-Pulido, L., Heuckmann, J. M., Toth, R., Macartney, T., Eppink, B., Kanaar, R., Ponting, C. P.,

Lilley, D. M. & Rouse, J. (2009). Coordination of structure-specific nucleases by human SLX4/BTBD12 is required for DNA repair. *Mol Cell,* 35(1):116-127.

Nag, D. K. & Cavallo, S. J. (2007). Effects of mutations in SGS1 and in genes functionally related to SGS1 on inverted repeat-stimulated spontaneous unequal sister-chromatid exchange in yeast. *BMC Mol Biol.,* 8:120.

Novac, O., Alvarez, D., Pearson, C. E., Price, G. B., Zannis-Hadjopoulos, M. (2002). The human cruciform-binding protein, CBP, is involved in DNA replication and associates *in vivo* with mammalian replication origins. *J Biol Chem.,* 277(13):11174-11183.

Novara, F., Beri, S., Bernardo, M. E., Bellazzi, R., Malovini, A., Ciccone, R., Cometa, A. M., Locatelli, F., Giorda, R. And Zuffardi, O. (2009). Different molecular mechanisms causing 9p21 deletions in acute lymphoblastic leukemia of childhood. *Hum Genet.,* 126:511-520.

Pan, X. & Leach, D. (2000). The roles of *mut*S, *sbc*CD and *rec*A in the propagation of TGG repeat in *Escherichia coli. Nucleic Acids Res,* 28: 3178–3184.

Pan, X. (2004). Self-maintenance of gene and occurrence of genetic diseases, *Science Press Beijing,* pp.1–403.

Pan, X. (2006). Mechanism of trinucleotide repeats instabilities: the necessities of repeat non-B secondary structure formation and the roles of cellular trans-acting factors. *J Genet Genomics,* 33: 1–11.

Pan, X., Ding, Y. F. & Shi, L. F. (2009). The roles of SbcCD and RNaseE in the transcription of GAA · TTC repeats in Escherichia coli. *DNA Repair* (Amst) , 8:1321–1327.

Pan, X., Liao, Y., Liu, Y., Chang, P., Liao, L., Yang, L. & Li, H. Q. (2010). Transcription of AAT · ATT Triplet Repeats in *Escherichia coli* Is Silenced by H-NS and IS1E Transposition. *PLoS ONE,* 5(12): e14271.

Pan, X., Chang, P., Long, J., Xiao, P., Li, H. Q. & Duan, F. (2011). The molecular maintenance of trinucleotide repeats AGG in *Escherichia coli* relies on both homologous recombination and UmuDC –catalyzed DNA translesion synthesis. *DNA Res.,* (submitted).

Parker, B. O. & Marinus, M. G. (1992). Repair of DNA heteroduplexes containing small heterologous sequences in *Escherichia coli. Proc Natl Acad Sci U S A.,* 89(5):1730-1734.

Patel, H. P., Lu, L., Blaszak, R. T. & Bissler, J. J. (2004). PKD1 intron 21: triplex DNA formation and effect on replication. *Nucleic Acids Res.,* 32(4):1460-1468.

Pearson, C. E., Zorbas, H., Price, G. B. & Zannis-Hadjopoulos, M. (1996). Inverted repeats, stem-loops, and cruciforms: significance for initiation of DNA replication. *J Cell Biochem.,* 63(1):1-22.

Pfohl-Leszkowicz A, Boiteux S, Laval J, Keith G, Dirheimer G. (1983). Enzymatic methylation of chemically alkylated DNA and poly(dG-dC) X poly(dG-dC) in B and Z forms. *Biochem Biophys Res Commun.,* 116(2):682-688.

Qin, Y. & Hurley, L. H. (2008). Structures, folding patterns, and functions of intramolecular DNA G-quadruplexes found in eukaryotic promoter regions. *Biochimie.,* 90(8):1149-1171.

Qiu, J. F. & Pan, X. (2008). The linkage of DNA replication, recombination and repair, *Prog. Biochem. Biophys.,* 35: 751–756.

Raghavan, S. C., Swanson, P. C., Wu, X., Hsieh, C. L. & Lieber, M. R. (2004). A non-B-DNA structure at the Bcl-2 major breakpoint region is cleaved by the RAG complex. *Nature*, 428, 88-93.

Raghavan, S. C., Chastain, P., Lee, J. S., Hegde, B. G., Houston, S., Langen, R., Hsieh, C. L., Haworth, I. S. & Lieber, M. R.(2005). Evidence for a triplex DNA conformation at the *bcl*-2 major breakpoint region of the t(14;18) translocation. *J Biol Chem.*, 280, 22749-22760.

Rao,S. R., Trivedi, S., Emmanuel, D., Merita, K., & Hynniewta, M. (2010). DNA repetitive sequences-types, distribution and function: A review. *J Cell and Mol Biol.*, 7(2) & 8(1): 1-11.

Rawal, P., Kummarasetti, V. B., Ravindran, J., Kumar, N., Halder, K., Sharma, R., Mukerji, M., Das, S. K. & Chowdhury, S. (2006). Genome-wide prediction of G4 DNA as regulatory motifs: role in *Escherichia coli* global regulation. *Genome Res.*, 16(5):644-55.

Razidlo, D. F. & Lahue, R. S. (2008). Mrc1, Tof1 and Csm3 inhibit CAG·CTG repeat instability by at least two mechanisms. *DNA Repair* (Amst), 7(4):633-640.

Ren, B., Kühn, J., Meslet-Cladiere, L., Briffotaux, J., Norais, C., Lavigne, R., Flament, D., Ladenstein, R. & Myllykallio, H. (2009). Structure and function of a novel endonuclease acting on branched DNA substrates. *EMBO J.*, 28(16):2479-2489.

Rich, A., Nordheim, A. & Azorin, F. (1983). Stabilization and detection of natural left-handed Z-DNA. *J Biomol Struct Dyn.*, 1(1):1-19.

Sannohe, Y. & Sugiyama, H. (2010). Overview of formation of G-quadruplex structures. *Curr Protoc Nucleic Acid Chem.*, Chapter 17:Unit 17.2.1-17.

Schwartz, M., Zlotorynski, E. & Kerem, B. (2006). The molecular basis of common and rare fragile sites. *Cancer Lett.*, 232(1):13-26.

Shah, S. N., Opresko, P. L., Meng, X., Lee, M. Y. & Eckert, K. A. (2010). DNA structure and the Werner protein modulate human DNA polymerase delta-dependent replication dynamics within the common fragile site FRA16D. *Nucleic Acids Res.*, 38(4):1149-1162.

Shimizu, M., Mori, T., Sakuri, T. & Shindo, H. (2000).Destablization of nucleosomes by an unusual DNA conformation adopted by poly(dA)poly(dT) tracts *in vivo*. *EMBO J.*,19(13):3358-3365.

Sinden, R. R., Zheng, G. X., Brankamp R. G. & Allen, K. N. (1991). On the deletion of inverted repeated DNA in *Escherichia coli*: effects of length, thermal stability, and cruciform formation *in vivo*. *Genetics*, 129, 991–1005.

Sinden, R. R., Hashem, V. I. & Rosche, W. A. (1999). DNA-directed mutations. Leading and lagging strand specificity. *Ann N Y Acad Sci.*, 870,173-189

Singh, P., Zheng, L., Chavez, V., Qiu, J. & Shen, B. (2007). Concerted action of exonuclease and Gap-dependent endonuclease activities of FEN-1 contributes to the resolution of triplet repeat sequences $(CTG)_n$- and $(GAA)_n$-derived secondary structures formed during maturation of Okazaki fragments. *J Biol Chem.*, 282(6):3465-3477.

Smith, R. (2008). Meeting DNA palindromes head-to-head. *Genes Dev.*, 22: 2612-2620.

Spiro, C., Pelletier, R., Rolfsmeier, M. L., Dixon, M. J., Lahue, R. S., Gupta, G., Park, M. S., Chen, X, Mariappan, S. V. & McMurray, C. T. (1999). Inhibition of FEN-1 processing by DNA secondary structure at trinucleotide repeats. *Mol Cell*, 4(6):1079-1085.

Stros, M., Muselíková-Polanská, E., Pospíšilová, S. & Strauss, F. (2004). High-affinity binding of tumor-suppressor protein p53 and HMGB1 to hemicatenated DNA loops. *Biochemistry.*, 43(22):7215-7225.

Suda,T., Mishma,Y., Takayanagi, K., et al. (1996). A novel activity of HMG domains promotion of the triple-stranded complex formation between DNA containing (GGA/TCC)$_{11}$ and d(GGA)$_{11}$ Oligonucleotides. *Nucleic Acids Res.*, 24, 4733-4740

Sun, H., Yabuki, A. & Maizels, N. (2001). A human nuclease specific for G4 DNA. *Proc Natl Acad Sci U S A.*, 98(22):12444-12449.

Sundararajan, R. & Freudenreich, C. H. (2011). Expanded CAG / CTG Repeat DNA Induces a checkpoint response that impacts cell Proliferation in Saccharomyces cerevisiae. *PLoS Genet.*, 7(3):e1001339.

Svozil, D., Kalina, J., Omelka, M. & Schneider, B. (2008). DNA conformations and their sequence preferences. *Nucleic Acids Res.*, 36(11):3690-3706.

Tsutakawa, S. E., Classen, S., Chapados, B. R., Arvai, A. S., Finger, L. D., Guenther, G., Tomlinson, C. G., Thompson, P., Sarker, A. H., Shen, B., Cooper, P. K., Grasby, J. A. & Tainer, J. A. (2011). Human Flap Endonuclease Structures, DNA Double-Base Flipping, and a Unified Understanding of the FEN1 Superfamily. *Cell.*, 145(2):198-211.

Urban, M., Joubert, N., Purse, B. W., Hocek, M. & Kuchta, R. D. (2010). Mechanisms by which human DNA primase chooses to polymerize a nucleoside triphosphate. *Biochemistry.*, 49(4):727-735.

Van Holde, K. & Zlatanova, J. (1994). Unusual DNA structures, chromatin and transcription. *Bioessays.*, 16(1):59-68.

Vasquez, K. M., Christensen, J., Li, L., Finch, R. A. & Glazer, P. M. (2002). Human XPA and RPA DNA repair proteins participate in specific recognition of triplex-induced helical distortions. *Proc Natl Acad Sci U S A.*, 99(9):5848-5853.

Voineagu, I., Freudenreich, C. H. & Mirkin, S. M. (2009).Checkpoint responses to unusualstructures formed by DNA repeats. *Mol. Carcinogenesis*, 48:309 - 318.

Voineagu, I., Surka, C. F., Shishkin, A. A., Krasilnikova, M. M. & Mirkin, S. M. (2009). Replisome stalling and stabilization at CGG repeats, which are responsible for chromosomal fragility. *Nat Struct Mol Biol.*, 16(2):226-228.

Wang, G. & Vasquez, K. M. (2004). Naturally occurring H-DNA-forming sequences are mutagenic in mammalian cells. *Proc Natl Acad Sci U S A.*, 101(37):13448-13453.

Wang, G. & Vasquez, K. M. (2006). Non-B DNA structure induced genetic instability. *Mutat Res.*, 598, 103-119

Wang, J. C. (1984). DNA supercoiling and its effects on the structure of DNA. *J Cell Sci.* Suppl., 1:21-29.

Wang, Y. W. & Griffith, J. D. (1996). Methylation of expanded CCG triplet repeat DNA from fragile X syndrome patients enhances nucleosome exclusion. *J Biol Chem.*, 271, 22937-22940.

Watanabe, Y. & Maekawa, M. (2010). Spatiotemporal regulation of DNA replication in the human genome and its association with genomic instability and disease. *Curr Med Chem.*, 17:222-233.

Watson, J. D. & Crick, F. H. C. (1953). Structure for deoxyribose nucleic acid, *Nature*, 171:737 - 738.

Weber, J. L. & Wong, C. (1993). Mutation of human short tandem repeats. *Hum Mol Genet.*, 2(8):1123-1128.

Wells, R. D. (2007). Non-B DNA conformations, mutagenesis and disease. *Trends Biochem Sci.*, 32(6): 271-278.

Wilkins, M. H. F., Stokes, A. R. & Wilson, H. R. (1953a). Molecular structure of DNA, *Nature*,171:738-740.

Wilkins, M. H. F., Seeds, W. E., Stokes, A. R. & Wilson, H. R. (1953b). Helical structure of crystalline DNA, *Nature*, 172: 759 - 762.

Wojciechowska, M., Napierala, M., Larson, J. E. & Wells, R. D. (2006). Non-B DNA conformations formed by long repeating tracts of Myotonic Dystrophy Type 1, Myotonic Dystrophy Type 2, and Friedreich's Ataxia genes, not the sequences per se, promote mutagenesis in flanking regions. *J Biol Chem.*, 281(34):24531-24543.

Yang, W. (2006). Poor base stacking at DNA lesions may initiate recognition by many repair proteins. *DNA Repair* (Amst), 5(6):654-666.

Zahra, R., Blackwood, J. K., Sales, J. & Leach, D. R. F. (2007). Proofreading and secondary structure processing determine the orientation dependence of CAG · CTG trinucleotide repeat instability in *Escherichia coli*. *Genetics.*, 276:27-41.

Zhao, J., Jain, A., Iyer, R. R., Modrich, P. L. & Vasquez, K. M. (2009). Mismatch repair and nucleotide excision repair proteins cooperate in the recognition of DNA interstrand crosslinks. *Nucleic Acids Res.*, 37(13):4420-4429.

Zhao, J., Bacolla, A., Wang, G. & Vasquez, K. M. (2010). Non-B DNA structure-induced genetic instability and evolution. *Cell Mol Life Sci.*, 67:43-62.

# 2

# ATP-Binding Cassette Properties of Recombination Mediator Protein RecF

Sergey Korolev
*Saint Louis University School of Medicine*
*USA*

## 1. Introduction

### 1.1 Recombinational repair

Homologous recombination (HR) is essential for genetic diversity and genome stability. The conserved RecA-like recombinases promote pairing and consequent exchange of fragments between two homologous DNA molecules during conjugation in bacteria and meiotic recombination in eukaryotes. HR is a main DNA repair pathway particularly important in case of large-scale DNA damages, including chromosome or double-stranded (ds) DNA breaks (DSBs) and long single-stranded (ss) DNA gaps (SSGs) (Cox, 1991; Kowalczykowski et al., 1994). The broken chain is paired with the intact DNA, which serves as a template for the synthesis of the damaged DNA. The same recombinases are also involved in the repair and origin-independent restart of stalled DNA replication, a frequently occurring event in every cell (Cox et al., 2000; Kowalczykowski, 2000; Kuzminov, 2001).

HR is initiated by the cooperative binding of RecA recombinase to ssDNA hundreds or thousands nucleotides long forming nucleoprotein filament, a so called presynaptic complex often designated as RecA*. The presynaptic complex can bind homologous dsDNA and exchange a DNA strands. RecA* has multiple activities beyond the strand invasion and exchange (Figure 1). Those include triggering DNA damage SOS response through stimulation of LexA autocleavage (Rehrauer et al., 1996) and activation of UmuD subunit of the error-prone DNA polymerase PolV important for translesion synthesis to bypass small-scale DNA errors (Jiang et al., 2009; Rajagopalan et al., 1992). RecA* was also suggested to stabilize and maintain stalled replication fork during DNA repair (Courcelle et al., 1997). Consequently, RecA binding to DNA is regulated at multiple levels (Cox, 2007).

### 1.2 Recombination mediator proteins

Transient ssDNA regions generated during replication are protected by ssDNA binding proteins like bacterial ssDNA binding (SSB) protein and eukaryotic replication protein A (RPA), which prevent recombinase binding. Under DNA damage conditions, ubiquitous recombination mediator proteins (RMPs) overcome inhibitory effect of SSB and initiate presynaptic complex formation (Fig. 1)(Beernink and Morrical, 1999; Symington, 2002). RMPs are not directly involved in the repair of specific DNA damages, but they regulate initiation of multiple DNA repair pathways and damage response signaling cascades (Courcelle, 2005; Kowalczykowski, 2005; Lee and Paull, 2005; Moynahan et al., 2001;

Williams et al., 2007). In addition to presynaptic complex formation, many RMPs also promote DNA annealing (Luisi-DeLuca and Kolodner, 1994; Sugiyama et al., 1998). The importance of RMPs is reflected by the fact that recombination and repair pathways are often named after specific RMPs, e. g. RecF, RecBC, Rad52 pathways. RMPs include phage UvsY (Sweezy and Morrical, 1999), prokaryotic RecBCD and RecFOR proteins (Fujii et al., 2006; Kolodner et al., 1985; Lloyd and Thomas, 1983; Wang and Smith, 1983), and numerous eukaryotic members (Symington, 2002). Mutations of human RMPs are associated with cancer predisposition, mental retardation, UV-sensitivity and premature aging (Ouyang et al., 2008; Powell et al., 2002; Tal et al., 2009; Thompson and Schild, 2002).

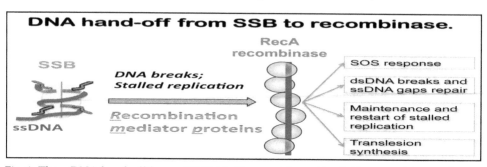

Fig. 1. The ssDNA hand-off from ssDNA binding proteins to RecA-like recombinase triggers multiple DNA damage response pathways important for DNA repair and origin-independent restart of stalled replication. Such DNA transaction is regulated by RMPs.

While ssDNA-binding proteins and RecA-like recombinases are well characterized, the mechanism of RMPs function remains poorly understood. Recent studies revealed a variety of RMPs structural domains. (Koroleva et al., 2007; Lee et al., 2004; Leiros et al., 2005; Makharashvili et al., 2004; Singleton et al., 2002; Yang et al., 2002). The diversity of RMPs structural domains reflects the plethora of different DNA damage response scenarios regulated by these proteins. The focus of this review is prokaryotic RMP RecF. Although a major bacterial recombination repair pathway is named after RecF, the mechanism of RecF activity and even its functional role remains one of the least understood and most controversial issues.

## 2. RecFOR recombination mediators

### 2.1 RecF pathway

The RecF was discovered as an alternative to RecBC pathway in genetic screens based on frequency of conjugation recombination in *E. coli*, and was found to be important for postreplication repair of extended SSGs (Horii and Clark, 1973; Lovett and Clark, 1983; Wang and Smith, 1984). Later, it was shown that *recF* mutants are even more hypersensitive to UV radiation than *RecBC*, that RecF pathway plays a major role in replication restart under UV damage conditions, and that RecF is involved in DSBs repair in the absence of RecBC and SbcBC (Clark, 1991; Courcelle, 2005; Courcelle et al., 1997; Ivancic-Bace et al., 2003; Kidane et al., 2004; Kusano et al., 1989; Whitby and Lloyd, 1995; Zahradka et al., 2006). Sequencing of new genomes revealed the ubiquitous nature of RecF pathway proteins found in most bacteria (Rocha et al., 2005), including the radiation resistant bacteria *Deinococcus radiodurans* (Bentchikou et al., 2010; Cox et al., 2010; Chang et al., 2010; Makarova et al.,

2001). RecF forms an epistatic group with RecO and RecR proteins (Asai and Kogoma, 1994; Courcelle et al., 1997; Courcelle and Hanawalt, 2003; Horii and Clark, 1973; Kolodner et al., 1985; Wang and Smith, 1984). All three proteins are equally important for recombinational repair in most genetic screens, although they do not form triple complex in solution. RecF and RecR genes are often located in DNA replication operons on chromosome, with the exception of extremophiles like *T. thermophiles* and *D. radiodurans* (Ream and Clark, 1983; Ream et al., 1980). In *E. coli*, RecF is co-transcribed with major subunits of replication machinery, DnaA and DnaN (Perez-Roger et al., 1991; Villarroya et al., 1998). RecF pathway proteins share either sequence or structural homology or functional similarities with eukaryotic proteins such as WRN, BLM, RAD52, and BRCA2, which are associated with cancer predisposition and premature aging when mutated (Karow et al., 2000; Kowalczykowski, 2005; Mohaghegh and Hickson, 2001; Yang et al., 2005).

Genetic studies demonstrated that RecF regulates several DNA repair and recombination pathways but is not directly involved in repair of specific DNA damage. For example, in *RecF* mutants DNA lesions are removed with the efficiencies comparable to wild-type cells, while the UV resistance is strongly compromised (Courcelle et al., 1999; Rothman and Clark, 1977). RecF-mediated loading of RecA on ssDNA is required for the maintenance of arrested replication forks, fot the protection and processing of DNA ends to permit DNA repair and replication restart at the site of disruption.

The regulatory role of RecF in replication restart is further supported by examples where RecF impairs cell survival, like in thymine starvation experiments (Nakayama et al., 1982). Another example is revealed by genetic studies of DNA helicases UvrD and Rep (Petit and Ehrlich, 2002). Mutants lacking both helicases are not viable and *RecF* mutations suppress the lethality of the *E. coli Rep/UvrD* double mutant. UvrD helicase disassembles RecA* filaments, the reaction opposite to that of RecFOR, while Rep helicase promotes replication through transcription sites (Boubakri et al., 2010; Centore and Sandler, 2007; Heller and Marians, 2005; Lane and Denhardt, 1975; Veaute et al., 2005). The frequent pausing of the replication fork can potentially stimulate RecF-mediated initiation of RecA* filament formation leading to illegitimate recombination in the absence of UvrD (Mahdi et al., 2006).

## 2.2 Mechanism of RecOR activities

The involvement of all three RecF, -O and -R proteins in HR initiation is well documented by genetic studies. However, the mechanism of their activities in the initiation process remains poorly understood, particularly with respect to RecF. RecO and RecR alone are sufficient to promote formation of the RecA filament on SSB-bound ssDNA (Cox, 2007; Umezu et al., 1993). RecO binds DNA and the C-terminal tail of SSB and these interactions are critical for RecOR function, at least in the absence of RecF (Inoue et al., 2011; Manfredi et al., 2010; Ryzhikov et al., 2011; Sakai and Cox, 2009; Umezu and Kolodner, 1994). In addition, RecO anneals complimentary ssDNA strands protected by cognate SSB (Kantake et al., 2002; Luisi-DeLuca and Kolodner, 1994), resembling the properties of the eukaryotic RMPs, Rad52 and BRCA2 (Grimme et al., 2010; Mazloum et al., 2007; Sugiyama et al., 1998). RecR binds either RecO or RecF (Makharashvili et al., 2009; Umezu and Kolodner, 1994; Webb et al., 1995, 1997). Although *E. coli* RecR does not bind DNA at submillimolar concentrations, it significantly affects DNA binding properties of both RecO and RecF (Kantake et al., 2002; Makharashvili et al., 2009; Webb et al., 1999). RecR inhibits DNA annealing properties of RecO, even though RecOR complex binds both ss- and dsDNA. In addition to initial loading of RecA, RecOR further stimulate homologous recombination by preventing the dissociation of RecA* filament from ssDNA in *E. coli* (Bork et al., 2001). Somewhat different properties were reported for *Bacillus subtilis* RecO, which does not require RecR for initiation of RecA* formation (Manfredi et al., 2008; Manfredi et al., 2010). Crystal structures of all three proteins and of the RecOR complex from *D. radiodurans* have been reported (Koroleva et al., 2007; Lee et al., 2004; Leiros et al., 2005; Makharashvili et al., 2004; Timmins et al., 2007). RecR structure resembles that of a DNA clamp-like tetramer (Lee et al., 2004). However, the role of a potential DNA clamp in RMPs-mediated reaction is unknown. Moreover, in the crystal structure of RecOR complex RecO occupies large portion of the clamp inner space. Such conformation makes it challenging to predict functionally relevant interaction of the complex with DNA. Another intriguing fact is that the crystal structure of RecO did not resemble any structural features of its functional eukaryotic analog Rad52 (Leiros et al., 2005; Makharashvili et al., 2004; Singleton et al., 2002), which supports two identical reactions.

## 2.3 Ambiguities of RecF function

In contrast to genetic data, initial biochemical studies did not reveal the function of RecF in recombination initiation (Umezu et al., 1993). RecF binds both ss- and dsDNA in the presence of ATP, and it is a weak DNA-dependent ATPase (Griffin and Kolodner, 1990; Madiraju and Clark, 1991, 1992). It interacts with RecR in the presence of ATP and DNA (Webb et al., 1999). Surprisingly however, RecF was initially shown to play an inhibitory role during RecOR-mediated loading of RecA on SSB-protected ssDNA (Umezu et al., 1993). The UV-sensitivity of *RecF* mutant can be suppressed by RecOR overexpression, suggesting that RecF plays a regulatory role (Sandler and Clark, 1994). In agreement with this hypothesis, RecF dramatically increases the efficiency of RecOR-mediated RecA loading at ds/ssDNA junctions with a 3′ ssDNA extension under specific conditions (Morimatsu and Kowalczykowski, 2003). RecF was suggested to recognize specific DNA junction structure to direct RecA loading at the boundary of SSGs. While initial experiments demonstrated such a preference (Hegde et al., 1996), later work did not support the binding preference of RecF to DNA junction (Webb et al., 1999). Purified RecF tends to gradually aggregate in solution (Webb et al., 1999). Apparently, nonspecific high molecular weight RecF aggregates interact with DNA resulting in the inhibitory effect of RecF or false positive interactions of RecF with specific DNA substrates (Hegde et al., 1996). In addition, RecFR complex limits the extension of RecA* beyond SSGs, the observation indirectly supporting RecF specificity towards boundaries of SSGs while in complex with other proteins (Webb et al., 1997).

RecF is co-transcribed with the replication initiation protein DnaA and with the β-clamp subunit of DNA polymerase III DnaN. However, its open reading frame is usually shifted by one or two nucleotides relatively to that of DnaN (Villarroya et al., 1998). *E. coli RecF* gene also has multiple rear codons. Thus, expression of RecF is likely to be down regulated at translational level. Consequently, there are only a few copies of RecF in an *E. coli* cell.

How RecF promotes recombination remains an open question. The ability of RecFR complex to limit extension of RecA* filament beyond the SSGs suggests that the RecFR complex may specifically interact with RecA*. However, no direct observation of such interactions has been reported so far. RecF also binds RecX protein (Lusetti et al., 2006). RecX is a negative regulator of presynaptic complex formation, which inhibits filament extension by binding to RecA. RecF scavenges RecX from solution through direct interaction, thus diminishing negative regulatory effect of RecX (Drees et al., 2004; Lusetti et al., 2006). Additional evidence of direct involvement of RecF in the initiation of RecA* filament formation was recently demonstrated in experiments with the SSB mutant lacking conserved C-terminus peptide. This SSB mutant inhibits RecOR-mediated recombination initiation, likely due to lack of interaction of SSB with RecO (Sakai and Cox, 2009). Surprisingly, RecF rescues the RecOR function with this SSB mutant, even on ssDNA plasmids without ds/ssDNA junction.

## 3. Structural studies of RecF

### 3.1 RecF is an ABC ATPase

The amino acid sequence of RecF contains three conserved motifs characteristic of ATP-binding cassette (ABC) ATPases: Walker A, Walker B, and a "signature" motif. Walker A, or P-loop, is a nucleotide binding site found in a variety of ATPases (Walker et al., 1982). Walker B motif provides acidic amino acids important for coordination of a water molecule and a metal ion during the hydrolysis of a triphosphate nucleotide bound to the Walker A motif. The signature motif is a unique feature of ABC ATPases, a diverse family of proteins ranging from membrane transporters to DNA-binding proteins (review in (Hopfner and Tainer, 2003). ATP-dependent dimerization is a common feature of this class of proteins. Signature motif residues interact with the nucleotide bound to an opposite monomer (Hopfner et al., 2000). This motif is important for both ATP-dependent dimerization and subsequent ATP hydrolysis. ABC ATPases are not motor proteins and utilize ATP binding and hydrolysis as a switch or sensor mechanism, regulating diverse signaling pathways and reactions.

DNA-binding ABC ATPases include DNA mismatch and nucleotide excision repair enzymes (Ban and Yang, 1998; Junop et al., 2001; Obmolova et al., 2000; Tessmer et al., 2008), structural maintenance of chromosome (SMC) proteins cohesin and condensin (Strunnikov, 1998), and DSBs repair enzyme Rad50 (Hirano et al., 1995). SMCs and Rad50 are characterized by the presence of a long coiled-coil structural domain inserted between N- and C-terminal halves of the globular head domain (Haering et al., 2002). RecF lacks a coiled-coil region, but it does exhibit an ATP-dependent DNA binding and a slow DNA-dependent ATP hydrolysis activity (Hegde et al., 1996; Madiraju and Clark, 1992; Webb et al., 1995). However, the SMC-like properties of RecF and their role in recombinational repair have not been addressed. Previously, only Walker A motif has been shown to be critical for RecF function (Sandler et al., 1992; Webb et al., 1999). All known ABC-type ATPases function as a heterooligomeric complexes in which a sequence of inter- and intra-molecular interactions is triggered by the ATP-dependent dimerization and the dimer-dependent ATP hydrolysis (Deardorff et al., 2007; Dorsett, 2011; Hopfner and Tainer, 2003; Junop et al., 2001; Moncalian et al., 2004; Smith et al., 2002). Thus, RecF may function in recombination initiation through a multistep pathway of protein-protein and DNA-protein interactions regulated by ATP-dependent RecF dimerization.

### 3.2 Structural similarity of RecF with Rad50 head domain

The diversity of ABC ATPases makes it difficult to predict to which subfamily RecF belongs to based on sequence comparison. RecF is a globular protein lacking long coiled-coil domains of Rad50 and SMC proteins. However, it does not have significant sequence similarity beyond three major motifs with globular DNA binding proteins like MutS. We crystalized and solved a high resolution structure of RecF from *D. radiodurans* (DrRecF) (Fig.

2) (Koroleva et al., 2007). The structure was solved with resolution of 1.6 Å using native and selenomethionine protein derivative crystals. The structure is comprised of two domains. The ATPase domain I is formed by two β-sheets wrapped around central α-helix A and is similar to the corresponding subdomain of the Rad50 head domain (Figure 2, right). Structures of nucleotide-binding domains are similar in all ABC ATPases. In contrast, structure of subdomain containing signature motif (Lobe II in Rad50) is highly diverse among even DNA binding ABC ATPases. However, all structural elements presented in RecF domain II are present in Rad50 Lobe II subdomain and these domains are structurally more similar than ATP-binding domains. The only difference is two long α-helixes of RecF which are connected at the apical part of this "arm-like" domain. In Rad50 analogous α-helixes are extended into an extremely long coiled-coil structure, absent in RecF. High degree of structural similarity unequivocally puts RecF in the same family together with Rad50 and SMC proteins. Therefore, RecF represents the only known globular protein with a structure highly homologous to that of the head domains of Rad50, cohesin and condensin.

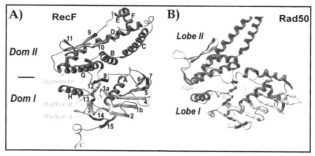

Fig. 2. Cartoon representation of **A**) RecF and **B**) Rad50 head domain structures. α-helices are shown in red and β-sheets in yellow. In RecF, α-helices are lettered and β -strands are numbered. Walker A, B, and signature motifs are highlighted in green and labeled. In RecF, ATP-binding domain is designated as Domain I and signature motif domain as Domain II. In Rad50 corresponding domains are referred as Lobe I and Lobe II subdomains.

### 3.3 The model of ATP-dependent dimer suggests mechanism of DNA binding

RecF was crystallized as a monomer. ATP-dependent dimer was modeled based on known intersubunit interactions conserved in ABC ATPases and, specifically, based on a known structure of Rad50 dimer (Fig. 3)(Hopfner et al., 2000). In all proteins of this family, a conserved serine of the signature motif interacts with a γ-phosphate group of ATP. The ATP bound to Walker A motif was modeled accordingly to its highly conserved conformation in all Walker A and B containing structures. These constrains unambiguously dictate a single conformation of the potential RecF dimer (Fig. 3A). The model suggests a potential DNA binding site located on the top of two nucleotide-binding domains, in a conformation similar to the proposed DNA binding site of Rad50 (Figs. 3B-D). The resulting RecF dimer forms a semi-clamp or a symmetrical crab-claw with two arms extending in the directions similar to those of coiled–coil regions of Rad50 dimer (Hopfner et al, 2001). The claw structure contains sufficient space to accommodate and cradle dsDNA. In this model, the majority of conserved residues map to the dimerization interface and pocket region of the claw, where DNA binding is expected to occur.

The proposed model explains an ATP-dependence of RecF DNA binding. First, it is an acidic protein with mostly negatively charged surface area. In the model of an ATP-dependent dimer, small patches of positively charged surface area are aligned on the top of the dimer, creating the extended basic surface area. Second, the arms of domain II form a deep cleft, sufficient to engulf a DNA helix. The constrains of a signature motif interaction with a γ-phosphate group of ATP does not allow to alter the distance between these arms in the model without significant structural clashes of surface exposed residues of the two monomers. Thus, the ATP-dependent dimerization leads to favorite juxtaposition of the surface charges and to surface complementarity, which stimulate DNA binding.

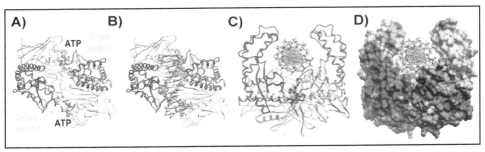

Fig. 3. A model of RecF dimer. **A)** Domains I and II of one RecF monomer are color-coded in yellow and orange, and of the other monomer in grey and blue. Signature motif residues are shown by stick representation in cyan and ATP by stick representation with nitrogen, oxygen, carbon and phosphate atoms are colored in blue, red, yellow and orange, correspondingly. **B)** The same dimer representation with bound dsDNA shown by stick representation in green. **C)** Orthogonal view of the dimer shown in B). **D)** Surface representation of DrRecF dimer in same orientation as in C) color-coded according to the surface electrostatic potential.

Proving ATP-dependent dimerization of RecF in solution was quite challenging due to poor solubility and a tendency of RecF to form nonspecific soluble aggregates (Webb et al., 1999). Initial attempts with size exclusion chromatography (SEC) yielded the monomeric form of E. coli RecF in the presence of ATP (Webb et al., 1999). The caveat of such experiment is in low protein solubility, when only solution with limited protein concentration can be run through column, and in a non-equilibrium nature of SEC, which may lead to dissociation of weak dimers. Later, it was shown that DrRecF nonspecifically interacts with the column resin even in a 1M KCl buffer (Koroleva et al., 2007). Therefore, a combination of SEC with static light scattering was utilized to determine the true molecular weight of eluted fractions. DrRecF does form an ATP-dependent dimer, though relatively unstable, which could dissociate on the column under non-equilibrium conditions at low protein concentration. The dimerization of wild type protein and specific mutants under equilibrium conditions was tested with a dynamic light scattering (DLS). DrRecF dimerizes only in the presence of ATP but not with ADP. Mutation of signature motif S276R resulted in lack of dimerization, as well as mutation of Walker motif A K39M, which prevents ATP binding. Walker A motif mutant K39R which binds, but does not hydrolyses ATP, forms dimer as well as mutants of Walker B motif D300N. Surprisingly, non-hydrolizable ATP analogs did not support dimerization in initial experiments, suggesting that RecF dimerization is highly sensitive to specific ATP-bound conformation. While DLS method is not suitable for quantitative analysis, it is highly sensitive to the presence of high molecular weight protein aggregates, and it was utilized to optimize RecF solution conditions for other experiments.

## 4. Functional significance of ABC-type ATPase properties of RecF

### 4.1 ATP-dependent dimerization is required for DNA binding

The DNA binding properties of RecF and their role in recombination initiation remain poorly understood and controversial. Different publications presented contradicting results of DNA junction recognition by RecF (Hegde et al., 1996; Webb et al., 1999). RecR was shown to stabilize ATP-dependent interaction of RecF with DNA. However, RecR also stimulated ATP hydrolysis, which theoretically should lead to destabilizing of RecF complex with DNA (Webb et al., 1995). Therefore, multiple complimentary equilibrium binding techniques were utilized to comprehensively address the relationship between dimerization, DNA binding and ATP binding and hydrolysis (Makharashvili et al., 2009). Quantitative characterization of RecF dimerization was performed using Föster (or Fluorescence) Resonance Energy Transfer (FRET) technique with a mixture of Cy3- and FAM(fluorescein)-labeled DrRecF (Fig. 4). The cysteine substitutions were introduced either at a topical part of domain II arm or at the C-terminal tail to crosslink DrRecF with fluorophores. The labeling of domain II interfered with DNA-binding (Makharashvili, 2009), indirectly confirming the dimer model presented in Fig. 3, where apical parts of domain II arms are situated close to each other in the dimer and the presence of bulky polar fluorophores may interfere with DNA binding. C-terminally labeled protein (A355C) was fully functional. Apparent dimerization constant of $L_d = 0.15 \pm 0.02$ µM was calculated from the plot of FRET signal versus DrRecF concentration (Fig. 4C). Alternatively, multiple data sets (Fig. 4B) were globally fitted into a two-step reaction model consisting of the ATP-binding and dimerization processes resulting in a dimerization constant of $L_d = 0.13 \pm 0.02$ µM and an ATP-binding constant of $K_d^{ATP} = 13 \pm 2$ µM.

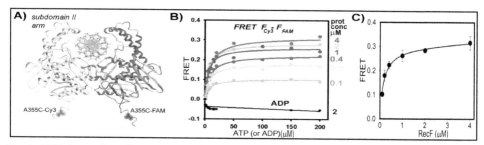

Fig. 4. ATP-dependent dimerization of DrRecF. **A)** Location of cysteines (A355C) are shown by magenta spheres on the model of DrRecF dimer with one monomer is colored in yellow and the other color-coded accordingly to its secondary structure elements with α-helixes in red and β-strands in green. The DNA is shown in cyan. **B)** Titration of labeled DrRecF by ATP. Different isotherms represent different concentration of DrRecF in solution (values are shown on the right). The black isotherm corresponds to titration of 2 µM DrRecF by ADP. **C)** A plot of maximal FRET signal versus DrRecF concentrations.

The DNA binding was first assayed using short FAM-labeled oligonucleotides with the fluorescent polarization anisotropy method (Fig. 5). To address initial DNA binding rate, reactions were performed for a relatively short time (10-15 min) and with the excess of ATP, taking an advantage of RecF being a slow ATPase (Fig. 6C, below). Alternatively, the rate of ATP hydrolysis was measured over 1 or 2 hours time upon titration of RecF by different DNA oligonucleotides (Fig. 6B). The binding of all DNA substrates was relatively weak with the apparent dissociation constants greater than 15 μM (Fig. 5). Neither a wild type DrRecF in the presence of ADP nor a signature motif mutant S279R in the presence of ATP were able to bind DNA (Fig. 5), suggesting that the ATP-dependent dimerization is essential for RecF interaction with all DNA substrates.

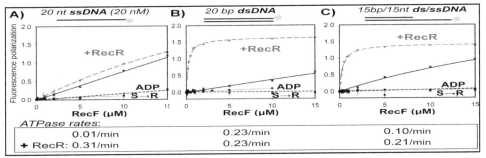

Fig. 5. ATP-dependent binding of DrRecF to different DNA substrates (top) and DNA-dependent ATP hydrolysis rates (bottom). DNA substrates are schematically represented above each plot with **A)** ssDNA, **B)** dsDNA and **C)** ds/ssDNA junction. Solid isotherms correspond to binding in the presence of ATP, dashed black – in the presence of ADP, dotted – to the binding of signature motif mutant S279R in the presence of ATP. Red isotherms correspond to DrRecF binding in the presence of ATP and 50 μM DrRecR. The maximum estimated ATP hydrolysis rates of DrRecF (Fig. 6A) are shown at the bottom with the top lane corresponding to reactions without DrRecR and the bottom – with RecR. DrRecF concentration is 10 μM, DNA- 20 nM, ATP – 2 mM.

## 4.2 RecR-dependent DNA specificity of RecF

DNA binding of DrRecF is drastically alters in the presence of DrRecR (red isotherms in Fig. 5). DrRecR significantly increases the affinity of DrRecF to dsDNA (Fig. 5B) with the estimated association binding constant at least two orders of magnitude stronger than without DrRecR. DrRecR does not alter DrRecF ssDNA binding according to the DNA binding assay. However, the ATPase assay clearly demonstrated interaction of DrRecR with DrRecF in the presence of ssDNA. ssDNA does not stimulate ATP hydrolysis by DrRecF, while the presence of both DrRecR and ssDNA results in strongest ATPase rate. This suggests that DrRecR stimulates the ATPase rate of DrRecF bound to ssDNA, potentially destabilizing dimerization and ssDNA binding. In case of dsDNA, maximum ATPase rates were similar with and without DrRecR. Therefore, DrRecR stabilizes DrRecF complex with dsDNA without increasing its ATPase rate. Due to this stabilization effect of RecR, we are able to measure DNA binding and dimerization of DrRecF in the presence of ATP analogs (Fig. 6B). Curiously, a weak dimerization is observed at highest DrRecF concentration even in the presence of ADP. Therefore, DrRecR selectively stimulates binding of DrRecF dimer to dsDNA, while potentially destabilizing DrRecF complex with ssDNA. Both dimerization and DNA binding reactions were also measured as a function of time to verify that under these conditions ATP hydrolysis does not significantly alter either interaction within first 10 minutes (Fig. 6C).

DrRecR is characterized by a weak DNA binding affinity in a millimolar range, while binding of *E. coli* RecR to DNA was not detected. DrRecR forms a tetrameric DNA clamp-like structure (Lee et al., 2004). This conformation is likely to be conserved for other RecR homologs since *E. coli* RecR is either a dimer or tetramer in solution (Umezu et al., 1993), and *H. influenzae* RecR also was crystallized in a similar tetrameric conformation (Koroleva, O., Baranova, E., Korolev, S. unpublished data). One way to explain the DNA-dependent interaction of RecR with RecF is through the binding of both proteins to a shared DNA substrate, as beads on a string. Moreover, since dimer to tetramer transition was proposed as a clamp loading mechanism (although not confirmed), the ATP-dependent dimerization of RecF may stimulate such loading of RecR clamp on DNA. To test the hypothesis of shared DNA substrate requirement for RecF interaction with RecR, the RecR-stimulated DNA binding of RecF and the ATPase rate were tested in the presence of different length dsDNA substrates. Surprisingly, 10 bp short oligonucleotide stimulates DrRecF interaction with DrRecR. Structural modeling suggests that RecF dimer can bind 12-15 bp long DNA, while RecR clamp may cover up to 8-12 bp. These results rule out the beads-on-a-string model of RecFR binding to dsDNA. Alternatively, RecR may interact with the domain II arms encircling RecF bound DNA in a model similar to that of Rad50/Mre11 complex (Hopfner et al., 2001; Lammens et al., 2011; Williams et al., 2011).

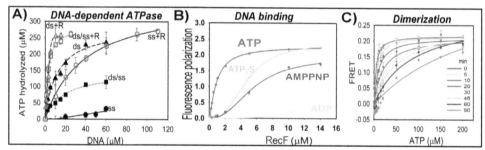

Fig. 6. **A)** ATP hydrolysis by DrRecF over 120 min was measured upon titration by different DNA substrates, with circles corresponding to ssDNA, triangles to dsDNA, and squares to ds/ssDNA. Red symbols correspond to titrations in the presence of 50 μM RecR. Concentration of RecF is 10 μM, and ATP 2 mM. **B)** dsDNA binding by RecFR in the presence of ATP analogs measured with the fluorescence polarization assay performed similarly to that in Fig. 5 with the following nucleotides: ATP (red), ATPγS (green), AMPPNP (blue), and ADP (cyan). **C)** Time dependence of RecF dimerization upon titration with ATP as measured by FRET of labeled RecF. Isotherms of different colors correspond to the FRET value at different time points shown on the right.

### 4.3 The lack of ss/dsDNA junction specificity

The steps of RecF interaction with DNA and RecR are schematically represented in Fig. 7. ATP binding stimulates RecF dimerization, essential for binding of all DNA substrates. The DNA-bound RecF dimer interacts with RecR, which either stabilizes the complex with dsDNA or destabilizes with ssDNA. Importantly, neither of the performed assays revealed any specificity of RecF and RecFR complex for ss/dsDNA junction. Both DNA binding and

ATPase rates had an average between ss- and dsDNA substrates values. Although all data were obtained with *D. radiodurans* proteins, RecF and RecR are highly homologous proteins. Moreover, *E. coli* RecR stimulates DNA binding of *D. radiodurans* RecF similarly to that of *D. radiodurans* RecR suggesting that DrRecF binds both Dr- and *E. coli* RecR proteins with similar affinities (Makharashvili, 2009). Therefore, the described above properties of *D. radiodurans* proteins are likely to be conserved for *E. coli* homologs. While DNA binding and ATPase assays did not reveal specificity of RecF towards DNA junction, functional studies clearly evidence the role of RecF at ss/dsDNA junction (Chow and Courcelle, 2004; Handa et al., 2009; McInerney and O'Donnell, 2007; Morimatsu and Kowalczykowski, 2003; Webb et al., 1997). The potential specificity of RecF to ds/ssDNA junction is likely to require additional protein partners of recombination initiation reaction including SSB, RecO and RecA. For example, RecR can be recruited to SSB-bound ssDNA while in complex with RecO (Ryzhikov et al., 2011). The increased local concentration of RecR on SSB-coated ssDNA may subsequently stimulate RecF interaction with the adjacent dsDNA region.

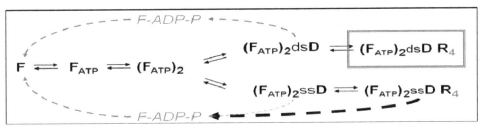

Fig. 7. Schematic representation of RecF interaction with ATP, ATP-dependent dimerization, DNA binding, and the effect of RecR on DNA binding and ATP hydrolysis. The complex formed on dsDNA in the presence of RecR (red box) is the most stable intermediate. In case of *D. radiodurans* homologs, RecF dimer interacts with RecR tetramer.

### 4.4 In vivo function of RecF conserved motifs

The role of RecF SMC motifs *in vivo* was initially addressed with *E. coli* RecF mutant cells transformed with RecF-containing vector (Koroleva et al., 2007). Only wild type RecF complemented the UV sensitivity of a *recF* cells. Mutations of Walker A, -B and signature motifs did not restore the UV resistance. Since the overexpression of RecF can potentially affect its function, similar mutants of *RecF* were constructed in chromosome (Michel-Marks et al., 2010). Importantly, different steps of RecFOR function were tested with each mutant. Those include the rate of DNA synthesis, degradation of nascent DNA, the presence of DNA intermediates, and cell survival upon UV irradiation. Mutants included Walker A motif K36M, deficient in ATP binding, a Walker A motif K36R and a Walker B D303N, which both retain ATP binding but are deficient in ATP hydrolysis, and two signature motif mutants S270R and Q273A, which prevent an ATP-dependent dimerization.

Following the UV-induced arrest of replication, the nascent DNA is partially degraded at the replication fork by RecQ helicase and RecJ nuclease and RecF limits such degradation (Courcelle and Hanawalt, 1999). The degree of nascent DNA degradation was measured with pulse labeling of growing cell culture with [$^{14}$C]thymine and [$^{3}$H]thymidine. Similarly to a null mutant(Courcelle and Hanawalt, 1999), approximately 50% of nascent DNA was degraded with all mutants with the exception of D303N, where degradation was less severe.

Therefore, all steps of the dynamic interactions of RecF with ATP and DNA are important for the very first step of RecFOR function in replication repair. The weak functionality of D303N can be explained by a potential residual ATPase activity of this mutant, as shown for other SMC proteins (Lammens et al., 2004). Experiments with ATP analogs (Fig. 6B) demonstrated that even minor conformational changes significantly affect RecF properties. Therefore, an alternative explanation may be that D303N mutant introduces the least conformational distortion at the ATP-binding site and may retain conformation of a wild type wild type dimer and DNA-binding activities better than K36R mutant.

The rate of DNA synthesis is reduced by approximately 90% immediately after UV irradiation, but is recovered to nearly initial rate within 100 min in wild-type cells. The overall accumulation of DNA is increased at that time approaching the level of unirradiated cells. In *recF* cells the initial reduction of DNA synthesis rate is similar, but there is no recovery. Like in the previous assay, all mutants with exception of D303N were similar to the null mutant. D303N mutant did support slight recovery of DNA replication rate, yet it was significantly weaker than that of a wild type. RecF is associated with appearance of specific replication intermediates during DNA damage, as visualized on two-dimensional agarose gel (Courcelle et al., 2003). In this assay, all mutants were equally deficient in accumulation of such intermediates similarly to the null mutant, although the detection level of this assay may not be sufficient to reveal weak activity of D303N mutant. Finally, the survival rate of cell culture after UV irradiation was assayed. D303N mutant was partially resistant, while all other mutants were as hypersensitive to UV irradiation as deletion of *recF*. These studies demonstrate that all steps of ATP binding, dimerization and hydrolysis by RecF are important to maintain stalled replication and to restart cell growth after DNA damage.

## 5. Conclusions

RecFOR proteins regulate RecA binding to ssDNA under DNA damage conditions. This reaction initiates a variety of DNA repair pathways including maintenance and restart of stalled replication. Correspondingly, recombinational repair is tightly regulated in cell. While the exact role and mechanism of RecF in these pathways remain controversial, the majority of known data suggest a regulatory function of RecF during initiation and subsequent steps of recombinational DNA repair. Intricate properties of the ATP-dependent interaction of RecF with DNA and of the DNA-dependent ATP hydrolysis as well as the dependence of these interactions on RecR strongly supports this hypothesis.

Regulatory function is further reinforced by the sequence and structural homology with the head domain of Rad50 and SMC proteins. Rad50 is involved in multiple steps of DNA damage response including initial detection of DSBs, triggering of cell signaling cascades, and in resection of dsDNA to create 3' ssDNA tail for recombinase binding (Nicolette et al., 2010). In bacteria, RecF is likely to be involved in multiple steps of replication restart as well, including initial detection of replication arrest. Neither Rad50 nor RecF specifically recognizes functionally relevant DNA substrates, blunt-end DNA and ss/dsDNA junction, correspondingly (de Jager et al., 2002). Rad50 functions in complex with other DNA binding proteins, including Mre11 nuclease, and protein-protein interactions regulate DNA binding and ATPase activities (Lammens et al., 2011; Lim et al., 2011; Williams et al., 2011). By analogy, we can speculate that ATP binding and hydrolysis may not simply control DNA binding and dissociation of RecF, but also regulate binding of RecF dimer to different protein partners. For example, the ability of short DNA fragments to promote RecR binding suggests that the DNA-dependent conformational changes of RecF are important for protein-protein recognition rather than simple binding to the shared DNA substrate.

It is important to note that RecF does not represent the exact analog of Rad50. It is a much smaller protein without long coiled-coil structures. RecF does not support DNA unwinding or resection, as well as additional adenylate kinase activity of Rad50 and SMC proteins (Bhaskara et al., 2007; Lammens and Hopfner, 2010). Instead, it is involved in the initiation of the presynaptic complex formation, the function performed by BRCA2 or Rad52 in eukaryotes (Moynahan et al., 2001; New et al., 1998; Shinohara and Ogawa, 1998; Sung, 1997; Yang et al., 2005). While Rad52 is rather unique protein (Singleton et al., 2002), structural and functional motifs of BRCA2 resemble that of RecFOR system (Yang et al., 2002). BRCA2 interacts with ssDNA through OB-fold domain, similarly to RecO, and has a putative dsDNA-binding domain. The latter function is likely to be performed by RecF, even with the lack of structural similarity.

RecF regulates RecQJ-dependent resection of nascent DNA at stalled replication fork (Courcelle and Hanawalt, 1999). This step occurs prior to RecA loading and initiation of SOS response. How RecF recognizes stalled replication remains unknown. It is tempting to speculate that RecF is a part of replisome (Kogoma, 1997) based on co-translation of RecF with replication initiation protein DnaA and polymerase subunit DnaN and on its early involvement in detection of replication arrest. However, no interactions of RecF with replication proteins have been identified so far. RecF may represent an alternative to PriA pathway of replication restart in case of arrested replication or postreplication repair (Sandler, 1996). Thus, it is important to find additional RecF-binding proteins. The detection of novel interactions is problematic due to low copy number of RecF in cells and poor solubility of purified RecF. The potential requirement of ATP- and DNA-dependent dimerization for RecF interaction with other proteins further complicates the search for interacting proteins.

The relationship of specific steps of ATP-dependent reactions with the DNA damage recognition and processing by RecF and Rad50 remains elusive. Since RecF is the smallest known DNA-biding ABC ATPase composed of the head domain only, it represents an excellent model system to address the role of allosteric regulations, governing function of this class of proteins. Importantly, both ATP binding and hydrolysis are likely to play an important mechanistic role in most of reactions (Fig. 7). For example, the first step of limiting degradation of nascent lagging DNA by RecQJ and loading of RecA may only require formation of a stable RecF dimer at DNA junction, while ATP hydrolysis and dimer dissociation may be important for the following steps. However, the involvement of all the conserved motifs to prevent degradation of nascent DNA suggests that both ATP binding and hydrolysis are important even for this initial step. Therefore, all steps of RecF function in DNA repair are likely to depend on dynamic interactions of RecF with ATP, DNA and DNA repair proteins. Delineating molecular basis and principles of these interactions is essential for understanding fundamental mechanisms of DNA repair, recombination and replication.

## 6. Acknowledgment

The research was supported by National Institutes of Health grant GM073837. The author is grateful for help during manuscript preparation to Mikhail Ryzhikov and Ekaterina Los. Significant portion of RecF studies were performed by graduate student Nodar Makharashivili.

## 7. References

Asai, T., and Kogoma, T. (1994). The RecF pathway of homologous recombination can mediate the initiation of DNA damage-inducible replication of the Escherichia coli chromosome. J Bacteriol 176, 7113-7114.

Ban, C., and Yang, W. (1998). Crystal structure and ATPase activity of MutL: implications for DNA repair and mutagenesis. Cell 95, 541-552.

Beernink, H.T., and Morrical, S.W. (1999). RMPs: recombination/replication mediator proteins. Trends Biochem Sci 24, 385-389.

Bentchikou, E., Servant, P., Coste, G., and Sommer, S. (2010). A major role of the RecFOR pathway in DNA double-strand-break repair through ESDSA in Deinococcus radiodurans. PLoS Genet 6, e1000774.

Bhaskara, V., Dupre, A., Lengsfeld, B., Hopkins, B.B., Chan, A., Lee, J.H., Zhang, X., Gautier, J., Zakian, V., and Paull, T.T. (2007). Rad50 adenylate kinase activity regulates DNA tethering by Mre11/Rad50 complexes. Mol Cell 25, 647-661.

Bork, J.M., Cox, M.M., and Inman, R.B. (2001). The RecOR proteins modulate RecA protein function at 5' ends of single- stranded DNA. Embo J 20, 7313-7322.

Boubakri, H., de Septenville, A.L., Viguera, E., and Michel, B. (2010). The helicases DinG, Rep and UvrD cooperate to promote replication across transcription units in vivo. Embo J 29, 145-157.

Chang, X., Yang, L., Zhao, Q., Fu, W., Chen, H., Qiu, Z., Chen, J.A., Hu, R., and Shu, W. (2010). Involvement of recF in 254 nm ultraviolet radiation resistance in Deinococcus radiodurans and Escherichia coli. Curr Microbiol 61, 458-464.

Centore, R.C., and Sandler, S.J. (2007). UvrD limits the number and intensities of RecA-green fluorescent protein structures in Escherichia coli K-12. J Bacteriol 189, 2915-2920.

Chow, K.H., and Courcelle, J. (2004). RecO acts with RecF and RecR to protect and maintain replication forks blocked by UV-induced DNA damage in Escherichia coli. J Biol Chem 279, 3492-3496.

Clark, A.J. (1991). rec genes and homologous recombination proteins in Escherichia coli. Biochimie 73, 523-532.

Courcelle, J. (2005). Recs preventing wrecks. Mutat Res 577, 217-227.

Courcelle, J., Carswell-Crumpton, C., and Hanawalt, P.C. (1997). recF and recR are required for the resumption of replication at DNA replication forks in Escherichia coli. Proc Natl Acad Sci U S A 94, 3714-3719.

Courcelle, J., Crowley, D.J., and Hanawalt, P.C. (1999). Recovery of DNA replication in UV-irradiated Escherichia coli requires both excision repair and recF protein function. J Bacteriol 181, 916-922.

Courcelle, J., Donaldson, J.R., Chow, K.H., and Courcelle, C.T. (2003). DNA damage-induced replication fork regression and processing in Escherichia coli. Science 299, 1064-1067.

Courcelle, J., and Hanawalt, P.C. (1999). RecQ and RecJ process blocked replication forks prior to the resumption of replication in UV-irradiated Escherichia coli. Mol Gen Genet 262, 543-551.

Courcelle, J., and Hanawalt, P.C. (2003). RecA-dependent recovery of arrested DNA replication forks. Annu Rev Genet 37, 611-646.

Cox, M.M. (1991). The RecA protein as a recombinational repair system. Mol Microbiol 5, 1295-1299.

Cox, M.M. (2007). Regulation of bacterial RecA protein function. Crit Rev Biochem Mol Biol 42, 41-63.

Cox, M.M., Goodman, M.F., Kreuzer, K.N., Sherratt, D.J., Sandler, S.J., and Marians, K.J. (2000). The importance of repairing stalled replication forks. Nature 404, 37-41.

Cox, M.M., Keck, J.L., and Battista, J.R. (2010). Rising from the Ashes: DNA Repair in Deinococcus radiodurans. PLoS Genet 6, e1000815.

de Jager, M., Wyman, C., van Gent, D.C., and Kanaar, R. (2002). DNA end-binding specificity of human Rad50/Mre11 is influenced by ATP. Nucleic Acids Res 30, 4425-4431.

Deardorff, M.A., Kaur, M., Yaeger, D., Rampuria, A., Korolev, S., Pie, J., Gil-Rodriguez, C., Arnedo, M., Loeys, B., Kline, A.D., et al. (2007). Mutations in cohesin complex members SMC3 and SMC1A cause a mild variant of cornelia de Lange syndrome with predominant mental retardation. Am J Hum Genet 80, 485-494.

Dorsett, D. (2011). Cohesin: genomic insights into controlling gene transcription and development. Curr Opin Genet Dev 21, 199-206.

Drees, J.C., Lusetti, S.L., and Cox, M.M. (2004). Inhibition of RecA protein by the Escherichia coli RecX protein: modulation by the RecA C terminus and filament functional state. J Biol Chem 279, 52991-52997.

Fujii, S., Isogawa, A., and Fuchs, R.P. (2006). RecFOR proteins are essential for Pol V-mediated translesion synthesis and mutagenesis. Embo J 25, 5754-5763.

Griffin, T.J.t., and Kolodner, R.D. (1990). Purification and preliminary characterization of the Escherichia coli K-12 recF protein. J Bacteriol 172, 6291-6299.

Grimme, J.M., Honda, M., Wright, R., Okuno, Y., Rothenberg, E., Mazin, A.V., Ha, T., and Spies, M. (2010). Human Rad52 binds and wraps single-stranded DNA and mediates annealing via two hRad52-ssDNA complexes. Nucleic Acids Res 38, 2917-2930.

Haering, C.H., Lowe, J., Hochwagen, A., and Nasmyth, K. (2002). Molecular architecture of SMC proteins and the yeast cohesin complex. Mol Cell 9, 773-788.

Handa, N., Morimatsu, K., Lovett, S.T., and Kowalczykowski, S.C. (2009). Reconstitution of initial steps of dsDNA break repair by the RecF pathway of E. coli. Genes Dev 23, 1234-1245.

Hegde, S.P., Rajagopalan, M., and Madiraju, M.V. (1996). Preferential binding of Escherichia coli RecF protein to gapped DNA in the presence of adenosine (gamma-thio) triphosphate. J Bacteriol 178, 184-190.

Heller, R.C., and Marians, K.J. (2005). Unwinding of the nascent lagging strand by Rep and PriA enables the direct restart of stalled replication forks. J Biol Chem 280, 34143-34151.

Hirano, T., Mitchison, T.J., and Swedlow, J.R. (1995). The SMC family: from chromosome condensation to dosage compensation. Curr Opin Cell Biol 7, 329-336.

Hopfner, K.P., Karcher, A., Craig, L., Woo, T.T., Carney, J.P., and Tainer, J.A. (2001). Structural biochemistry and interaction architecture of the DNA double-strand break repair Mre11 nuclease and Rad50-ATPase. Cell 105, 473-485.

Hopfner, K.P., Karcher, A., Shin, D.S., Craig, L., Arthur, L.M., Carney, J.P., and Tainer, J.A. (2000). Structural biology of Rad50 ATPase: ATP-driven conformational control in DNA double-strand break repair and the ABC-ATPase superfamily. Cell *101*, 789-800.

Hopfner, K.P., and Tainer, J.A. (2003). Rad50/SMC proteins and ABC transporters: unifying concepts from high-resolution structures. Curr Opin Struct Biol *13*, 249-255.

Horii, Z., and Clark, A.J. (1973). Genetic analysis of the recF pathway to genetic recombination in Escherichia coli K12: isolation and characterization of mutants. J Mol Biol *80*, 327-344.

Inoue, J., Nagae, T., Mishima, M., Ito, Y., Shibata, T., and Mikawa, T. (2011). A mechanism for single-stranded DNA-binding protein (SSB) displacement from single-stranded DNA upon SSB-RecO interaction. J Biol Chem *286*, 6720-6732.

Ivancic-Bace, I., Peharec, P., Moslavac, S., Skrobot, N., Salaj-Smic, E., and Brcic-Kostic, K. (2003). RecFOR Function Is Required for DNA Repair and Recombination in a RecA Loading-Deficient recB Mutant of Escherichia coli. Genetics *163*, 485-494.

Jiang, Q., Karata, K., Woodgate, R., Cox, M.M., and Goodman, M.F. (2009). The active form of DNA polymerase V is UmuD'(2)C-RecA-ATP. Nature *460*, 359-363.

Junop, M.S., Obmolova, G., Rausch, K., Hsieh, P., and Yang, W. (2001). Composite active site of an ABC ATPase: MutS uses ATP to verify mismatch recognition and authorize DNA repair. Mol Cell *7*, 1-12.

Kantake, N., Madiraju, M.V., Sugiyama, T., and Kowalczykowski, S.C. (2002). Escherichia coli RecO protein anneals ssDNA complexed with its cognate ssDNA-binding protein: A common step in genetic recombination. Proc Natl Acad Sci U S A *99*, 15327-15332.

Karow, J.K., Wu, L., and Hickson, I.D. (2000). RecQ family helicases: roles in cancer and aging. Curr Opin Genet Dev *10*, 32-38.

Kidane, D., Sanchez, H., Alonso, J.C., and Graumann, P.L. (2004). Visualization of DNA double-strand break repair in live bacteria reveals dynamic recruitment of Bacillus subtilis RecF, RecO and RecN proteins to distinct sites on the nucleoids. Mol Microbiol *52*, 1627-1639.

Kogoma, T. (1997). Is RecF a DNA replication protein? Proc Natl Acad Sci U S A *94*, 3483-3484.

Kolodner, R., Fishel, R.A., and Howard, M. (1985). Genetic recombination of bacterial plasmid DNA: effect of RecF pathway mutations on plasmid recombination in Escherichia coli. J Bacteriol *163*, 1060-1066.

Koroleva, O., Makharashvili, N., Courcelle, C.T., Courcelle, J., and Korolev, S. (2007). Structural conservation of RecF and Rad50: implications for DNA recognition and RecF function. Embo J *26*, 867-877.

Kowalczykowski, S.C. (2000). Initiation of genetic recombination and recombination-dependent replication. Trends Biochem Sci *25*, 156-165.

Kowalczykowski, S.C. (2005). Cancer: catalyst of a catalyst. Nature *433*, 591-592.

Kowalczykowski, S.C., Dixon, D.A., Eggleston, A.K., Lauder, S.D., and Rehrauer, W.M. (1994). Biochemistry of homologous recombination in Escherichia coli. Microbiol Rev 58, 401-465.

Kusano, K., Nakayama, K., and Nakayama, H. (1989). Plasmid-mediated lethality and plasmid multimer formation in an Escherichia coli recBC sbcBC mutant. Involvement of RecF recombination pathway genes. J Mol Biol 209, 623-634.

Kuzminov, A. (2001). DNA replication meets genetic exchange: chromosomal damage and its repair by homologous recombination. Proc Natl Acad Sci U S A 98, 8461-8468.

Lammens, A., and Hopfner, K.P. (2010). Structural basis for adenylate kinase activity in ABC ATPases. J Mol Biol 401, 265-273.

Lammens, A., Schele, A., and Hopfner, K.P. (2004). Structural biochemistry of ATP-driven dimerization and DNA-stimulated activation of SMC ATPases. Curr Biol 14, 1778-1782.

Lammens, K., Bemeleit, D.J., Mockel, C., Clausing, E., Schele, A., Hartung, S., Schiller, C.B., Lucas, M., Angermuller, C., Soding, J., et al. (2011). The Mre11:Rad50 Structure Shows an ATP-Dependent Molecular Clamp in DNA Double-Strand Break Repair. Cell 145, 54-66.

Lane, H.E., and Denhardt, D.T. (1975). The rep mutation. IV. Slower movement of replication forks in Escherichia coli rep strains. J Mol Biol 97, 99-112.

Lee, B.I., Kim, K.H., Park, S.J., Eom, S.H., Song, H.K., and Suh, S.W. (2004). Ring-shaped architecture of RecR: implications for its role in homologous recombinational DNA repair. Embo J 23, 2029-2038.

Lee, J.H., and Paull, T.T. (2005). ATM activation by DNA double-strand breaks through the Mre11-Rad50-Nbs1 complex. Science 308, 551-554.

Leiros, I., Timmins, J., Hall, D.R., and McSweeney, S. (2005). Crystal structure and DNA-binding analysis of RecO from Deinococcus radiodurans. Embo J 24, 906-918.

Lim, H.S., Kim, J.S., Park, Y.B., Gwon, G.H., and Cho, Y. (2011). Crystal structure of the Mre11-Rad50-ATP{gamma}S complex: understanding the interplay between Mre11 and Rad50. Genes Dev. 25, 1091-1104

Lloyd, R.G., and Thomas, A. (1983). On the nature of the RecBC and RecF pathways of conjugal recombination in Escherichia coli. Mol Gen Genet 190, 156-161.

Lovett, S.T., and Clark, A.J. (1983). Genetic analysis of regulation of the RecF pathway of recombination in Escherichia coli K-12. J Bacteriol 153, 1471-1478.

Luisi-DeLuca, C., and Kolodner, R. (1994). Purification and characterization of the Escherichia coli RecO protein. Renaturation of complementary single-stranded DNA molecules catalyzed by the RecO protein. J Mol Biol 236, 124-138.

Lusetti, S.L., Hobbs, M.D., Stohl, E.A., Chitteni-Pattu, S., Inman, R.B., Seifert, H.S., and Cox, M.M. (2006). The RecF protein antagonizes RecX function via direct interaction. Mol Cell 21, 41-50.

Madiraju, M.V., and Clark, A.J. (1991). Effect of RecF protein on reactions catalyzed by RecA protein. Nucleic Acids Res 19, 6295-6300.

Madiraju, M.V., and Clark, A.J. (1992). Evidence for ATP binding and double-stranded DNA binding by Escherichia coli RecF protein. J Bacteriol *174*, 7705-7710.

Mahdi, A.A., Buckman, C., Harris, L., and Lloyd, R.G. (2006). Rep and PriA helicase activities prevent RecA from provoking unnecessary recombination during replication fork repair. Genes Dev *20*, 2135-2147.

Makarova, K.S., Aravind, L., Wolf, Y.I., Tatusov, R.L., Minton, K.W., Koonin, E.V., and Daly, M.J. (2001). Genome of the extremely radiation-resistant bacterium Deinococcus radiodurans viewed from the perspective of comparative genomics. Microbiol Mol Biol Rev *65*, 44-79.

Makharashvili, N. (2009). What RecFOR are for: Structure-function studies of recombination mediator proteins. In Biochemistry (Saint Louis, Missouri, USA, Saint Louis University), pp. 152.

Makharashvili, N., Koroleva, O., Bera, S., Grandgenett, D.P., and Korolev, S. (2004). A novel structure of DNA repair protein RecO from Deinococcus radiodurans. Structure (Camb) *12*, 1881-1889.

Makharashvili, N., Mi, T., Koroleva, O., and Korolev, S. (2009). RecR-mediated modulation of RecF dimer specificity for single- and double-stranded DNA. J Biol Chem *284*, 1425-1434.

Manfredi, C., Carrasco, B., Ayora, S., and Alonso, J.C. (2008). Bacillus subtilis RecO nucleates RecA onto SsbA-coated single-stranded DNA. J Biol Chem *283*, 24837-24847.

Manfredi, C., Suzuki, Y., Yadav, T., Takeyasu, K., and Alonso, J.C. (2010). RecO-mediated DNA homology search and annealing is facilitated by SsbA. Nucleic Acids Res *38*, 6920-6929.

Mazloum, N., Zhou, Q., and Holloman, W.K. (2007). DNA binding, annealing, and strand exchange activities of Brh2 protein from Ustilago maydis. Biochemistry *46*, 7163-7173.

McInerney, P., and O'Donnell, M. (2007). Replisome fate upon encountering a leading strand block and clearance from DNA by recombination proteins. J Biol Chem *282*, 25903-25916.

Michel-Marks, E., Courcelle, C.T., Korolev, S., and Courcelle, J. (2010). ATP binding, ATP hydrolysis, and protein dimerization are required for RecF to catalyze an early step in the processing and recovery of replication forks disrupted by DNA damage. J Mol Biol *401*, 579-589.

Mohaghegh, P., and Hickson, I.D. (2001). DNA helicase deficiencies associated with cancer predisposition and premature ageing disorders. Hum Mol Genet *10*, 741-746.

Moncalian, G., Lengsfeld, B., Bhaskara, V., Hopfner, K.P., Karcher, A., Alden, E., Tainer, J.A., and Paull, T.T. (2004). The rad50 signature motif: essential to ATP binding and biological function. J Mol Biol *335*, 937-951.

Morimatsu, K., and Kowalczykowski, S.C. (2003). RecFOR Proteins Load RecA Protein onto Gapped DNA to Accelerate DNA Strand Exchange. A Universal Step of Recombinational Repair. Mol Cell *11*, 1337-1347.

Moynahan, M.E., Pierce, A.J., and Jasin, M. (2001). BRCA2 is required for homology-directed repair of chromosomal breaks. Mol Cell 7, 263-272.

Nakayama, H., Nakayama, K., Nakayama, R., and Nakayama, Y. (1982). Recombination-deficient mutations and thymineless death in Escherichia coli K12: reciprocal effects of recBC and recF and indifference of recA mutations. Can J Microbiol 28, 425-430.

New, J.H., Sugiyama, T., Zaitseva, E., and Kowalczykowski, S.C. (1998). Rad52 protein stimulates DNA strand exchange by Rad51 and replication protein A. Nature 391, 407-410.

Nicolette, M.L., Lee, K., Guo, Z., Rani, M., Chow, J.M., Lee, S.E., and Paull, T.T. (2010). Mre11-Rad50-Xrs2 and Sae2 promote 5' strand resection of DNA double-strand breaks. Nat Struct Mol Biol 17, 1478-1485.

Obmolova, G., Ban, C., Hsieh, P., and Yang, W. (2000). Crystal structures of mismatch repair protein MutS and its complex with a substrate DNA. Nature 407, 703-710.

Ouyang, K.J., Woo, L.L., and Ellis, N.A. (2008). Homologous recombination and maintenance of genome integrity: cancer and aging through the prism of human RecQ helicases. Mech Ageing Dev 129, 425-440.

Perez-Roger, I., Garcia-Sogo, M., Navarro-Avino, J.P., Lopez-Acedo, C., Macian, F., and Armengod, M.E. (1991). Positive and negative regulatory elements in the dnaA-dnaN-recF operon of Escherichia coli. Biochimie 73, 329-334.

Petit, M.A., and Ehrlich, D. (2002). Essential bacterial helicases that counteract the toxicity of recombination proteins. EMBO J 21, 3137-3147.

Powell, S.N., Willers, H., and Xia, F. (2002). BRCA2 keeps Rad51 in line. High-fidelity homologous recombination prevents breast and ovarian cancer? Mol Cell 10, 1262-1263.

Rajagopalan, M., Lu, C., Woodgate, R., O'Donnell, M., Goodman, M.F., and Echols, H. (1992). Activity of the purified mutagenesis proteins UmuC, UmuD', and RecA in replicative bypass of an abasic DNA lesion by DNA polymerase III. Proc Natl Acad Sci U S A 89, 10777-10781.

Ream, L.W., and Clark, A.J. (1983). Cloning and deletion mapping of the recF dnaN region of the Escherichia coli chromosome. Plasmid 10, 101-110.

Ream, L.W., Margossian, L., Clark, A.J., Hansen, F.G., and von Meyenburg, K. (1980). Genetic and physical mapping of recF in Escherichia coli K-12. Mol Gen Genet 180, 115-121.

Rehrauer, W.M., Lavery, P.E., Palmer, E.L., Singh, R.N., and Kowalczykowski, S.C. (1996). Interaction of Escherichia coli RecA protein with LexA repressor. I. LexA repressor cleavage is competitive with binding of a secondary DNA molecule. J Biol Chem 271, 23865-23873.

Rocha, E.P., Cornet, E., and Michel, B. (2005). Comparative and evolutionary analysis of the bacterial homologous recombination systems. PLoS Genet 1, e15.

Rothman, R.H., and Clark, A.J. (1977). The dependence of postreplication repair on uvrB in a recF mutant of Escherichia coli K-12. Mol Gen Genet 155, 279-286.

Ryzhikov, M., Koroleva, O., Postnov, D., Tran, A., and Korolev, S. (2011). Mechanism of RecO recruitment to DNA by single-stranded DNA binding protein. Nucleic Acids Res.

Sakai, A., and Cox, M.M. (2009). RecFOR and RecOR as distinct RecA loading pathways. J Biol Chem *284*, 3264-3272.

Sandler, S.J. (1996). Overlapping functions for recF and priA in cell viability and UV-inducible SOS expression are distinguished by dnaC809 in Escherichia coli K-12. Mol Microbiol *19*, 871-880.

Sandler, S.J., Chackerian, B., Li, J.T., and Clark, A.J. (1992). Sequence and complementation analysis of recF genes from Escherichia coli, Salmonella typhimurium, Pseudomonas putida and Bacillus subtilis: evidence for an essential phosphate binding loop. Nucleic Acids Res *20*, 839-845.

Sandler, S.J., and Clark, A.J. (1994). RecOR suppression of recF mutant phenotypes in Escherichia coli K-12. J Bacteriol *176*, 3661-3672.

Shinohara, A., and Ogawa, T. (1998). Stimulation by Rad52 of yeast Rad51-mediated recombination. Nature *391*, 404-407.

Singleton, M.R., Wentzell, L.M., Liu, Y., West, S.C., and Wigley, D.B. (2002). Structure of the single-strand annealing domain of human RAD52 protein. Proc Natl Acad Sci U S A *99*, 13492-13497.

Smith, P.C., Karpowich, N., Millen, L., Moody, J.E., Rosen, J., Thomas, P.J., and Hunt, J.F. (2002). ATP binding to the motor domain from an ABC transporter drives formation of a nucleotide sandwich dimer. Mol Cell *10*, 139-149.

Strunnikov, A.V. (1998). SMC proteins and chromosome structure. Trends Cell Biol *8*, 454-459.

Sugiyama, T., New, J.H., and Kowalczykowski, S.C. (1998). DNA annealing by RAD52 protein is stimulated by specific interaction with the complex of replication protein A and single-stranded DNA. Proc Natl Acad Sci U S A *95*, 6049-6054.

Sung, P. (1997). Function of yeast Rad52 protein as a mediator between replication protein A and the Rad51 recombinase. J Biol Chem *272*, 28194-28197.

Sweezy, M.A., and Morrical, S.W. (1999). Biochemical interactions within a ternary complex of the bacteriophage T4 recombination proteins uvsY and gp32 bound to single-stranded DNA. Biochemistry *38*, 936-944.

Symington, L.S. (2002). Role of RAD52 epistasis group genes in homologous recombination and double-strand break repair. Microbiol Mol Biol Rev *66*, 630-670, table of contents.

Tal, A., Arbel-Goren, R., and Stavans, J. (2009). Cancer-associated mutations in BRC domains of BRCA2 affect homologous recombination induced by Rad51. J Mol Biol *393*, 1007-1012.

Tessmer, I., Yang, Y., Zhai, J., Du, C., Hsieh, P., Hingorani, M.M., and Erie, D.A. (2008). Mechanism of MutS searching for DNA mismatches and signaling repair. J Biol Chem *283*, 36646-36654.

Thompson, L.H., and Schild, D. (2002). Recombinational DNA repair and human disease. Mutat Res *509*, 49-78.

Timmins, J., Leiros, I., and McSweeney, S. (2007). Crystal structure and mutational study of RecOR provide insight into its mode of DNA binding. Embo J 26, 3260-3271.

Umezu, K., Chi, N.W., and Kolodner, R.D. (1993). Biochemical interaction of the Escherichia coli RecF, RecO, and RecR proteins with RecA protein and single-stranded DNA binding protein. Proc Natl Acad Sci U S A 90, 3875-3879.

Umezu, K., and Kolodner, R.D. (1994). Protein interactions in genetic recombination in Escherichia coli. Interactions involving RecO and RecR overcome the inhibition of RecA by single-stranded DNA-binding protein. J Biol Chem 269, 30005-30013.

Veaute, X., Delmas, S., Selva, M., Jeusset, J., Le Cam, E., Matic, I., Fabre, F., and Petit, M.A. (2005). UvrD helicase, unlike Rep helicase, dismantles RecA nucleoprotein filaments in Escherichia coli. EMBO J 24, 180-189.

Villarroya, M., Perez-Roger, I., Macian, F., and Armengod, M.E. (1998). Stationary phase induction of dnaN and recF, two genes of Escherichia coli involved in DNA replication and repair. EMBO J 17, 1829-1837.

Walker, J.E., Saraste, M., Runswick, M.J., and Gay, N.J. (1982). Distantly related sequences in the alpha- and beta-subunits of ATP synthase, myosin, kinases and other ATP-requiring enzymes and a common nucleotide binding fold. Embo J 1, 945-951.

Wang, T.C., and Smith, K.C. (1983). Mechanisms for recF-dependent and recB-dependent pathways of postreplication repair in UV-irradiated Escherichia coli uvrB. J Bacteriol 156, 1093-1098.

Wang, T.V., and Smith, K.C. (1984). recF-dependent and recF recB-independent DNA gap-filling repair processes transfer dimer-containing parental strands to daughter strands in Escherichia coli K-12 uvrB. J Bacteriol 158, 727-729.

Webb, B.L., Cox, M.M., and Inman, R.B. (1995). An interaction between the Escherichia coli RecF and RecR proteins dependent on ATP and double-stranded DNA. J Biol Chem 270, 31397-31404.

Webb, B.L., Cox, M.M., and Inman, R.B. (1997). Recombinational DNA repair: the RecF and RecR proteins limit the extension of RecA filaments beyond single-strand DNA gaps. Cell 91, 347-356.

Webb, B.L., Cox, M.M., and Inman, R.B. (1999). ATP hydrolysis and DNA binding by the Escherichia coli RecF protein. J Biol Chem 274, 15367-15374.

Whitby, M.C., and Lloyd, R.G. (1995). Altered SOS induction associated with mutations in recF, recO and recR. Mol Gen Genet 246, 174-179.

Williams, G.J., Williams, R.S., Williams, J.S., Moncalian, G., Arvai, A.S., Limbo, O., Guenther, G., Sildas, S., Hammel, M., Russell, P., et al. (2011). ABC ATPase signature helices in Rad50 link nucleotide state to Mre11 interface for DNA repair. Nat Struct Mol Biol 18, 423-431.

Williams, R.S., Williams, J.S., and Tainer, J.A. (2007). Mre11-Rad50-Nbs1 is a keystone complex connecting DNA repair machinery, double-strand break signaling, and the chromatin template. Biochem Cell Biol 85, 509-520.

Yang, H., Jeffrey, P.D., Miller, J., Kinnucan, E., Sun, Y., Thoma, N.H., Zheng, N., Chen, P.L., Lee, W.H., and Pavletich, N.P. (2002). BRCA2 function in DNA binding and recombination from a BRCA2-DSS1-ssDNA structure. Science *297*, 1837-1848.

Yang, H., Li, Q., Fan, J., Holloman, W.K., and Pavletich, N.P. (2005). The BRCA2 homologue Brh2 nucleates RAD51 filament formation at a dsDNA-ssDNA junction. Nature *433*, 653-657.

Zahradka, K., Simic, S., Buljubasic, M., Petranovic, M., Dermic, D., and Zahradka, D. (2006). sbcB15 And DeltasbcB mutations activate two types of recf recombination pathways in Escherichia coli. J Bacteriol *188*, 7562-7571.

# 3

# DNA Double-Strand Break Repair Through Non-Homologous End-Joining: Recruitment and Assembly of the Players

Radhika Pankaj Kamdar[1,2] and Yoshihisa Matsumoto[1]
[1]*Research Laboratory for Nuclear Reactors, Tokyo Institute of Technology, Tokyo*
[2]*Department of Human Genetics, Emory University, Atlanta, Georgia*
[1]*Japan*
[2]*USA*

## 1. Introduction

DNA, this vitally important genetic macromolecule, is under constant assault via endogenous and exogenous agents which cause damage to DNA and thus to cells leading to genomic instability. The primary endogenous cause of DNA damage is caused during continuous replication of DNA at the S phase of the cell cycle effecting spontaneous mutations. Other endogenous DNA damaging agents are reactive oxygen species (ROS) produced as metabolic byproducts. Additionally, breaks are introduced to DNA in the process of recombination, *e.g.*, V(D)J recombination in immune systems and meiotic recombination in reproductive organs. The exogenous DNA damaging agents are ionizing radiations and chemical compounds, which are intercalated into major or minor grooves of DNA strand or form chemical bond with bases.

DNA damages include base elimination, modification, cross-linking and strand break. Strand break includes single-strand break (SSB) and double-strand break (DSB). Among these various types of DNA damages, DSB is considered most fatal. Hence healing DSB is vital to circumvent genomic instability encompassing chromosomal aberrations, translocations and tumorigenesis. Eukaryotes have evolved two major pathways to repair DSBs, *i.e.*, homologous recombination (HR) and non-homologous end-joining (NHEJ). This chapter will review the mechanisms of the latter, especially how the players are recruited to the sites of DSBs and are assembled into multi-protein repair machinery.

## 2. DNA double-strand break repair through non-homologous end-joining pathway

### 2.1 Homologous Recombination and Non-Homologous End-Joining

HR is a reaction wherein the genetic material is exchanged between two similar or identical strands of DNA. In the repair of DSB through HR, undamaged DNA serves as a template to reconsitute the original sequence across the break. On the other hand, NHEJ is the direct rejoining of the broken DNA ends without much regard for homology at these ends.

Therefore, NHEJ may sometimes incur nucleotide deletions or insertions at the junction or joining with incorrect partner, leading to chromosomal abberations like duplications, inversions or translocations. Hence it is considered that NHEJ is less accurate than HR but, nevertheless, important especially in vertebrates.

In HR, the template should be found in homologous chromosome or in sister chromatid. Organisms like budding yeast can avail homologous chromosome as the template. However, vertebrate can utilize only sister chromatid, but not homologous chromosome, as the template for HR and, therefore, the repair of DSB through HR is limited to late S and G2 phases. The majolity of the cells reside in G0 or G1 phases in vertebrate body, where only NHEJ can operate.

Additionally, only small portions of the genome in vertebrate are encoding protein or functional RNA and other portions are intervening or repetitive sequences. These regions may have important roles in the structural maintenance of the genome, proper replication/segregation of the genome or spatiotemporal regulation of the genen expression. Nevertheless, small deletion or insertion of nucleotides might be tolerated in most portion of the vertebrate genome.

Finally, whereas HR is utilized in meiotic recombination in reproductive organs, NHEJ is utilized in V(D)J recombination in immune system to establish diversity of immunogloblins and T cell receptors. Thus, genetic defect in either one of NHEJ components results not only in elevated sensitivity toward radiation and radiomimetic agents but also in immunodeficiency.

## 2.2 Processes of NHEJ

NHEJ process may be divided into three steps, i.e., (i) detection, (ii) processing and (iii) ligation of DSB ends (Fig.1). The detection and ligation steps comprises the core reaction while the processing step is required only when the ends are not readily ligatable. In the detection step, Ku protein, heterodimer consisting of Ku70 and Ku86 (also known as Ku80), first binds to the ends of double-stranded DNA and then recruits DNA-PK catalytic subunit (DNA-PKcs). The complex consisting of Ku70, Ku86 and DNA-PKcs is termed DNA-dependent protein kinase (DNA-PK). Upon binding of DNA-PKcs to DNA ends, it exerts kinase catalytic activity to phosphorylate substrate proteins. Thus, DNA-PK is considered the molecular sensor of DSB, triggering the signalling cascade. At the final ligation step, DNA ligase IV in a tight association with XRCC4 catalyzes the reaction to join the two DNA ends. XRCC4-like factor, XLF, which is also known as Cernunnos, is also essential at this step, especially when two ends are not compatible. Thus, six polypeptides, i.e., Ku70, Ku86, DNA-PKcs, DNA ligase IV, XRCC4 and XLF are core components of NHEJ. Processing step might involve a number of enzymes depending on the shape of each DNA end and compatibility of two ends to be ligated. Presumed processing enzymes contain Artemis, DNA polymerase $\mu/\lambda$, polynucleotide kinase/phosphatase (PNKP), Aprataxin (APTX) and Aprataxin and PNKP-like factor (APLF, also known as PALF, C2orf13 or Xip1).

## 2.3 Components of NHEJ
### 2.3.1 Ku

Ku protein was initially found as the antigen of autoantibody in a patient of polymyositis-scleroderma overlap syndrome (Mimori et al., 1981). Biochemical approach, including immunoprecipitation of [32P]orthophosphate or [35S]methionine-labeled cell extract and

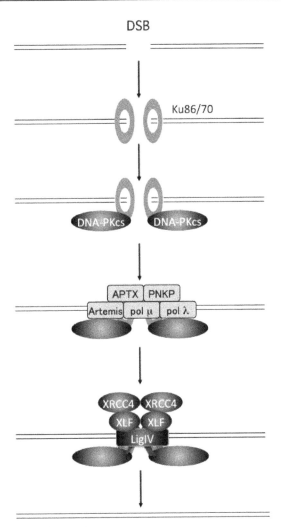

Fig. 1. Repair of DSB through NHEJ.

immunoaffinity purification, lead to identify Ku as a DNA-binding protein made up of two subunits of 70,000Da and 80,000Da, respectively, which are now known as Ku70 and Ku80 (or Ku86) (Mimori et al., 1986). It is also estimated that Ku is an abundant protein, existing as 400,000 copies in logarithmically growing HeLa cells. Protein-DNA interaction studies, including footprint analysis, led to the finding that Ku binds to the ends of double-stranded DNA without requirement for specific sequence (Mimori & Hardin, 1986). Because of this striking property, possible role of Ku in DNA repair or in transposition was suspected. In early 1990s, Ku was found to be an essential component of DNA-PK (Dvir et al., 1992, 1993; Gottlieb and Jackson, 1993). It was also found that Ku80 is equivalent to XRCC5 (X-ray repair cross complementing) gene product, which is missing in X-ray sensitive rodent cell lines including xrs-5, -6, XR-V9B and XR-V15B (Taccioli et al., 1994; Smider et al., 1994).

These cell lines also exhibit defect in V(D)J recombination, indicating the role of Ku in this process. Ku80 knockout mice showed immunodeficiency and radiosensitivity, like *scid* mice (below) and also exhibited growth defect; body weight was 40-60% of age-matched control (Nussenzweig et al., 1996; Zhu et al., 1996). Ku70 knockout mice also showed immunodeficiency, radiosensitivity and growth defect. However, immunological defect in Ku70 knockout mice was less severe than in Ku80 knockout mice, as it shows partial production and differentiation of T cells (Gu et al., 1997; Ouyang et al., 1997). Ku is also implied to play critical roles in telomere capping in mammalian cells (Hsu et al., 2000).

Homologues of Ku proteins were identified not only in mammals but also in other eukaryota including budding yeast, where it is referred to as HDF1 and HDF2 (high- affinity DNA binding factor), or Yku70 and Yku80, respectively (Feldmann & Winnacker, 1993; Milne et al., 1996; Feldmann et al., 1996; Boulton & Jackson, 1996). Yeast Ku is shown to play important roles in NHEJ, telomere maintenance and silencing (Boulton & Jackson, 1996, 1998; Porter et al., 1996).

Ku70 and Ku80 show low but significant sequence similarity, indicating common evolutional origin, and share a similar structural configuration (Dynan & Yoo, 1998; Gell & Jackson, 1999). Expectedly, "single" Ku orthologue was identified in bacteria and in bacteriophage (Weller et al., 2002; d'Adda di Fagagna et al., 2003). As revealed by X-ray crystallography, Ku 70 and Ku80 fold to form an asymmetric ring shaped structure forming an aperture large enough to let DNA thread through it; thus playing a crucial role in DSB recognition (Walker et al., 2001). The core of Ku required to form dimer and aperture is conserved among all Ku orthologues. Both of Ku70 and Ku80 bear von Willebrant factor A domain, which may be essential for heterodimer formation. The C-termial of Ku70 bears SAP domain, which may mediate DNA binding, and the C-terminal of Ku80 bears a conserved motif to interact with DNA-PKcs (Gell & Jackson, 1999; Falck et al.., 2005).

Ku translocates along DNA in an ATP-independent manner, allowing several dimers to bind on a single DNA molecule (Zhang and Yaneva, 1992; Bliss and Lane, 1997). Ku was identified also as a ssDNA dependent ATPase stimulating the DNA polymerase α primase activity (Vishwanatha and Baril, 1990; Cao et al., 1994) and as an ATP dependent DNA helicase II (HDH II) (Tuteja et al., 1994). Recent study demonstrated that Ku has 5'-RP/AP lyase activity, nicking 3'-side of abasic site (Roberts et al., 2010). Thus, Ku might exert multiple functions, not only binding to DSBs but also activating damage signal via DNA-PKcs and processing DSB ends removing the obstacle for ligation.

### 2.3.2 DNA-PKcs

DNA-PK activity was first found as an activity to phosphorylate Hsp90 in the presence of double-stranded DNA in the extracts of HeLa cell, rabbit reticulocyte, Xenopus egg and sea urchin egg (Walker et al., 1985). DNA-PK was purified from Hela cell nuclei as a 300-350 kDa protein, which is now called DNA-PKcs for DNA-PK catalytic subunit (Carter et al., 1990; Lees-Miller et al., 1990). Later it was found that Ku is an essential component of DNA-PK and that DNA-PK requires binding to DNA ends to be activated (Dvir et al., 1992, 1993; Gottlieb and Jackson, 1993). Following the finding that XRCC5 is equivalent to Ku80, DNA-PKcs is found to correspond to XRCC7, which is deficient in *scid* (severe combined immunodeficiency) mouse (Kirchgessner et al., 1995; Blunt et al., 1995; Peterson et al., 1995), lacking mature B and T cells due to a defect in V(D)J recombination (Bosma et al., 1983; Lieber et al., 1988; Fulop & Phillips, 1990; Biederman et al., 1991). Scid due to defect in DNA-

PKcs is also found in horse (Wiler et al., 1995; Shin and Meek, 1997) and in dog (Meek et al., 2001). M059J, a human glioma cell line, defective in DNA-PKcs, also showed radiosensitivity with defective DSB repair (Lees-Miller et al., 1995). Recently, DNA-PKcs missense mutation was identified in human radiosensitive T- B- severe combined immunodeficiency (TB-SCID) (van der Burg et al., 2009). Cells from the patient exhibit normal DNA-PK activity but may have defect in Artemis activation (below).

Cloning of gene revealed that DNA-PKcs is a 4,127 amino acid polypeptide, one of the largest molecules in the cell (Hartley et al., 1995). The carboxy-terminal between amino acid residues 3719 – 4127 compose the catalytic domain that is categolized into phosphatidylinositol-3 kinase and like kinase (PIKK) family (Hartley et al., 1995; Poltoratsky et al., 1995). PIKK family include ataxia-telangiectasia mutated (ATM) (Savitsky et al., 1995) and ATM- and Rad3-related (ATR) (Cimprich et al., 1996), both of which are protein kinases with roles in DNA repair and cell cycle checkpoint as sensors of DNA damages. Although orthologues of ATM and ATR can be found in fruit fly, nematoda, plants and yeast, DNA-PKcs has been found only in vertebrate, some arthropods (Dore et al., 2004) and dictyostelium (Hudson et al., 2005).

*In vitro* studies had revealed that DNA-PK can phosphorylate a number of nuclear, DNA binding proteins with supposed functions in transcription, replication, recombination and repair (Lees-Miller et al., 1992). The sites phosphorylated by DNA-PK were identified as serine and threonine that are immediately followed by a glutamine on the linear sequence; SQ/TQ (Lees-Miller et al., 1992), although there are a considerable number of exceptions reported. The protein phosphorylation by DNA-PK should be essential for NHEJ, as catalytically inactive form of DNA-PKcs can restore at most partial NHEJ activity to DNA-PKcs deficient cells (Kurimasa et al., 1999). However, it is presently unclear what is/are the *in vivo* phosphorylation target(s) essential for DNA repair.

Recent studies have shed light on the phosphorylation of DNA-PKcs itself. At least 16 sites of autophosphorylation have been identified (Chan et al., 2002; Douglas et al., 2002; Ding et al., 2003). Most of them are clustered within 2023 - 2056 (PQR cluster), 2609 – 2647 (ABCDE cluster) and 2671 – 2677. Some of them may be phosphorylated by ATM or ATR *in cellulo* (Chen et al., 2007; Yajima et al., 2006) It has been demonstrated that, *in vitro*, autophosphorylation of DNA-PK leads to loss of kinase activity and dissociation from Ku (Chan et al., 1996). It should be also noted that substitution of serines and threonines within ABCDE cluster with alanine results in greater radiation sensitivity than DNA-PKcs null cells and also in reduced rates of HR. Thus, autophosphorylation, especially within ABCDE cluster might regulate DNA-PK activity negatively or switch repair pathway from NHEJ to HR.

### 2.3.3 XRCC4-DNA ligase IV

XRCC4 was isolated and cloned from a human cDNA sequence whose expression in the XR-1 cells, derived from Chinese Hamster ovary and phenotypically similar to *scid* and *xrs*, conferred normal V(D)J recombination ability and also DSB repair activity (Li et al., 1995). Biochemical studies lead to finding that it is associated with DNA ligase IV (Critchlow et al., 1997; Grawunder et al., 1997). Mutations in DNA ligase IV gene have been identified in radiosensitive leukemia patient (Badie et al, 1995; Riballo et al., 1999) and in patients exhibiting developmental delay and immunodeficiency, which is called ligase IV syndrome (O'Driscoll et al., 2001). Although mutation in XRCC4 gene has not been found in humans, there are some polymorphisms associated with colorectal cancer and childfood leukemia (Bau et al., 2010; Wu et al., 2010). Disruption of either XRCC4 or DNA Ligase IV gene in mice leads to embryonic lethality with a primary defect in neurogenesis and severe neuronal apoptosis (Barnes et al., 1998; Frank et al., 1998; Gao et al., 1998). Mutants of *DNL4* and *LIF1* genes, the yeast orthologue of human DNA Ligase IV and XRCC4, respectively, exhibited a phenotype similar to that of HDF1 and 2 mutants, indicating its role in recombination and repair (Wilson et al., 1997; Teo and Jackson, 1997, 2000).

XRCC4-DNA Ligase IV is a critical complex formed *in vivo* (Critchlow et al., 1997; Grawunder et al., 1997) for the ligation of the broken DNA ends via NHEJ pathway. The presence of XRCC4 stabilize and activates DNA Ligase IV (Grawunder et al., 1997; Bryans et al., 1999) by stimulating its adenylation which is the first chemical step in ligation (Modesti et al., 1999). XRCC4 forms a homodimer and associates with a polypeptide at the C-terminus of DNA Ligase IV (Critchlow et al., 1997; Junop et al., 2000; Sibanda et al., 2001). This interaction is mapped to the central coiled coil domain of XRCC4 and the inter BRCT linker region at the C-terminus of DNA Ligase IV. This region within DNA Ligase IV, termed as the XRCC4-interacting region (XIR) was deemed necessary and sufficient for XRCC4-Ligase IV interaction (Grawunder et al., 1998). Recently a high resolution crystal structure of human XRCC4 bound to the C-terminal tandem BRCT repeat of DNA Ligase IV was reported. It revealed an extensive binding interface formed by helix-loop-helix structure within the inter-BRCT linker region of Ligase IV, as well as significant interactions involving the second BRCT domain that induces a kink in the tail region of XRCC4 (Wu et al., 2009). This interaction was demonstrated as essential to stabilize the interaction between the XIR of DNA Ligase IV and XRCC4, while the first BRCT domain was considerably dispensable.

### 2.3.4 XLF/ cernunnos

Although above five factors had been identified by 1998, there were indications of the existence of additional factor essential for mammalian NHEJ. First, 2BN cell line, which is derived from radiosensitive and immunodeficient patient, showed defective NHEJ but all the known NHEJ components were normal. Second, in 2001, NEJ1/LIF2 was identified as a new essential factor of NHEJ in budding yeast (Kegel et al., 2001; Valencia et al., 2001; Ooi et al., 2001; Frank-Vaillant & Marcand, 2001).

XLF was identified in the yeast two hybrid screen for XRCC4 inteacting protein (Ahnesorg et al., 2006) and named XRCC4-like factor, as it was predicted to have 3D structure similar to that of XRCC4. It is also identified as Cernunnos missing in patients with growth retardation, microcephaly, immunodeficiency, increased cellular sensitivity to ionizing radiation and a defective V(D)J recombination (Buck et al, 2006). It is a 33kDa protein with 299 amino acid residues. NHEJ deficient 2BN cells lacked XLF due to a frameshift mutation (Ahnesorg, 2006). XLF was found to be a genuine homologue of Nej1p from budding yeast (Callebaut et al., 2006). XLF was also shown to be conserved across evolution (Hentges et al., 2006) and to be a paralogue of XRCC4 (Callebaut et al., 2006).

Chromatographic analyses established XLF existing as dimer and crystallographic studies demonstrated its interaction through globular head-to-head domain with that of XRCC4 (Andres et al., 2007; Li et al., 2008). Three-dimensional X-ray scattering characterized a tetramer formation of XRCC4, while the XRCC4-XLF interaction was still mediated through glogular head domains which rendered it suitable for DNA alignment and Ligase IV function (Hammel et al., 2010). XLF possesses DNA binding activity dependent on the length of DNA (Lu et al., 2007a) and ability to ligate mismatched and non-cohesive ends (Tsai et al., 2007).

## 2.3.5 Processing enzymes

Pathologic and physiologic breaks create incompatible DNA ends which are not as easy to rejoin as those created *in vitro* by restriction enzyme digestion. It requires removal of excess DNA and fill-in of gaps and overhangs in order to make them compatible for the DNA ligase activity.

Artemis was identified as the causative gene for human RS-SCID (Moshous et al., 2001). Artemis forms a complex with DNA-PKcs and expresses 5' to 3' exonuclease activity and endonuclease and endonuclease activity at the junction of single-stranded and double-stranded DNA (Ma et al., 2002). Although, the signal joint formation during V(D)J recombination does not require Artemis or DNA-PKcs for joining, all of the components of NHEJ including Artemis are required for coding ends. Artemis in association with DNA-PKcs is deemed necessary for the opening of hairpin structures (Lu et al., 2007b). Artemis is phosphorylated both by DNA-PKcs and ATM (Poinsignon et al., 2004; Zhang et al., 2004).

Polymerases μ and λ belong to pol X family and might fill gaps and 5'overhangs (Ramadan et al., 2003). Polynucleotide kinase/phosphatase (PNKP) adds phosphate group to 5'-hydroxyl end and also removes phosphate group from 3'-phosphorylated end (Koch et al., 2004; Clements et al., 2004;Whitehouse et al., 2001). Aprataxin (APTX) is initially identified as the product of the gene defective in genetic disorder early-onset ataxia with oculomoter apraxia (Date et al., 2001) and later shown to remove AMP from abortive intermediates of ligation (Ahel et al., 2009). PNK- and APTX-like FHA protein (PALF, also known as APLF, C2orf13 or Xip1) has AP endonuclease activity (Kanno et al., 2007; Iles et al., 2007). Recent study showed that APLF also has histone chaperone activity (Mehrotra et al., 2011) and that it co-operates with PARP-3, which is newly found as a DSB sensor (Rulten et al., 2011). It might be noted that all of these factors bears BRCT or FHA domain as module to bind phosphorylated proteins. Polymerases μ and λ possess BRCT domain. PNKP, APTX and PALF possess FHA domain, which is structurally similar to each other and known to interact with CKII-phosphorylated XRCC1 or XRCC4 (see below).

## 2.4 Alternative NHEJ pathways

Apart from the classical NHEJ model, there are also studies by several groups highlighting NHEJ as a more sophisticated and complex mechanism involving a cross–talk between pathways including proteins other than DNA-PKcs, Ku, XRCC4-DNA Ligase IV.

### 2.4.1 ATM dependent pathway

Human genetic disorder, Ataxia Telangiectasia (AT) is caused by mutation in the ATM (Ataxia Telangiectasia mutated) gene and is characterized by chromosomal instability, immunodeficiency, radiosensitivity, defective cell cycle checkpoint activation and predisposition to cancer indicating its responsibility in genome surveillance (Jorgensen and Shiloh, 1996). ATM deficiency causes early embryonic lethality in Ku or DNA-PKcs deficient mice, thus providing NHEJ an independent role for the DNA-PK holoenzyme (Sekiguchi et al., 2001). ATM and Artemis, together with NBS1, Mre11 and 53BP1 function in a sub-pathway that repairs approximately 10% of DSBs, probably requiring end-processing (Riballo et al., 2004). Another study suggested three parallel, but mutually crosstalking, pathways of NHEJ, *i.e.*, core pathway mediated by DNA-PKcs and Ku, ATM-Artemis pathway and 53BP1 pathway, all of which finally converge on XRCC4-DNA Ligase IV (Iwabuchi et al., 2006).

### 2.4.2 Back-up NHEJ pathway

Repair in IR-induced DSBs in higher eukaryotes is mainly dominated by NHEJ which is faster as compared to other mechanisms. However, it is severely compromised in case of defects in DNA-PKcs, Ku and DNA Ligase IV (DiBiase et al., 2000; Wang et al., 2001). An array of biochemical and genetic studies have shown that despite the prevalence of DNA-PK dependent pathway, cells deficient in either of its components are still able to rejoin a majority of DSBs, operating with slower kinetics, using an alternative pathway (Nevaldine et al., 1997; Wang et al., 2003). Chicken DT40 cells defective in HR rejoin IR induced DSBs with kinetics similar to those of other cells with much lower levels of HR. Nevertheless, rejoining of DSBs with slow kinetics is associated with incorrect DNA end-joining which is incompatible with the mechanism of HR (Löbrich et al., 1995). These observations led to the model that DNA DSBs are rejoined by two pathways, one of which is DNA-PK dependent (D-NHEJ) and an alternative pathway termed as Back-up (B-NHEJ) pathway (Wang et al., 2003) possibly prone to erroneous re-joining and utilization of microhomologies (Roth DB, 1986). Further investigations ascertained the role of DNA-PK in the functional co-ordination of D-NHEJ and B-NHEJ, suggesting that the binding of inactive DNA-PK to DNA ends not only blocks the D-NHEJ but also interferes with the function of B-NHEJ (Perrault et al., 2004). The DNA-PK and Ku complex is believed to recruit other repair proteins like XRCC4-DNA Ligase IV complex and stimulate the ligation of DNA ends (Ramsden and Gellert, 1998) in D-NHEJ pathway.

DNA Ligase IV deficient mouse embryonic fibroblasts retained significant DNA end-joining activity which was reduced upto 80% by knocking down DNA Ligase III. Thus DNA Ligase III was identified as a vital component of B-NHEJ (Wang et al., 2005). PARP-1 was initially pointed to bind to DSBs with a higher efficacy than to SSBs (Weinfeld et al., 1997) and with a greater affinity than that of DNA-PKcs (D'Silva et al., 1999). It has also been shown to interact with both the subunits of DNA-PK (Galande and Kohwi-Shigematsu, 1999; Ariumi et al., 1999) catalyzing their poly(ADP-ribosyl)ation (Li et al., 2004; Ruscetti et al., 1998). Using chemically potent producer of DSBs, calicheamicin γ1, a new mechanism was identified operating independently but complementing the classical NHEJ pathway. Proteins such as, PARP-1, XRCC1 and DNA Ligase III, which were believed to be otherwise involved in Base Excision Repair (Caldecott, 2003) and SSB repair (Caldecott, 2001) surmised a new mechanism encompassing synapsis and end-joining activity.

Above mentioned studies evidently illustrate alternative DNA end-joining pathways to contribute in the repair of DSBs in order to maintain the genomic integrity when D-NHEJ is compromised. However, due to their low fidelity, they are directly implicated in genomic instability (Ferguson et al., 2000), aberrant coding and signal joint formation during V(D)J recombination (Taccioli et al., 1993; Bogue et al., 1997) as well as formation of soft tissue sarcomas (Sharpless et al., 2001) that potentially leads to cancer.

## 3. Recruitment and assembly of NHEJ factors at DSB

The key players of NHEJ are named, but the mechanism of their recruitment and hierarchy of assembly on the DNA DSB is not yet well clarified. Many proteins in the HR pathway, *e.g.*, Nbs1-Mre11-Rad50, BRCA1 and Rad51, exhibit local accumulation after DSB induction, forming microscopically visible structures, termed ionizing radiation-induced foci (IRIF) (Maser et al., 1997). Such change in the localization of HR proteins has been observed also in partial volume irradiation (Nelms et al., 1998) and laser micro-irradiation experiments (Kim et al., 2002). As the distribution of these proteins after irradiation, at least partially, overlapped with irradiated area or DSBs, visualized by DNA end labeling or immunofluorescence analysis of γ-H2AX, these phenomena are believed to reflect the accumulation of these proteins around DSB sites. In the case of NHEJ proteins, however, IRIF has been observed only for autophosphorylated form of DNA-PKcs (Chan et al., 2002). Recently, there are increasing number of studies using laser micro-irradiation demonstrating the accumulation of NHEJ molecules in irradiated area. Another approach to examine the association of DNA repair proteins with damaged DNA is sequential extraction with increasing concentration of detergent or salt.

## 3.1 Recruitment of XRCC4 to chromatin DNA in response to ionizing radiation

We employed sequential extraction with detergent-containing buffer to examine the binding of XRCC4 to DSB (Kamdar and Matsumoto, 2010). The retention of XRCC4 to subcellular fraction consisting of chromatin DNA and other nuclear matrix structures increased in response to irradiation. Micrococcal nuclease enzyme which specifically cleaves the chromatin DNA into smaller nucleosomal fragments revealed that XRCC4 is tethered to chromatin DNA after irradiation.

Through quantitative analyses, it was estimated that only one or few XRCC4 molecules might be recruited to each DNA end at the DSB site. This can be speculated based on the stoichiometric results depicting a complex consisting of two XRCC4 molecules forming a dimer and one Ligase IV molecule (Junop et al., 2000). The accumulation of XRCC4 on the damaged chromatin is very rapid and sensitive as the response after radiation is observed in $\leq 0.1$hr and is stable until at least 4 hrs. This phenomenon is in parallel to the appearance of phosphorylation of H2AX which is observed as foci until the DSBs are repaired and then their disappearance from the resealed DNA (Svetlova et al., 2010). XRCC4 could be retained on the damaged chromatin as long as the repair complex carries out the rejoining of the DNA ends which pivotally includes ligation by XRCC4-DNA Ligase IV. In addition, the residence of XRCC4 on chromatin might be very transient, particularly after the irradiation with small and conventional doses. These observations can reasonably explain why it has been difficult to capture the movement of NHEJ enzymes to DSB sites.

Using a similar approach, the movement of NHEJ molecules in response to DSB induction by neocarzinostatin or bleomycin was reported (Drouet et al., 2005). Conversely, there are several differences between the results of the two studies. First, they observed that DNase I treatment released DNA-PKcs and Ku but not XRCC4 and DNA Ligase IV, leading to the idea that XRCC4 and DNA Ligase IV were bound to nuclear matrix or other structures rather than chromatin itself. In the present study, XRCC4 retained after buffer extraction could be released by micrococcal nuclease treatment, indicating its binding to chromatin DNA. Second, they mentioned that the movement of NHEJ molecules could be observed only after high doses of irradiation in their study. The present study has demonstrated small but significant increase in the chromatin binding of XRCC4 even after physiologically relevant dose, *i.e.*, 2Gy, of irradiation.

## 3.2 Phosphorylation of XRCC4

Several studies have shown that DNA-PK can phosphorylate XRCC4 *in vitro*, decreasing its interaction with DNA, although the significance of this phenomenon is presently unclear (Critchlow et al., 1997; Leber et al., 1998; Modesti et al., 1999). Moreover, our research group demonstrated XRCC4 phosphorylation in living cells, which was induced by ionizing radiation in a manner dependent on DNA-PKcs (Matsumoto et al., 2000), indicating that XRCC4 is an *in vivo* and not merely an *in vitro*, substrate of DNA-PK. However, the presence of DNA-PK did not seem as a pre-requisite for XRCC4 recruitment to chromatin as demonstrated by siRNA and specific kinase inhibitors against DNA-PKcs.

DNA-PK is autophosphorylated and leads to the phosphorylation events on the target proteins. An earlier study also detected XRCC4 on DNA ends in a phosphorylated form dependent on DNA-PK. However, phosphorylation was deemed dispensable for XRCC4-DNA Ligase IV loading at DNA ends since stable complexes involving DNA-PK and the ligation complex were recovered in the presence of wortmannin which is a PIKK inhibitor (Calsou et.al 2003). A recent study using laser irradiation demonstrated XRCC4 accumulation in irradiated area, which also did not require DNA-PKcs (Mari et al., 2006; Yano et al., 2008). All these observations in aggregate thus lead to the unanswered question as to what mechanism is involved in XRCC4 recruitment to damaged chromatin DNA.

Then, what is the importance of the phosphorylation of XRCC4, if any? It has been awaited to find the biological consequence of XRCC4 phosphorylation by DNA-PK through the identification and elimination of the phosphorylation site(s). Several groups, employing mass spectrometry, identified Ser260 and Ser318 as the major phosphorylation sites in XRCC4 by DNA-PK *in vitro* (Lee et al., 2002; Yu et al., 2003; Lee et al., 2003; Wang et al., 2004). However, it is presently unclear whether these sites are phosphorylated in living cells, especially, in response to DNA damage. Furthermore, the mutants lacking these phosphorylation sites appeared fully competent in the restoration of radioresistance and V(D)J recombination in CHO-derived XRCC4-deficient XR-1 cells and also exhibited normal activity in DNA joining reaction in cell-free system, leading to the conclusion that XRCC4 phosphorylation by DNA-PK was unnecessary for these functions (Lee et.al 2003; Yu et.al 2003). However, our group recently identified four additional phosphorylation sites in XRCC4 by DNA-PK and found that at least three of them would be important for DSB repair, because disruption of these sites resulted in elevated radiosensitivity (Sharma, Matsumoto et al., unpublished results).

### 3.3 Recruitment dynamics of NHEJ complex on damaged chromatin

XRCC4 associates in a tight complex with DNA Ligase IV. XRCC4 is essential for the stability of ligase IV in mammalian cells (Bryans et.al 1999). It also initiates the chemical reaction of ligation reaction by bringing about the adenylation on Ligase IV to rejoin the DNA. Radiation induced modification, i.e phosphorylation of XRCC4 is also observed in the cells harboring the ligase IV gene. Although, it is evident from the above reports that phosphorylation is not a necessary phenomenon required for XRCC4 recruitment to chromatin, it occurs as a modification induced in response to radiation. These observations lead to two possible hierarchies; (a) ionizing radiation induces phosphorylation on DNA-PKcs which then in turn phosphorylates XRCC4 and the phosphorylated form is recruited to DSBs or (b) ionizing radiation stimulates XRCC4 recruitment to DSBs, chaperoned by other factors like ligase IV, and also recruitment of DNA-PKcs independently and then the kinase would bring about the phosphorylation events. However, since current evidences render phosphorylation dispensable for recruitment of XRCC4, the second mechanism may seem more plausible.

Moreover, movement of DNA-PKcs to chromatin DNA is also diminished in the absence of DNA Ligase IV and Ku. In addition, structural and crystallographic studies have displayed that the interaction between XRCC4 dimer and DNA Ligase IV is via the linker region on ligase IV between the tandem BRCT domains (Grawunder et al., 1998). A recent high resolution crystallographic study has revealed an extensive DNA Ligase IV binding interface for XRCC4 forming a helix-loop-helix structure forming a clamp within the inter-BRCT linker region. This loop buries and packs against a large hydrophobic surface of XRCC4, thus inducing a kink in the tail region of XRCC4, thereby involving numerous interactions between the BRCT2 domain of ligase IV and XRCC4 which are expected to play a major role in the interactions between the two proteins (Wu et.al., 2009). Mutational analysis in several of these hydrophobic residues would give a better insight in the mode of interaction alterating the conformation of both the molecules for recruitment on DNA ends.

XLF or Cernnunos is also considered a vital component of the ligation complex to reseal the DNA ends. XLF has been demonstrated to interact with XRCC4 via the globular head domains at the amino-terminal region of both the proteins forming a heterodimeric structure (Andres et al., 2007). The response to ionizing radiation could thus be expected to be similar to that evoked in XRCC4. Conversely, the protein was not found to be tethered to chromatin even after extraction with a high detergent concentration. Contrasting to that observed in case of XRCC4, XLF accumulation was neither rapid or transient nor sensitive to be observed at conventional radiation dose. This leads to the possibility that XLF association to XRCC4 is highly unstable and does not directly adhere to chromatin structures. A parallel observation was drawn by another study wherein they demonstrated that XRCC4 was dispensable for XLF recruitment to DSBs, although it could act as a stabilizing factor and cause a dynamic exchange between the free and bound protein once XLF is recruited on the DNA free ends (Yano et.al., 2008). Very recent study indicated that 10 amino acid region at the C-terminal of XLF is essential for interaction with Ku and for recruitment to DSB (Yano et al., 2011).

Intriguingly, transgenetically expressed XLF protein demonstrated a similar trend, except that the retention was observed in the subcellular nucleosolic fraction, alleged as tethered to chromatin. This disparity in the observation can be attributed to the difference in behaviour between endogenous and exogenously expressed molecules.

Owing to the recruitment of XRCC4 during the inhibition of phosphorylation by the kinases, a possible speculation leads to the idea that either or both of ligase IV and XLF molecules could play a role as a chaperone responsible for the recruitment of XRCC4 to damaged chromatin.

Live cell imaging studies have demonstrated that Ku recruits XLF and is also likely to mediate the XLF-DNA interaction (Yano et al., 2008). Therefore, the vital component of NHEJ, Ku might be mediating the interaction between XRCC4 and DSB via DNA Ligase IV or also between Ligase IV and DSB via XRCC4, though ligase IV possess a DNA-binding region at the N-terminus.

Another very intriguing analysis has exhibited that PARP-3, whose function was previously unknown, accumulates APLF (Aprataxin-like factor) to the site of DSBs which in turn supports the retention of XRCC4-DNA Ligase IV on the chromatin (Rulten et al., 2011).

Another possibility is that XRCC4 moves to a DSB site autonomously due to its intrinsic DNA end-binding activity (Modesti et al., 1999). Furthermore, XRCC4 was shown to interact with polynucleotide kinase (PNK) (Koch et al., 2004) or aprataxin (APTX) (Clements et al., 2004), depending on the phosphorylation by casein kinase II. Unexpectedly, unphosphorylated XRCC4 interacts with PNKP, although with a lower affinity, but CKII mediated XRCC4 phosphorylation inhibited the PNKP activity (Mani et al., 2010). In addition, XRCC4 has been shown to undergo monoubiquitination (Foster et al., 2006) and SUMOylation (Yurchenko et al., 2006), the former of which was shown to be DNA damage-inducible. The role of such posttranslational modifications on the chromatin-recruitment of XRCC4 is of another interest.

Additionally, studies by several groups have suggested that NHEJ is more sophisticated than thought initially and involves many proteins other than DNA-PKcs, Ku, XRCC4-DNA ligase IV, XLF/Cernunnos. In order to investigate into the entirety of the complex compounding several molecules from NHEJ and particularly from other repair or physiological pathways: XRCC4 associated complex bound to chromatin, supposedly at the last step of resealing the DNA nicks and gaps, can be isolated and analysed.

One of the other speculations is that the unwinding may be carried out by Ku since it possesses helicase activity in an ATP dependent manner (Blier et al., 1993) and is supposedly the earliest protein in repair hierarchy. Certain studies have shown a functional interaction between the Ku heterodimer and WRN (Karmakar et al., 2002) emphasizing its significance in DNA repair and metabolism pathways. The exonuclease but not the helicase activity of WRN is stimulated by physical interaction with XRCC4-ligase IV (Kusumoto et al., 2008).

ATM and Artemis, together with Nbs1, Mre11 and 53BP1, function in a subpathway of NHEJ that repairs approximately 10% of DSBs, probably those require DNA end processing (Riballo et. al, 2004). Another study suggested three parallel, but mutually crosstalking, pathways of NHEJ, *i.e.*, core pathway mediated by DNA-PKcs and Ku, ATM-Artemis pathway and 53BP1 pathway, all of which finally converge on XRCC4-DNA ligase IV (Iwabuchi et al., 2006). Recent studies indicated the requirement of chromatin remodeling factors, like ALC1 and ACF1, for the recruitment of NHEJ molecules to DSB (Ahel et al., 2009; Lan et al., 2010).

It will be of interest to investigate whether all of the above mentioned proteins play some role in the recruitment of XRCC4-DNA Ligase IV to DSB sites or, conversely, are recruited to DSB sites through interaction with XRCC4. This entire conglomerate of proteins has yet to reveal complex mechanisms and cross-talk between other repair and cellular pathways.

These questions may be addressed by examining the chromatin-recruitment of deletion or point mutants of XRCC4 and by applying siRNA or inhibitors of the above listed molecules in experimental systems. They could then be optimized for use as adjuvants in radiotherapy.

Proteomic analysis is one of the vital instruments to examine any kinase network involving *in vivo* substrates. Such modern technologies have helped to understand that the DNA damage repair response is much sophisticated and complicated than anticipated earlier. It connects NHEJ with chromatin remodelling as well as transcription processes which are also pivotal to cellular functions; thereby aspiring to investigate the cross-talks involved in the repair mechanics.

## 3.4 Future perspectives

There have been several studies including ours, demonstrating various mechanisms for the dynamics and assembly of the repair machinery on the damaged DNA site in response to various forms of endogenous and exogenous stress. A certain study also suggests that the DNA damage response does not require the DNA damage but the stable association of the repair factors for a prolonged period of time with chromatin which is likely a critical step in triggering, amplifying and maintaining the DNA damage response signal (Soutoglou and Misteli, 2008). It will thus be interesting to investigate the capricious questions as to what are the exact signalling mechanisms to trigger the DSB repair response or the role of several macromolecules involved in different cellular processes. Thus, the assembly of non-homologous end joining protein complex at DSB was not as simple as thought in classical models and further studies are warranted to fully elucidate the processes. Another important aspect, not clarified, is to understand the hierarchy and mechanism of the disassembly of the repair machinery, involved in NHEJ or from cross-talk pathways, from the site of refurbished DNA. Finally, understanding the mechanisms of DNA repair at molecular levels might bring us a new approach to be applied in cancer radiotherapy or chemotherapy.

## 4. Conclusions

DSB repair through NHEJ has been considered rather simple reaction, basically comprised of six core factors, Ku70, Ku80, DNA-PKcs, XRCC4, DNA Ligase IV and XLF. However, the mechanism how these molecules are recruited to DSBs and assembled into repair machinery is not fully understood. It has been difficult even to observe the recruitement of NHEJ molecules by immunofluorescence or simple labeling with fluorescent proteins. However, laser microirradiation technique combined with fluorescent protein and biochemical fractionation enabled us to capture the binding of NHEJ factors to DSBs. NHEJ would involve a number of processing enzymes, whose function or regulation is largely unclear. Additionally, most recent study shed light on the importance of chromatin remodeling prior to the binding of Ku. Obviously, further studies are warranted to elucidate this complexity.

## 5. Acknowledgements

Our work was supported in part by Grant-in-Aid for Scientific Research from the Ministry of Education, Culture, Sport, Science and Technology of Japan to Y.M.  Y.M. also received supports from Foundation for Promotion of Cancer Research, Sato Memorial Foundation for Cancer Research, Public Trust Haraguchi Memorial Cancer Research Fund and Osaka Cancer Research Foundation.  A part of the work is the result of "Initiating Events in the Recognition and Repair of DNA Double-Strand Breaks" carried out under the Strategic Promotion Program for Basic Nuclear Research, the Ministry of Education, Culture, Sport, Science and Technology, Japan.

## 6. References

Ahel, D., Horejsi, Z., Wiechens, N., Polo, S.E., Garcia-Wilson, E., Ahel, I., Flynn, H., Skehel, M., West, S.C., Jackson, S.P., Owen-Hughes, T. & Boulton, S.J. (2009) Poly(ADP-ribose)-dependent regulation of DNA repair by the chromatin remodeling enzye ALC1. *Science*, Vol. 325, pp. 1240-1243.

Ahel, I., Rass, U., El-Khamisy, S., Katyal, S., Clements, P., McKinnon, P., Caldecott, K. & West, S. (2006) The neurodegenerative disease protein aprataxin resolves abortive DNA ligation intermediates. *Nature*, Vol. 443, pp.713-716.

Ahnesorg, P., Smith, P. & Jackson, S. (2006) XLF interacts with the XRCC4-DNA ligase IV complex to promote DNA nonhomologous end-joining. *Cell*, Vol. 124, pp. 301-313.

Andegeko, Y., Moyal, L., Mittelman, L., Tsarfaty, I., Shiloh, Y. & Rotman, G. (2001) Nuclear retention of ATM at sites of DNA double strand breaks. *Journal of Biological Chemistry*, Vol. 276, pp. 38224-38230.

Anderson, C.W. & Lees-Miller, S.P. (1992) The nuclear serine/threonine protein kinase DNA-PK. *Critical Reviews on Eukaryotic Gene Expression*, Vol. 2., No.4, pp. 283-314.

Andres, S.N.; Modesti, M.; Tsai, C.J.; Chu, G. & Junop, M.S. (2007) Crystal structure of human XLF: a twist in nonhomologous DNA end-joining. *Molecular Cell*, Vol. 28, pp. 1093-1101.

Ariumi, Y.; Masutani, M.; Copeland, T.D.; Mimori, T.; Sugimura,T.; Shimotohno, K.; Ueda, K.; Hatanaka, M.; and Noda, M. (1999) Suppression of the poly(ADP-ribose) polymerase activity by DNA-dependent protein kinase in vitro. *Oncogene*, Vol. 18, pp. 4616-4625.

Badie, C.; Iliakis, G.; Foray, N.; Alsbeih, G.; Pantellias, G.E.; Okayasu, R.; Cheong, N.; Russell, N.S.; Begg, A.C.; Arlett, C.F. & Malaise, E.P. (1995) Defective repair of DNA double-strand breaks and chromosome damage in fibroblast from a radiosensitive leukemia patient. Cancer Research, Vol. 55, pp. 1232-1234.

Barnes, D., Stamp, G., Rosewell, I., Denzel, A. & Lindahl, T. (1998) Targeted disruption of the gene encoding DNA ligase IV leads to lethality in embryonic mice. *Current Biology*, Vol. 8, pp. 1395-1398.

Bau, D., Yang, M., Tsou, Y., Lin, S., Wu, C., Hsieh, H., Wang, R., Tsai, C., Chang, W., Hsieh, H., Sun, S. & Tsai, R. (2010) Colorectal cancer and genetic polymorphism of DNA double-strand break repair gene XRCC4 in Taiwan. *Anticancer Research*, Vol. 30, pp. 2727-2730.

Biederman, K.A.; Sun, J.; Giaccia, A.J.; Tosto, L.M. & Brown, J.M. (1991) scid mutation in mice confers hypersensitivity to ionizing radiation and a deficiency in DNA double-strand break repair. *Proceeding of National Academy of Science United States of America*, Vol. 88, pp. 1394-1397.

Blier, P.R.; Griffith, A.J.; Craft, J.; and Hardin, J.A. (1993) Binding of Ku protein to DNA. Measurement of affinity for ends and demonstration of binding to nicks. *J Biol. Chem.*, Vol. 268, pp. 7594-7601.

Bliss, T. & Lane, D. (1997) Ku selectively transfers between DNA molecules with homologous ends. *Journal of Biological Chemistry*, Vol. 272, pp. 5765-5773.

Blunt, T.; Finnie,N.; Taccioli, G.; Smith, G.; Demengeot, J.; Gottlieb, T.; Mizuta, R.; Varghese, A.; Alt, F.; Jeggo, P. & Jackson, S.P. (1995) Defective DNA-dependent protein kinase activity is linked to V(D)J recombination and DNA repair defects associated with the murine scid mutation. *Cell*, Vol. 80, pp. 813-823.

Bogue, M.A.; Wang, C.; Zhu, C. and Roth, D.B. (1997). V(D)J recombination in Ku86-deficient mice: distinct effects on coding, signal, and hybrid joint formation. *Immunity*, Vol. 7, pp. 37-47.

Bosma, G.; Custer, R. & Bosma, M. (1983) A severe combined immunodeficiency mutation in the mouse. *Nature*, Vol. 301, pp. 527-530.

Boulton, S.J. & Jackson, S.P. (1996) Identification of a *Saccharomyces cerevisiae* Ku80 homologue: roles in DNA double strand break rejoining and in telomeric maintenance. *Nucleic Acids Research*, Vol. 24, pp. 4639-4648.

Boulton, S.J. & Jackson, S.P. (1998) Components of the Ku-dependent non-homologous end-joining pathway are involved in telomeric length maintenance and teleomeric silencing. EMBO Journal, Vol. 17, pp. 1819-1828.

Bryans, M., Valenzano, M. & Stamato, T. (1999). Absence of DNA ligase IV protein in XR-1 cells: evidence for stabilization by XRCC4. *Mutation Research*, Vol. 433, pp. 53-58.

Buck, D., Malivert, L., de Chasseval, R., Barraud, A., Fondanèche, M., Sanal, O., Plebani, A., Stéphan, J., Hufnagel, M., le Deist, F., Fischer, A., Durandy, A., de Villartay, J. & Revy, P. (2006) Cernunnos, a novel nonhomologous end-joining factor, is mutated in human immunodeficiency with microcephaly. *Cell*, Vol. 124, pp. 287-299.

Caldecott, K.W. (2001) Mammalian DNA single-strand break repair: an X-ra(y)ted affair. *Bioessays*, Vol. 23, pp. 447-455.

Caldecott, K.W. (2003) XRCC1 and DNA strand break repair. *DNA Repair (Amst)*, Vol. 2, pp. 955-969.

Callebaut, I., Malivert, L., Fischer, A., Mornon, J., Revy, P. & de Villartay, J. (2006) Cernunnos interacts with the XRCC4 x DNA-ligase IV complex and is homologous to the yeast nonhomologous end-joining factor Nej1. *Journal of Biological Chemistry*, Vol. 281, pp. 13857-13860.

Calsou, P., Delteil, C., Frit, P., Drouet, J. & Salles,B. (2003) Coordinated assembly of Ku and p460 subunits of the DNA-dependent protein kinase on DNA ends is necessary for XRCC4-ligase IV recruitment. *Journal of Molecular Biology*, Vol.326, pp.93-103.

Cao, Q., Pitt, S., Leszyk, J. & Baril, E. (1994) DNA-dependent ATPase from HeLa cells is related to human Ku autoantigen. *Biochemistry*, Vol. 33, pp. 8548-8557.

Carter, T.; Vancurova, I.; Sun, I.; Lou, W. & DeLeon, S. (1990) A DNA-activated protein kinase from HeLa cell nuclei. *Molecular and Cellular Biology*, Vol. 10, pp. 6460-6471.

Chan, D.W. & Lees-Miller, S.P. (1996) The DNA-dependent protein kinase is inactivated by autophosphorylation of the catalytic subunit. *Journal of Biological Chemistry*, Vol. 271, pp. 8936-8941.

Chan, D.W., Chen, B.P., Prithivirajsingh, S., Kurimasa, A., Story, M.D., Qin, J. & Chen, D.J. (2002) Autophosphorylation of the DNA-dependent protein kinase catalytic subunit is required for rejoining of DNA double-strand breaks. *Genes & Development*, Vol. 16, pp. 2333-2338.

Chen, B.P.-C.; Uematsu, N.; Kobayashi, J.; Lerenthal, Y.; Krempler, A.; Yajima, H.; Lobrich, M.; Shiloh, Y. & Chenk D.J. (2007) ATM is essential for DNA-PKcs phosphorylation at T2609 cluster upon DNA double-strand break. *Journal of Biological Chemistry*, Vol. 282, pp. 6582-6587.

Chen, L., Trujillo, K., Sung, P. & Tomkinson, A. (2000). Interactions of the DNA ligase IV-XRCC4 complex with DNA ends and the DNA-dependent protein kinase. *Journal of Biological Chemistry*, Vol. 275, pp. 26196-26205.

Clements, P., Breslin, C., Deeks, E., Byrd, P., .Ju, L., Bieganowski, P., Brenner, C., Moreira, M., Taylor, A. & Caldecott, K. (2004). The ataxia-oculomotor apraxia 1 gene product has a role distinct from ATM and interacts with the DNA strand break repair proteins XRCC1 and XRCC4. *DNA Repair (Amst.)*, Vol. 3, pp. 1493-1502.

Cimprich,K.A.; Shin, T.B.; Keith, C.T. & Schreiber,S.L. (1996) cDNA cloning and gene mapping of a candicated human cell cycle checkpoint protein. *Proceeding of National Academy of Science United States of America*, Vol. 93, pp. 2850-2855.

Critchlow, S.; Bowater, R. & Jackson, S. (1997) Mammalian DNA double-strand break repair protein XRCC4 interacts with DNA ligase IV. *Current Biology*, Vol. 7, pp. 588-598.

d'Adda di Fagagna, F.; Weller, G.R.; Doherty, A.J. & Jackson, S.P. (2003) The Gam protein of bacteriophage Mu is an orthologue of eukaryotic Ku. *EMBO Reports*, Vol. 4, pp. 47-52.

Dai, Y., Kysela, B., Hanakahi, K., Manolis, K., Riballo, E., Stumm, M., Harville, T. O., West, S., C., Oettinger, M. A., and Jeggo, P. A. (2003) Nonhomologous end joining and V(D)J recombination require an additional factor. *Proceeding of National Academy of Science United States of America*, Vol. 100, pp. 2462-2467.

Date,H., Onodera,O., Tanaka,H., Iwabuchi,K., Uekawa,K., Igarashi,S., Koike,R., Hiroi,T., Yuasa,T., Awaya, Y., Sakai, T., Takahashi, T., Nagatomo, H., Sekijima, Y., Kawachi, I., Takiyama, Y., Nishizawa, M., Fukuhara, N., Saito, K., Sugano, S. & Tsuji, S. (2001) Early-onset ataxia with ocular motor apraxia and hypoalbuminemia is caused by mutations in a new HIT superfamily gene. *Nature Genetics*, Vol. 29, pp. 184–188.

DiBiase, S.J.; Zeng, Z.C.; Chen, R.; Hyslop, T.; Curran, Jr., W.J. and Iliakis, G. (2000). DNA-dependent protein kinase stimulates an independently active, nonhomologous, end-joining apparatus. *Cancer Res.*, Vol. 60, pp. 1245-1253.

Ding, Q.; Reddy, Y.; Wang, W.; Woods, T.; Douglas, P.; Ramsden, D.; Lees-Miller, S.P. & Meek, K. (2003) Autophosphorylation of the catalytic subunit of the DNA-dependent protein kinase is required for efficient end processing during DNA double-strand break repair. *Molecular and Cellular Biology*, Vol. 23, pp. 5836-5848.

D'Silva, I.; Pelletier, J.D.; Lagueux, J.; D'Amours, D.; Chaudhry, M.A.; Weinfeld, M.; Lees-Miller, S.P. and Poirier, G. G. (1999) Relative affinities of poly(ADP-ribose) polymerase and DNA-dependent protein kinase for DNA strand interruptions. *Biochim. Biophys. Acta.*, Vol. 1430, pp. 119-126.

Dvir, A.; Peterson, S.R.; Knuth, M.W.; Lu, H. & Dynan,W.S. (1992) Ku autoantigen is the regulatory component of a template-associated protein kinase that phosphorylates RNA polymerase II. *Proceeding of National Academy of Science United States of America*, Vol. 89, pp. 11920-11924.

Dvir, A.; Stein, L.Y.; Calore, B.L. & Dynan,W.S. (1993) Purification and characterization of a template-associated protein kinase that phosphorylates RNA polymerase II. *Journal of Biological Chemistry*, Vol. 268, pp.10440-10447.

Dore, A.S.; Drake, A.C., Brewerton, S.C. & Blundell, T.L. Identification of DNA-PK in the arthropods. Evidence for the ancient ancestry of vertebrate non-homologous end-joining. DNA Repair, Vol. 3., pp. 33-41.

Douglas, P., Sapkota, G., Morrice, N., Yu, Y., Goodarzi, A., Merkle, D., Meek, K., Alessi, D. & Lees-Miller, S. (2002) Identification of in vitro and in vivo phosphorylation sites in the catalytic subunit of the DNA-dependent protein kinase. *Biochemical Journal*, Vol. 368, pp.243-251.

Drouet, J., Delteil, C., Lefrancois, J., Concannon, P., Salles, B. & Calsou, P. (2005) DNA-dependent protein kinase and XRCC4-DNA ligase IV mobilization in the cell in response to DNA double strand breaks. *Journal of Biological Chemistry*, Vol. 280 pp. 7060-7069.

Dynan, W.S. & Yoo, S. (1998) Interaction of Ku protein and DNA-dependent protein kinase catalytic subunit with nucleic acids. *Nucleic Acids Research*, Vol. 26, pp. 1551-1559.

Falck, J.; Coates, J. & Jackson, S.P. (2005) Conserved modes of recruitment of ATM, ATR and DNA-PKcs to sites of DNA damage. *Nature*, Vol. 434, pp. 605-611.

Feldmann, H.; Driller, L.; Meier, B.; Mages, G.; Kellermann, J. & Winnacker,E.L. (1996) HDF2, the second subunit of the Ku homologue from *Saccharomyces cerevisiae*. *Journal of Biological Chemistry*, Vol. 271, pp. 27765-27769.

Feldmann, H. & Winnacker, E.L. (1993) A putative homologue of the human autoantigen Ku from Saccharomyces cerevisiae. *Journal of Biological Chemistry*, Vol. 268, pp. 12895-12900.

Ferguson, D.O.; Sekiguchi, J.M.; Chang, S.; Frank, K.M.; Gao, Y.; DePinho, R.A. and Alt, F.W. (2000) The nonhomologous end-joining pathway of DNA repair is required for genomic stability and the suppression of translocations. *Proc. Natl. Acad. Sci. U. S. A*, Vol. 97, pp. 6630-6633.

Foster, R., Nnakwe, C., Woo, L. & Frank, K. (2006) Monoubiquitination of the nonhomologous end joining protein XRCC4. *Biochemical & Biophysical Research Communications*, Vol. 341, pp. 175-183.

Frank, K., Sekiguchi, J., Seidl, K., Swat, W., Rathbun, G., Cheng, H., Davidson, L., Kangaloo, L. & Alt, F. (1998) Late embryonic lethality and impaired V(D)J recombination in mice lacking DNA ligase IV. *Nature*, Vol. 396, pp. 173-177.

Frank-Vaillant, M. & Marcand, S. (2001) NHEJ regulation by mating type is exercised through a novel protein, Lif2p, essential to the Ligase IV pathway. *Genes & Development*, Vol. 15, pp. 3005-3012.

Fulop, G.M. & Phillips, R.A. (1990). The scid mutation in mice causes a general defect in DNA repair. *Nature*, Vol. 347, pp. 479-482.

Gao, Y., Sun, Y., Frank, K., Dikkes, P., Fujiwara, Y., Seidl, K., Sekiguchi, J., Rathbun, G., Swat, W., Wang, J. *et al.* (1998) A critical role for DNA end-joining proteins in both lymphogenesis and neurogenesis. *Cell*, Vol. 95, pp. 891-902.

Galande, S. and Kohwi-Shigematsu, T. (1999) Poly(ADP-ribose) polymerase and Ku autoantigen form a complex and synergistically bind to matrix attachment sequences. *J Biol. Chem.*, Vol. 274, pp. 20521-20528.

Gell, D. & Jackson, S.P. (1999) Mapping of protein-protein interactions within the DNA-dependent protein kinase complex. *Nucleic Acids Research*, Vol. 27, pp.3494-3502.

Gottlieb, T.M. & Jackson,S.P. (1993) The DNA-dependent protein kinase: requirement for DNA ends and association with Ku antigen. *Cell*, Vol. 72, pp.131-142.

Grawunder, U., Wilm, M., Wu, X., Kulesza, P., Wilson, T., Mann, M. & Lieber, M. (1997) Activity of DNA ligase IV stimulated by complex formation with XRCC4 protein in mammalian cells. *Nature*, Vol. 388, pp. 492-495

Grawunder, U., Zimmer, D. & Leiber, M. (1998) DNA ligase IV binds to XRCC4 via a motif located between rather than within its BRCT domains. *Current Biology*, Vol. 8, pp. 873-876.

Gu, Y., Seidl, K., Rathbun, G., Zhu, C., Manis, J., der Van, S., Davidson, L., Cheng, H., Sekiguchi, J., Frank, K., Stanhope-Baker, P., Schlissel, M., Roth, D. & Alt, F. (1997) Growth retardation and leaky SCID phenotype of Ku70-deficient mice. *Immunity*, Vol. 7, pp. 653-665.

Hammel, M., Yu, Y., Fang, S., Lees-Miller, S. & Tainer, J. (2010) XLF regulates filament architecture of the XRCC4 ·ligase IV complex. *Structure*, Vol. 18, pp. 1431-1442.

Hartley, K.; Gell, D.; Smith, C.; Zhang, H.; Divecha, N.; Connelly, M.; Admon, A.; Lees-Miller, S.; Anderson, C. & Jackson, S. (1995) DNA-dependent protein kinase catalytic subunit: a relative of phosphatidylinositol 3-kinase and the ataxia telangiectasia gene product. *Cell*, Vol. 82, pp. 849-856.

Hartley, K.O.; Gell, D.; Smith, G.C.; Zhang, H.; Divecha, N.; Connelly, M.A.; Admon, A.; Lees-Miller, S.P.; Anderson, C.W. and Jackson, S.P. (1995). DNA-dependent protein kinase catalytic subunit: a relative of phosphatidylinositol 3-kinase and the ataxia telangiectasia gene product. *Cell*, Vol. 82, pp. 849-856.

Hentges, P., Ahnesorg, P., Pitcher, R., Bruce, C., Kysela, B., Green, A., Bianchi, J., Wilson, T., Jackson, S. & Doherty, A. (2006). Evolutionary and functional conservation of the DNA non-homologous end-joining protein, XLF/Cernunnos. *Journal of Biological Chemistry*, Vol. 281, pp. 37517-37526.

Herrmann, G.; Lindahl, T. & Schar, P. (1998) Saccharomyces cerevisiae LIF1: a function involved in DNA double-strand break repair related to mammalian XRCC4. EMBO Journal, Vol. 17, pp. 4188-4198.

Hsu, H., Gilley, D., Galande, S., Hande, M., Allen, B., Kim, S., Li, G., Campisi, J., Kohwi-Shigematsu,T. & Chen, D. (2000) Ku acts in a unique way at the mammalian telomere to prevent end joining. *Genes & Development*, Vol. 14, pp. 2807-2812.

Hudson, J.J.; Hsu, D.W.; Guo, K.; Zhukovskaya, N.; Liu, P.H.; Williams, J.G.; Pears, C.J. & Lakin, N.D. (2005) DNA-PKcs-dependent signaling of DNA damage in Dyctyostelium discodeum. *Current Biology*, Vol. 15, pp. 1880-1885.

Iles, N.; Rulten, S.; El-Khamisy, S.F. & Caldecott, K.W. (2007) APLF (C2orf13) is a novel human protein involved in the cellular response to chromosomal DNA strand breaks. *Molecular and Cellular Biology*, Vol. 27, pp. 3793-3803.

Iwabuchi, K., Hashimoto, M., Matsui, T., Kurihara, T., Shimizu, H., Adachi, N., Ishiai, M., Yamamoto, K.-i., Tauchi, H., Takata, M., Koyama, H. and Date, T. (2006) 53BP1 contributes to survival of cells irradiated with X-ray during G1 without Ku70 or Artemis. *Genes to Cells*, Vol. 11, pp. 935-948.

Jorgensen, T.J. and Shiloh, Y. (1996) The ATM gene and the radiobiology of ataxia-telangiectasia. *Int. J Radiat. Biol.*,Vol. 69, pp. 527-537.

Junop, M., Modesti, M., Guarne, A., Ghirlando, R., Gellert, M. & Yang, W. (2000) Crystal structure of the Xrcc4 DNA repair protein and implications for end joining. *EMBO Journal*, Vol. 19, pp. 5962-5970.

Kamdar, R. & Matsumoto, Y. (2010) Radiation-induced XRCC4 association with chromatin DNA analyzed by biochemical fractionation. *Journal of Radiation Research*, Vol. 51, pp. 303-313.

Kanno, S.; Kuzuoka, H.; Sasao, S.; Hong, Z.; Lan, L.; Nakajima, S. & Yasui, A. (2007) A novel human AP endonuclease with conserved zinc-finger-like motifs involved in DNA strand break responses. EMBO Journal, Vol. 26, pp. 2094-2103.

Karmakar, P., Snowden, C.M.; Ramsden, D.A. and Bohr, V.A. (2002) Ku heterodimer binds to both ends of the Werner protein and functional interaction occurs at the Werner N-terminus. *Nucleic Acids Res.*, Vol. 30, pp. 3583-3591.

Kegel, A.; Sjostrand, J.O.O. & Astrom, S.U. (2001) Nej1p, a cell type-specific regulator of nonhomologous end joining in yeast. *Current Biology*, Vol. 11, pp. 1611-1617.

Kienker, L.J.; Shin, E.K. & Meek, K. (2000) Both V(D)J recombination and radioresistance require DNA-PK kinase activity, though minimal levels suffice for V(D)J recombination. *Nucleic Acids Research*, Vol. 28, pp. 2752-2761.

Kim, J. S., Krasieva, T. B., LaMorte, V., Taylor, A. M. R. and Yokomori, K. (2002) Specific recruitment of human cohesin to laser-induce DNA damage. *Journal of Biological Chemistry*, Vol. 277, pp. 45149-45153.

Kirchgessner, C.; Patil, C.; Evans, J.; Cuomo, C.; Fried, L.; Carter, T.; Oettinger, M. & Brown, M. (1995) DNA-dependent kinase (p350) as a candidate gene for the murine SCID defect. *Science*, Vol. 267, pp. 1178-1183.

Koch, C., Agyei, R., Galicia, S., Metalnikov, P., O'Donnell, P., Starostine, A., Weinfeld, M. & Durocher, D. (2004). Xrcc4 physically links DNA end processing by polynucleotide kinase to DNA ligation by DNA ligase IV. *EMBO Journal*, Vol. 23, pp.3874-3885.

Kurimasa, A.; Kumano, S.; Boubnov, N.; Story, M.; Tung, C.; Peterson, S. & Chen, D. (1999) Requirement for the kinase activity of human DNA-dependent protein kinase catalytic subunit in DNA strand break rejoining. *Molecular and Cellular Biology*, Vol. 19, pp. 3877-3884.

Kusumoto, R., Dawut, L., Marchetti, C., Wan, L., Vindigni, A., Ramsden, D. & Bohr, V. (2008) Werner protein cooperates with the XRCC4-DNA ligase IV complex in end-processing. *Biochemistry*, Vol. 47, pp. 7548-7556.

Lan, L.; Ui, A.; Nakajima, S.; Hatakeyama, K.; Hoshi, M.; Watanabe, R.; Janicki, S.M.; Ogiwara, H.; Kohno, T.; Kannno, S. & Yasui, A. (2010) The ACF1 complex is required for DNA double-strand break repair in human cells. *Molecular Cell*, Vol. 40, pp. 976-987.

Leber, R.;.Wise, T.W.; Mizuta, R. and Meek,K.(1998) The XRCC4 gene product is a target for and interacts with the DNA-dependent protein kinase. *J Biol. Chem.*, Vol. 273, pp. 1794-1801.

Lee, K.-J., Dong, X., Wang, J., Takeda, Y., & Dynan, W.S. (2002) Identification of human autoantibodies to the DNA ligase IV/XRCC4 and mapping of an autoimmune epitope to a potential regulatory region. *Journal of Immunology*, Vol. 169, pp. 3413-3421.

Lee, K.-J., Jovanovic, M., Udayakumar, D., Bladen, C.L., & Dynan, W.S. (2003) Identification of DNA-PKcs phosphorylation sites in XRCC4 and effects of mutation at these sites on DNA end joining in a cell-free system. *DNA Repair*, Vol. 3, pp. 267-276.

Lees-Miller, S.P.; Chen,Y.-R. & Anderson,C.W. (1990) Human cells contain a DNA-activated protein kinase that phosphorylates simian virus 40 T antigen, mouse p53, and the human Ku autoantigen. *Molecular and Cellular Biology*, Vol. 10, pp. 6472-6481.

Lees-Miller, S., Sakaguchi, K., Ullrich, S., Appella, E. & Anderson, C. (1992) Human DNA-activated protein kinase phosphorylates serines 15 and 37 in the amino-terminal transactivation domain of human p53. *Molecular and Cellular Biology*, Vol. 12, pp. 5041-5049.

Lees-Miller, S.; Godbout, R.; Chan, D.; Weinfeld, M.; Day III, R.; Barron, G. and Allalunis-Turner, J. (1995). Absence of p350 subunit of DNA-activated protein kinase from a radiosensitive human cell line. *Science*, Vol. 267, pp. 1183-1185.

Li, B.; Navarro, S.; Kasahara, N. and Comai, L. (2004) Identification and biochemical characterization of a Werner's syndrome protein complex with Ku70/80 and poly(ADP-ribose) polymerase-1. *J Biol. Chem.*, Vol. 279, pp. 13659-13667.

Li, Y., Chirgadze, D., Bolanos-Garcia, V., Sibanda, B., Davies, O., Ahnesorg, P., Jackson, S. and Blundell, T. (2008) Crystal structure of human XLF/Cernunnos reveals unexpected differences from XRCC4 with implications for NHEJ. *EMBO Journal*, Vol. 27, pp. 290-300.

Li, Z.; Otevrel, T.; Gao, Y.; Cheng, H.; Seed, B.; Stamato, T.; Taccioli, G. & Alt, F. (1995) The XRCC4 gene encodes a novel protein involved in DNA double-strand break repair and V(D)J recombination. *Cell*, Vol., 83, pp. 1079-1089.

Lieber,M.R.; Hesse, J.E.; Lewis, S.; Bosma, G.C.; Rosenberg, N.; Mizuuchi, K.; Bosma, M.J. & Gellert,M. (1988). The defect in murine severe combined immune deficiency: joining of signal sequences but not coding segments in V(D)J recombination. *Cell*, Vol. 55, pp. 7-16.

Löbrich, M.; Rydberg, B. and Cooper, P.K. (1995) Repair of x-ray-induced DNA double-strand breaks in specific Not I restriction fragments in human fibroblasts: joining of correct and incorrect ends. *Proc. Natl. Acad. Sci. U. S. A.*, Vol. 92, pp.12050-12054.

Lu, H., Pannicke, U., Schwarz, K. & Lieber, M. (2007a) Length-dependent binding of human XLF to DNA and stimulation of XRCC4.DNA ligase IV activity. *Journal of Biological Chemistry*, Vol. 282, pp. 11155-11162.

Lu, H., Schwarz, K. & Lieber, M. (2007b) Extent to which hairpin opening by the Artemis:DNA-PKcs complex can contribute to junctional diversity in V(D)J recombination. *Nucleic Acids Research.*, Vol. 35, pp. 6917-6923.

Ma, Y.; Pannicke, U.; Schwarz, K. & Lieber, M.R. (2002) Hairpin opening and overhang processing by an Artemis/DNA-dependent protein kinase complex in nonhomologous end joining and V(D)J recombination. *Cell*, Vol. 108, pp. 781-794.

Mani, R., Yu, Y., Fang, S., Lu, M., Fanta, M., Zolner, A., Tahbaz, N., Ramsden, D., Litchfield, D., Lees-Miller, S., & Weinfeld, M. (2010) Dual modes of interaction between XRCC4 and polynucleotide kinase/phosphatase: implications for nonhomologous end joining. *J.Biol.Chem.*, Vol. 26, pp. 37619-37629.

Mari, P., Florea, B., Persengiev, S., Verkaik, N., Bruggenwirth, H., Modesti, M., Giglia-Mari, G., Bezstarosti, K., Demmers, J., Luider, T. & van Gent, D.C. (2006) Dynamic assembly of end-joining complexes requires interaction between Ku70/80 and XRCC4. *Proceeding of National Academy of Science United States of America*, Vol. 103, pp. 18597-18602.

Maser, R.S.; Monsen, K.J.; Nelms, B.E. & Petrini, J.H.J. (1997) hMre11 and hRad50 nuclear foci are induced during normal cellular response to DNA double-strand breaks. *Molecular and Cellular Biology*, Vol. 17, pp. 6087-6096.

Matsumoto, Y.; Suzuki, N.; Namba, N.; Umeda, N.; Ma, X.-J.; Morita, A.; Tomita, M.; Enomoto, A.; Serizawa, S.; Hirano, K.; Sakai, K.; Yasuda, H. & Hosoi, Y. (2000) Cleavage and phosphorylation of XRCC4 protein induced by X-irradiation. *FEBS Letters*, Vol. 478, pp. 67-71.

Matsuoka, S., Ballif, B., Smogorzewska, A., McDonald III, E., Hurov, K., Luo, J., Bakalarski, C., Zhao, Z., Solimini, N., Lerenthal, Y., Shiloh, Y., Gygi, S.P. & Elledge, S.J. (2007) ATM and ATR substrate analysis reveals extensive protein networks responsive to DNA damage. *Science,* Vol. 316, pp. 1160-1166.

McElhinny, N., Snowden, C., McCarville, J. & Ramsden, D. (2000). Ku recruits the XRCC4-ligase IV complex to DNA ends. *Molecular and Cellular Biology,* Vol. 20, pp.2996-3003.

Meek, K.; Kienker, L.; Dallas, C.; Wang, W.; Dark, M.J.; Venta, P.J.; Huie, M.L.; Hirschhorn, R. & Bell, T. (2000) SCID in Jack Russell terriers: a new animal model of DNA-PKcs deficiency. *Journal of Immunology,* Vol. 167, pp. 2142-2150.

Mehrotra, P.V.; Ahel, D.; Ryan, D.P.; Weston, R.; Wiechens, N.; Kraehenbuehl, R.; Owen-Hughes, T. & Ahel, I. (2011) DNA repair factor APLF is a histone chaperone. Molecular Cell, Vol. 41, pp. 46-55.

Milne, G.T.; Jin, S.; Shannon, K.B. & Weaver,D.T. (1996) Mutations in two Ku homologs define a DNA end-joining repair pathway in *Saccharomyces cerevisiae. Molecular and Cellular Biology,* Vol. 16, pp. 4189-4198.

Mimori, T.; Akizuki, M.; Yamagata, H.; Inada, S.; Yoshida, S. & Homma, M. (1981) Characterization of a high molecular weight acidic nuclear protein recognized by autoantibodies in sera from patients with polymyositis-scleroderma overlap. *Journal of Clinical Investigations,* Vol. 68, pp. 611-620.

Mimori, T.; Hardin, J.A. & Steitz, J.A. (1986) Characterization of the DNA-binding protein antigen Ku recognized by autoantibodies from patients with rheumatic disorders. *Journal of Biological Chemistry,* Vol. 261, pp.2274-2278.

Mimori, T. & Hardin, J.A. (1986) Mechanism of interaction between Ku protein and DNA. *Journal of Biological Chemistry,* Vol. 261, pp.10375-10379.

Modesti, M., Hesse, J. & Gellert, M. (1999). DNA binding of Xrcc4 protein is associated with V(D)J recombination but not with stimulation of DNA ligase IV activity. *EMBO Journal,* Vol. 18, pp. 2008-2018.

Moshous, D, Callebaut, I., de Chasseval, R., Corneo, B., Cavazzana-Calvo, M., le Deist, F., Tezcan, I., Sanal, O., Bertrand, Y., Philippe, N., Fischer, A. & de Villartay, J.P. (2001) Artemis, a novel DNA double-strand break repair/V(D)J recombination protein, is mutated in human severe combined immune deficiency. *Cell,* Vol. 105, pp.177-186.

Nelms, B.E.; Maser, R.S.; MacKay, J.F.; Lagally, M.G. & Petrini, J.H.J. (1998) In situ visualization of DNA double-strand break repair in human fibroblasts. *Science,* Vol. 280, pp. 590-592.

Nevaldine, B.; Longo, J.A. and Hahn, P.J. (1997). The scid defect results in much slower repair of DNA double-strand breaks but not high levels of residual breaks. *Radiat. Res.,* Vol. 147, pp. 535-540.

Nussenzweig, A.; Chen, C.; de Costa Soares, V.; Sanchez, M.; Sokol, K.; Nussenzweig, M.C. & Li, G.C. (1996) Requirement for Ku80 in growth and immunoglobulin V(D)J recombination. *Nature,* Vol. 382, pp. 551-555.

O'Driscoll, M.; Cerosaletti, K.M.; Girard, P.M.; Dai, Y.; Stumm, M.; Kysela, B.; Hirsch, B.; Gennery, A.; Palmer, S.E.; Seidel, J.; Gatti, R.A.; Varon, R.; Oettinger, M.A.; Neitzel, H.; Jeggo, P.A. & Concannon, P. (2001) DNA ligase IV mutations identified in patients exhibiting developmental delay and immunodeficiency. *Molecular Cell,* Vol. 8, No.6, (December 2001), pp. 1175-1185.

Ooi, S.L.; Shoemaker, D.D. & Boeke, J.D. (2001) A DNA microarray-based genetic screen for nonhomologous end-joining mutants in *Saccharomyces cerevisiae*. *Science*, Vol. 294, pp. 2552-2556.

Ouyang, H., Nussenzweig, A., Kurimasa, A., Soares, V., Li, X., Cordon-Cardo, C., Li, W., Cheong, N., Nussenzweig, M., Iliakis, G., Chen, D. & Li, G. (1997). Ku70 is required for DNA repair but not for T cell antigen receptor gene recombination In vivo. *Journal of Experimental Medicine*, Vol. 186, pp. 921-929.

Perrault, R., Wang, H., Wang, M., Rosidi, B. & Iliakis, G. (2004) Backup pathways of NHEJ are suppressed by DNA-PK. *Journal of Cellular Biochemistry*, Vol. 92, pp. 781-794.

Peterson, S., Kurimasa, A., Oshimura, M., Dynan, W., Bradbury, E. & Chen, D. (1995) Loss of the catalytic subunit of the DNA-dependent protein kinase in DNA double-strand-break-repair mutant mammalian cells. *Proceeding of National Academy of Science United States of America*, Vol. 92, pp. 3171-3174.

Poinsignon, C.; De, C.R.; Soubeyrand, S.; Moshous, D.; Fischer, A.; Hache, R.J.and de Villartay, J.P. (2004). Phosphorylation of Artemis following irradiation-induced DNA damage. *Eur. J Immunol.*, Vol. 34, pp. 3146-3155.

Porter, S.; Greenwell, P.; Ritchie, K. & Petes, T. (1996) The DNA-binding protein Hdf1p (a putative Ku homologue) is required for maintaining normal telomere length in Saccharomyces cerevisiae. *Nucleic Acids Research*, Vol. 24, pp. 582-585.

Poltoratsky,V.; Shi,X.; York, J.; Lieber, M. & Carter, T. (1995) Human DNA-activated protein kinase (DNA-PK) is homologous to phosphatidylinositol kinases. *Journal of Immunology*, Vol. 155, pp. 4529-4533.

Ramadan, K.; Maga, G.; Shevelev, I.V.; Villani, G.; Blanco, L. and Hubscher, U. (2003). Human DNA polymerase lambda possesses terminal deoxyribonucleotidyl transferase activity and can elongate RNA primers: implications for novel functions. *J Mol. Biol.*, Vol. 328, pp. 63-72.

Ramsden, D. & Gellert, M. (1998) Ku protein stimulates DNA end joining by mammalian DNA ligases: a direct role for Ku in repair of DNA double-strand breaks. *EMBO Journal*, Vol. 17, pp. 609-614.

Riballo, E., Kuhne, M., Rief, N., Doherty, A., Smith, G., Recio, M., Reis, C., Dahm, K., Fricke, A., Krempler, A., Parker, A., Jackson, S., Gennery, A., Jeggo, P. & Lobrich, M. (2004). A pathway of double-strand break rejoining dependent upon ATM, Artemis, and proteins locating to gamma-H2AX foci. *Molecular Cell*, Vol. 16, pp. 715-724.

Riballo, E.; Critchlow, S.E.; Teo, S.H.; Doherty, A.J.; Priestley, A.; Broughton, B.; Kysela, B.; Beamish, H.; Plowman N.; Arlett, C.F.; Lehmann, A.R.; Jackson, S.P. & Jeggo, P.A. (1999) Identification of a defect inDNA ligase IV in a radiosensitive leukemia patient. *Current Biology*, Vol. 9, pp. 699-702.

Roberts, S.A.; Strande, N.; Burkhalter, M.D.; Strom, C.; Havener, J.M.; Hasty, P. & Ramsden, D.A. (2008) Ku is a 5'-dRP/AP lyase that excises nucleotide damage near broken ends. *Nature*, Vol. 464, pp. 1214-1217.

Roth, D.B. and Wilson, J.H. (1986) Nonhomologous recombination in mammalian cells: role for short sequence homologies in the joining reaction. *Mol. Cell Biol.*, Vol. 6, pp. 4295-4304.

Rulten, S., Fisher, A., Robert, I., Zuma, M., Rouleau, M., Ju, L., Poirier, G., Reina-San-Martin, B. & Caldecott, K. (2011) PARP-3 and APLF function together to accelerate nonhomologous end-joining. *Molecular Cell,* Vol. 41, pp. 33-45.

Ruscetti, T.; Lehnert, B.E.; Halbrook, J.; Le, T.H.; Hoekstra, M.F.; Chen, D.J.and Peterson, S.R. (1998). Stimulation of the DNA-dependent protein kinase by poly(ADP-ribose) polymerase. *J Biol. Chem.,* Vol. 273, pp. 14461-14467.

Savitsky, K.; Bar-Shira, A.; Gilad, S.; Rotman, G.; Ziv, Y.; Vanagaite, L.; Tagle, D.A.; Smith, S.; Uziel, T.; Sfez, S.; Ashkenazi, M.; Pecker, I.; Frydman, M.; Harnik, R.; Patanjali, S.R.; Simmons, A.; Clines, G.A.; Sartiel, A.; Jaspers, N.G.J.; Taylor, A.M.R.; Arlett, C.F.; Miki, T.; Weissmn, S.M.; Lovett, M.; Collins, F.S. & Shiloh,Y. (1995) A single ataxia telangiectasia gene with a product similar to PI-3 kinase. *Science,* Vol. 268, pp. 1749-1753.

Sekiguchi, J., Ferguson, D., Chen, H., Yang, E., Earle, J., Frank, K., Whitlow, S., Gu, Y., Xu, Y., Nussenzweig, A. & Alt, F.W. (2001) Genetic interactions between ATM and the nonhomologous end-joining factors in genomic stability and development. *Proceeding of National Academy of Science United States of America,* Vol. 98, 3243-3248.

Sharpless, N.E.; Ferguson, D.O.; O'Hagan, R.C.; Castrillon, D.H.; Lee, C.; Farazi, P.A.; Alson, S.; Fleming, J.; Morton, C.C.; Frank, K.; Chin, L.; Alt, F.W and DePinho, R.A. (2001) Impaired nonhomologous end-joining provokes soft tissue sarcomas harboring chromosomal translocations, amplifications, and deletions. *Mol. Cell,* Vol. 8, pp. 1187-1196.

Shin, E., Perryman, L. & Meek, K. (1997). A kinase negative mutation of DNAPKcs in equine SCID results in defective coding and signal joint formation. *Journal of Immunology,* Vol. 158, pp. 565-3589.

Sibanda, B., Critchlow, S., Begun, J., Pei, X., Jackson, S., Blundell, T. & Pellegrini, L. (2001) Crystal structure of an Xrcc4-DNA ligase IV complex. *Nature Structural Bioloty,* Vol. 8, pp. 1015-1019.

Smider, V.; Rathmell, W.K.; Lieber,M.R. & Chu,G. (1994) Restoration of X-ray resistance and V(D)J recombination in mutant cells by Ku cDNA. *Science,* Vol. 266, pp.288-291.

Soutoglou, E. & Misteli, T. (2008) Activation of the cellular DNA damage response in the absence of DNA lesions. *Science,* Vol. 320, pp. 1507-1510.

Svetlova, M.; Solovjeva, L. & Tomilin, N. (2010) Mechanism of elimination of phosphorylated histone H2AX from chromatin after repair of DNA double-strand breaks. *Mutation Research,* Vol. 685, 54-60.

Taccioli, G.E.; Rathbun, G.; Oltz, E.; Stamato, T.; Jeggo, P.A. and Alt, F.W. (1993). Impairment of V(D)J recombination in double-strand break repair mutants. *Science,* Vol. 260, pp. 207-210.

Taccioli, G.E.; Gottlieb, T.M.; Blunt, T.; Priestley, A.; Demengeot, J.; Mizuta, R.; Lehmann, A.R.; Alt, F.W.; Jackson, S.P. & Jeggo, P.A. (1994) Ku80: product of the *XRCC5* gene and its role in DNA repair and V(D)J recombination. *Science,* Vol. 265, pp. 1442-1445.

Teo, S. & Jackson, S. (1997). Identification of Saccharomyces cerevisiae DNA ligase IV: involvement in DNA double-strand break repair. *EMBO Journal,* Vol. 16, pp. 4788-4795.

Teo, S. & Jackson, S. (2000). Lif1p targets the DNA ligase Lig4p to sites of DNA double-strand breaks. *Current Biology,* Vol. 10, pp. 165-168.

Tsai,C., Kim, S. & Chu, G. (2007). Cernunnos/XLF promotes the ligation of mismatched and noncohesive DNA ends. *Proceeding of National Academy of Science United States of America*, Vol. 104, pp. 7851-7856.

Tuteja, N.; Tuteja, R.; Ochem, A.; Taneja, P.; Huang, N.W.; Simoncsits, A.; Susic, S.; Rahman, K.; Marusic, L.; Chen, J.; Zhang, J.; Wang, S; Pongor, S. & Falaschi, A. (1994) Human DNA helicase II: a novel DNA unwinding enzyme identified as the Ku autoantigen. *EMBO Journal*, Vol .13, pp.4991-5001.

Valencia, M.; Bentele, M.; Vaze, M.B.; Herrmann, G.; Kraus, E.; Lee, S.E.; Schar, P. & Haber, J.E. (2001) NEJ1 controls non-homologous end joining in *Saccharomyces cerevisiae*. *Nature*, Vol. 414, pp. 666-669.

van der Burg, M.; IJspeert, H.; Verkaik, N.S.; Turul, T.; Wiegant, W.W.; Morotomi-Yano, K.; Mari, P.-O.; Tezcan, I.; Chen, D.J.; Zdzienicka, M.Z.; van Dongen, J.J.M.; van Gent, D.C. (2009) A DNA-PK mutation in a radiosensitive T-B- SCID patients inhibits Artemis activation and nonhomologous end-joining. *Journal of Clinical Investigations*, Vol. 119, pp.91-98.

Vishwanatha, J. & Baril, E. (1990). Single-stranded-DNA-dependent ATPase from HeLa cells that stimulates DNA polymerase alpha-primase activity: purification and characterization of the ATPase. *Biochemistry*, Vol. 29, pp. 8753-8759.

Walker, A.I.; Hunt, T.; Jackson, R.J. & Anderson, C.W. (1985) Double-stranded DNA induces the phosphorylation of several proteins including the 90 000 mol. wt. heat-shock protein in animal cell extracts. *EMBO Journal*, Vol. 4., pp. 139-145.

Walker, J.R.; Corpina, R.A. & Goldberg, J. (2001) Structure of the Ku heterodimer bound to DNA and its implication for double-strand break repair. *Nature*, Vol. 412, pp. 607-614.

Wang, H., Zeng, Z., Perrault, A., Cheng, X., Qin, W. & Iliakis, G. (2001). Genetic evidence for the involvement of DNA ligase IV in the DNA-PK-dependent pathway of non-homologous end joining in mammalian cells. *Nucleic Acids Research*, Vol. 29, pp. 1653-1660.

Wang, H., Perrault, R., Takeda, Y., Qin, W., Wang, H. & Iliakis,G. (2003). Biochemical evidence for Ku-independent backup pathways of NHEJ. *Nucleic Acids Research*, Vol. 31, pp. 5377-5388.

Wang, Y., Nnakwe, C., Lane, W., Modesti, M. & Frank, K. (2004). Phosphorylation and regulation of DNA ligase IV stability by DNA-dependent protein kinase. *Journal of Biological Chemistry*, Vol. 279, pp. 37282-37290.

Wang, H., Rosidi, B., Perrault, R., Wang, M., Zhang, L., Windhofer, F. & Iliakis, G. (2005). DNA ligase III as a candidate component of backup pathways of nonhomologous end joining. *Cancer Research*, Vol. 65, pp. 4020-4030.

Weinfeld, M., Chaudhry, M, D'Amours, D., Pelletier, J., Poirier, G., Povirk, L. & Lees-Miller, S. (1997) Interaction of DNA-dependent protein kinase and poly(ADP-ribose) polymerase with radiation-induced DNA strand breaks. *Radiation Research*, Vol. 148, pp. 22-28.

Weller, G.R.; Kysela, B.; Roy, R.; Tonkin, L.M.; Scanlan, E.; Della, M.; Devine, S.K.; Day, J.P.; Wilkinson, A.; d'Adda di Fagagna, F.; Devine, K.M.; Bowarter, R.P.; Jeggo, P.A.; Jackson, S.P. & Doherty, A.J. (2002) Identification of a DNA nonhomologous end-joining complex in bacteria. *Science*, Vol. 297, pp. 1686-1689.

Whitehouse, C.J.; Taylor, R.M.; Thistlethwaite, A.; Zhang, H.; Karimi-Busheri, F.; Lasko, D.D.; Weinfeld, M.and Caldecott, K.W.(2001) XRCC1 stimulates human polynucleotide kinase activity at damaged DNA termini and accelerates DNA single-strand break repair. *Cell*, Vol. 104, pp. 107-117.

Wiler, R.; Leber, R.; Moore, B.B.; VanDyk, L.F.; Perryman, L.E. & Meek, K. (1995) Equine severe combined immunodeficiency: a defect in V(D)J recombination and DNA-dependent protein kinase activity. *Proceeding of National Academy of Science United States of America*, Vol. 92, pp. 11485-11489.

Wilson, T.; Grawunder, U. & Lieber, M. (1997) Yeast DNA ligase IV mediates non-homologous DNA end joining. *Nature*, Vol. 388, pp. 495-498.

Wu, K., Wang, C., Yang, Y., Peng, C., Lin, W., Tsai, F., Lin, D. & Bau D. (2010). Significant association of XRCC4 single nucleotide polymorphisms with childhood leukemia in Taiwan. *Anticancer Research*, Vol. 30, pp. 529-533.

Wu, P., Frit, P., Malivert, L., Revy, P., Biard, D., Salles, B. & Calsou, P. (2007) Interplay between Cernunnos-XLF and nonhomologous end-joining proteins at DNA ends in the cell. *Journal of Biological Chemistry*, Vol. 282, pp. 31937-31943.

Wu, P., Frit, P., Meesala, S., Dauvillier, S., Modesti, M., Andres, S., Huang, Y., Sekiguchi, J. Calsou, P., Salles, B. & Junop, M. (2009) Structural and functional interaction between the human DNA repair proteins DNA ligase IV and XRCC4. *Molecular and Cellular Biology*, Vol. 29, pp. 3163-3172.

Yajima, H.; Lee, K.J. & Chen, B.P. (2006) ATR-dependent phosphorylation of DNA-dependent protein kinase catalytic subunit in response to UV-induced replication stress. *Molecular & Cellular Biology*, Vol. 26, pp. 7520-7528.

Yaneva, M., Wen, J., Ayala, A. & Cook, R. (1989). cDNA-derived amino acid sequence of the 86-kDa subunit of the Ku antigen. *Journal of Biological Chemistry*, Vol. 264, pp. 13407-13411.

Yano, K., Morotomi-Yano, K., Wang, S., Uematsu, N., Lee, K., Asaithamby, A., Weterings, E. & Chen, D. (2008). Ku recruits XLF to DNA double-strand breaks. *EMBO Reports*, Vol. 9, pp. 91-96.

Yano, K.; Morotomi-Yano, K.; Lee, K.J. & Chen, D.J. (2011) Functional significance of the interaction with Ku in DNA double-strand recognition of XLF. FEBS Letters, Vol. 585, pp. 841-846.

Yu, Y., Wang, W., Ding, Q., Ye R., Chen, D., Merkle, D., Schriemer, D., Meek, K. & Lees-Miller, S.P. (2003) DNA-PK phosphorylation sites in XRCC4 are not required for survival after radiation or for V(D)J recombination. *DNA Repair*, Vol. 2, pp. 1239-1252.

Yurchenko, V., Xue, Z., & Sadofsky, M. (2006) SUMO modification of human XRCC4 regulates its localization and function in DNA double-strand break repair. *Molecular & Cellular Biology*, Vol. 26, pp. 1786-1794.

Zhang, W.W. and Yaneva, M. (1992). On the mechanisms of Ku protein binding to DNA. *Biochem. Biophys. Res. Commun.*, Vol. 186, pp. 574-579.

Zhang, X., Succi, J., Feng, Z., Prithivirajsingh, S., Story, M. & R.J.Legerski. (2004) Artemis is a phosphorylation target of ATM and ATR and is involved in the G2/M DNA damage checkpoint response. *Molecular & Cellular Biology*, Vol. 24, pp. 9207-9220.

Zhu, C.; Bogue, M.A.; Lim, D.-S.; Hasty, P. & Roth, D.B. (1996) Ku86-deficient mice exhibit severe combined immunodeficiency and defective processing of V(D)J recombination intermediates. *Cell*, Vol. 86, pp. 379-389.

# 4

# DNA Damage Recognition for Mammalian Global Genome Nucleotide Excision Repair

Kaoru Sugasawa

*Biosignal Research Center, Organization of Advanced Science and Technology*
*Kobe University*
*Japan*

## 1. Introduction

As a blueprint for genetic information, the structural and functional integrity of DNA must be maintained during cell division and gamete formation. However, this fundamental principle is threatened continuously by the vulnerability of DNA itself and/or by assaults from endogenously produced agents, such as reactive oxygen species and other metabolites, as well as various environmental agents including ultraviolet light (UV), ionizing radiation and chemical compounds (Friedberg et al., 2006). Among the DNA components, bases in particular are frequently the targets for such insults. Because DNA replication and transcription rely on the formation of specific base pairs, even a subtle change in the base structures can compromise faithful propagation and the expression of genetic information. For instance, replicative DNA polymerases, which exhibit very high intrinsic fidelity, are often blocked at sites where template bases are modified, which can lead to replication fork collapse and consequent chromosomal aberrations and/or cell death. This problem is overcome, at least partly, by translesion DNA synthesis, which is an error-prone process (Friedberg et al., 2005). To minimize the risk of mutagenesis, it is crucial for growing cells to detect and to remove damaged bases as much as possible before replication forks collide with them.

Nucleotide excision repair (NER) is a major DNA repair pathway that can eliminate an extremely broad spectrum of base damage. The NER substrates include dipyrimidinic UV photolesions, such as cyclobutane pyrimidine dimers (CPDs) and pyrimidine-pyrimidone (6-4) photoproducts (6-4PPs), intrastrand crosslinks caused by bifunctional alkylating agents (e.g., cisplatin), and bulky base adducts induced by numerous chemical carcinogens (Gillet & Schärer, 2006). The common feature shared by all of these insults does not reside in their chemical structure, but rather in the accompanying distortions of the otherwise regular DNA helical structure. Two subpathways are associated with mammalian NER: global genome NER (GG-NER) and transcription-coupled NER (TC-NER). GG-NER is a general pathway that operates throughout the genome. It minimizes the collision of replication forks with damaged bases and, thereby, contributes to the maintenance of genome integrity (Gillet & Schärer, 2006). TC-NER is specialized to remove transcription-blocking lesions from the template DNA strands, which ensures rapid recovery of transcriptional activity and thus averts apoptosis (Hanawalt & Spivak, 2008). In humans, hereditary defects in NER are associated with several autosomal recessive disorders, including xeroderma pigmentosum (XP), Cockayne syndrome (CS) and trichothiodystrophy (TTD) (Bootsma et al., 2001). The clinical hallmarks exhibited by patients with XP, which include marked photosensitivity and a predisposition to skin cancer, explicitly indicate that the impaired repair of UV-induced DNA photolesions promotes mutagenesis and carcinogenesis in the skin. Numerous genetic complementation groups have been identified for the above diseases, including 8 for XP (XP-A through -G, and variant), 2 for CS (CS-A and -B) and 1 for TTD (TTD-A). Cloning of the responsible genes has revealed that all of them encode proteins involved in the NER pathway. The notable exception is the XP variant (*XPV*) gene encoding DNA polymerase η that is involved in translesion DNA synthesis, but not in NER. Another important milestone in elucidating the NER mechanism has been the establishment of the cell-free system, which faithfully recapitulates the NER reaction with human whole cell extracts. Together, genetic and biochemical studies have successfully identified more than 30 polypeptides that are involved in mammalian GG-NER (Fig. 1). A fundamental challenge for GG-NER is that the cells must detect a small number of injured bases among the vast excess of normal bases comprising the huge genome. Although the complete network of mechanisms has not yet been entirely uncovered, recent studies have revealed some of the sophisticated molecular mechanisms that accomplish this difficult task, which involves concerted actions of multiple protein factors. This chapter overviews the latest progress in our understanding of the damage-recognition mechanism for mammalian GG-NER.

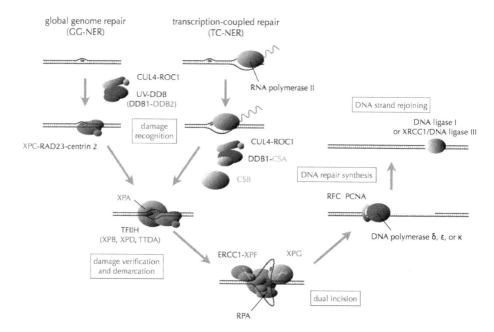

Fig. 1. Model of the mammalian NER mechanism. The 2 subpathways, GG-NER and TC-NER, differ in their strategies for initial damage recognition, but converge into a common process. The disease-related gene products are indicated by letters in different colors: XP, red; CS, green; TTD, blue.

## 2. Primary damage sensors for the initiation of GG-NER

One of the remarkable characteristics of GG-NER resides in its extremely broad substrate specificity, which encompasses UV-induced photolesions and other bulky base adducts that can be induced by numerous chemical compounds (Gillet & Schärer, 2006). These GG-NER substrates are associated with considerable levels of DNA helical distortion. This situation is in marked contrast to substrates for base excision repair (BER), such as uracils and oxidative base lesions, which are supposed to induce only marginal structural distortions. Initial damage detection for BER is accomplished by a set of DNA glycosylases, each of which exhibits a certain (partially overlapping) range of substrate specificity. In contrast, a virtually infinite spectrum of helix-distorting insults can be handled by the unified molecular machinery in GG-NER. In addition, GG-NER must survey the huge genome continuously and discriminate a small number of injured bases from normal bases with very high efficiency and accuracy. Recent biochemical studies have uncovered some of the sophisticated molecular mechanisms that achieve this difficult task.

### 2.1 Indirect sensing of DNA damage by XPC

The *XPC* gene was isolated from a cDNA expression library (Legerski & Peterson, 1992) that corrected the UV sensitivity of fibroblasts from patients with XP-C. Cells lacking XPC are incompetent for GG-NER, but TC-NER functions normally (Venema et al., 1990). By using the cell-free NER system, a protein factor that is missing in XP-C cells was purified from HeLa cell extracts (Masutani et al., 1994; Shivji et al., 1994). This biochemical approach revealed that the XPC protein forms a stable complex in vivo with 1 of the 2 human homologues of *Saccharomyces cerevisiae* Rad23p (designated RAD23A and RAD23B). Depletion of RAD23 markedly destabilized the XPC protein, thereby compromising GG-NER function (Ng et al., 2003; Okuda et al., 2004). Another component of the XPC complex, centrin-2 (Araki et al., 2001; Nishi et al., 2005), belongs to the calmodulin superfamily of small calcium-binding proteins containing 4 conserved EF-hand motifs. A subpopulation of centrin-2 localizes to the centrosomes and plays a vital role in cell cycle regulation (Lutz et al., 2001; Salisbury et al., 2002). Centrin-2 also binds to an α-helix near the C-terminus of XPC: this interaction potentiates the DNA-binding activity of the complex (Bunick et al., 2006; Nishi et al., 2005; Popescu et al., 2003; Thompson et al., 2006).

The XPC protein complex has been known to be associated with DNA-binding activity since it was first purified (Masutani et al., 1994; Shivji et al., 1994), although its preference for damaged DNA was discovered sometime later (Batty et al., 2000; Sugasawa et al., 1998). With conventional electrophoretic mobility shift assays (EMSAs) and DNase I footprint analyses with defined DNA substrates, we demonstrated that XPC prefers to associate with sites containing a helix-distorting lesion, such as 6-4PP or *N*-(guanin-8-yl) *N*-acetyl-2-amino-fluorene (dG-AAF) adduct (Sugasawa et al., 1998; Sugasawa et al., 2001). However, the addition of an appropriate competitor DNA was necessary to reveal the damage specificity, by preventing XPC from binding to the undamaged part of the DNA. Several physicochemical approaches subsequently were undertaken to assess the affinities of XPC for various DNA structures in more dynamic states (Hey et al., 2002; Roche et al., 2008; Trego & Turchi, 2006).

Involvement of the XPC complex in the very early stages of NER was first proposed on the basis of the results obtained with the cell-free NER system (Sugasawa et al., 1998). In this system, 2 plasmid DNAs containing AAF adducts were preincubated separately with different sets of NER factors, for which either XP cell extracts or purified recombinant proteins were used. After the 2 mixtures were combined and missing NER factors, if any, were supplemented, the initial repair rates of the 2 damaged DNA substrates were compared directly in one reaction. Damaged DNA preincubated in the presence of XPC was always repaired preferentially compared to DNA preincubated in its absence. Because a similar repair bias was not observed with other NER factors, these findings strongly suggest that XPC initiates in vitro NER, and its binding to damaged DNA is sufficient to recruit the whole repair machinery.

Several subsequent studies have supported this model. Local UV irradiation through micropore membrane filters has been used to visualize the recruitment of NER factors in cultured cells to the sites of DNA damage. Use of this method revealed that XPC accumulates at subnuclear UV-damaged areas, even when any other XP genes were mutated (Volker et al., 2001). Conversely, none of the other NER-related XP proteins (except for DDB2; see below) was recruited to the sites of DNA damage in XPC-deficient cells, consistent with the role of XPC as the initiator of GG-NER. Through the use of paramagnetic beads immobilized with a damaged DNA substrate, more refined biochemical studies were undertaken to determine the order of arrival and departure of individual NER proteins at the lesion: these studies also concluded that XPC arrives first (Riedl et al., 2003). It should be noted that only GG-NER is impaired in XP-C (and also XP-E) cells, unlike other NER-deficient XP cells, in which both GG-NER and TC-NER are affected. Considering that TC-NER is supposed to be triggered by RNA polymerase II stumbling at damaged bases on the template DNA strand, it could be assumed that the 2 NER subpathways vary only in their strategies for initial damage recognition and eventually merge into a common process.

Because XPC appeared to bind specifically to various lesions that did not share any common chemical structure, it was of great interest to understand which feature of DNA determined its binding specificity. To examine this, using EMSA, we tested XPC binding with various DNA substrates containing a defined lesion and/or artificial structure (Sugasawa et al., 2001; Sugasawa et al., 2002). XPC was able to recognize and to bind DNA duplexes containing a partially single-stranded region, such as bubble and loop structures, even though these substrates contained only base mismatches, but no chemical modifications. Further analyses using various oligonucleotides as competitors revealed that XPC was targeted preferentially to a branched DNA structure containing a double-stranded region attached to a single-stranded 3'-overhang. On the basis of these results, it might be better to refer to XPC as a *structure-specific DNA-binding factor*, rather than as a damage recognition factor.

The binding of XPC to sites of DNA damage seems to depend solely on the extent of local unwinding of the DNA duplex caused by a given lesion: typically, XPC showed very little affinity for sites of CPD, because of the subtle DNA helical distortion associated with this lesion. In contrast, the presence of 1 or 2 mismatched bases opposite the photodimer significantly enhanced binding by XPC (Sugasawa et al., 2001). Accordingly, this biochemical feature of XPC may provide an important molecular basis for the substrate specificity of GG-NER, including an infinite range of helix-distorting lesions, but not a number of nonbulky lesions, such as oxidized and deaminated bases.

More recently, a structural study corroborated this DNA-binding mode of XPC (Min & Pavletich, 2007). The *S. cerevisiae* NER protein Rad4p is presumed to be the counterpart of mammalian XPC: both proteins share several conserved structural domains in their C-terminal regions, including the transglutaminase-homology domain (TGD) and 3 consecutive β-hairpin domains (designated BHD1, BHD2, and BHD3). The X-ray crystal structure was solved with the C-terminal region of Rad4p bound to a short DNA duplex containing a CPD (which was placed within 3-base mismatches to enhance recognition by Rad4p). Consistent with the results of our footprint analyses with XPC, the results showed that Rad4p binds asymmetrically to the damaged DNA: it interacts with an 11-base pair segment of DNA duplex on the 3' side of CPD, mainly through TGD and BHD1, leaving the other double-stranded part on the 5' side of the lesion completely free. In the closer vicinity of the lesion, BHD3 is inserted into the major groove, such that BHD2 and BHD3 appear to pinch the phosphate-sugar backbone of the undamaged strand. As a result, 2 "normal" bases on the undamaged DNA strand are flipped out and held by BHD2-BHD3, while the CPD is also flipped out structurally disordered, and devoid of any contact with the protein (Fig. 2). The Rad4p binding results in a ~42° bend of DNA, as observed by our scanning force microscopy with the XPC-DNA complex (Janićijević et al., 2003). In conclusion, XPC/Rad4p appears to function as a versatile damage-recognition factor that senses the presence of oscillating normal bases within the DNA duplex.

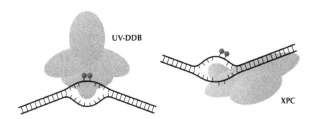

Fig. 2. Different binding modes of UV-DDB and XPC to damaged DNA sites. The β-hairpin of UV-DDB on the DDB2 β-propeller is inserted between the two strands of the DNA, so that DDB2 interacts directly with the damaged nucleotides flipped out of the DNA duplex. In contrast, XPC interacts with normal bases on the undamaged DNA strand without any contact with the damaged bases.

## 2.2 UV-DDB facilitates detection of UV-induced photolesions

In accordance with the proposed function of XPC as the initiator of GG-NER, most of the DNA lesions that are subject to GG-NER in vivo are recognized by XPC in vitro. However, CPD serves as a noticeable exception. Like other GG-NER substrates, CPDs are not removed from the global genome in *XPC*-deficient cells, although XPC by itself cannot find this type of insult (as described above). From this apparent discrepancy, it can be assumed that a certain factor (other than XPC) is responsible for the initial detection of CPDs, whereas XPC must be involved in later steps.

UV-damaged DNA-binding protein complex (UV-DDB) was first discovered as a factor that bound UV-damaged DNA with high affinity and specificity. The factor responsible for this binding activity was purified and revealed as a complex consisting of 2 subunits, designated DDB1 and DDB2, respectively (for a review, see Tang & Chu, 2002). It was later demonstrated that mutations in the *DDB2* gene constitute the XP genetic complementation group E (Rapić-Otrin et al., 2003). Recent studies have redefined DDB1 as an adaptor protein that mediates interactions between the CUL4-ROC1 ubiquitin ligase complex and a member of the substrate-recruiting subunit family, called DDB1-CUL4 associating factor (DCAF)

(Angers et al., 2006; Lee & Zhou, 2007). The DNA-binding specificity of purified UV-DDB has been characterized extensively (Fujiwara et al., 1999; Payne & Chu, 1994; Reardon et al., 1993; Treiber et al., 1992; Wittschieben et al., 2005). Concerning UV-induced photolesions, UV-DDB exhibits extraordinarily high affinity and specificity for 6-4PPs, although it also binds CPDs moderately. Although binding to chemical-induced base adducts seems not to be pronounced, abasic sites are relatively good substrates for UV-DDB.

Despite the above biochemical characteristics that explicitly point to roles in UV-damage recognition, the impact of defects in UV-DDB on NER has remained enigmatic. Cells from patients with XP-E have defects in GG-NER, but not in TC-NER. However, in contrast to XP-C, cells from patients with XP-E are proficient in removal of 6-4PPs from the global genome, while repair of CPDs seems to be affected profoundly (Hwang et al., 1999; Tang et al., 2000). As a result, among the NER-deficient XP groups, XP-E cells show the highest levels of residual UV-induced unscheduled DNA synthesis (>50% of normal cells) and resistance to killing by UV (Tang & Chu, 2002).

Unlike other XP-related gene products, DDB2 reportedly accumulates to local UV-damaged areas within the nucleus, even in the absence of XPC (Wakasugi et al., 2002), although XPC can relocate to sites containing UV-induced DNA damage in a DDB2-independent manner (Moser et al., 2005). Although UV-DDB and XPC appear to be recruited independently, UV irradiation always induces a mixture of various sorts of DNA injuries, including 6-4PPs, CPDs, and other less frequent insults. To solve this problem, elegant experiments have been undertaken, in which 6-4PPs were erased soon after local UV irradiation with the aid of an ectopically expressed 6-4PP photolyase (Fitch et al., 2003). Under these conditions where the remaining photolesions were mostly CPDs, DDB2-dependent recruitment of XPC became evident. These results clearly indicate that differential pathways are used for the deployment of XPC to sites of UV damage, depending on the type of lesions.

Considering the role for UV-DDB in CPD repair and its much stronger binding to 6-4PPs, one could assume that UV-DDB plays a role in the detection and repair of 6-4PPs. However, 6-4PPs are rapidly removed from the global genome even in the absence of DDB2 (most likely through direct recognition by XPC), so that stimulation by UV-DDB, if any, cannot be clearly discerned. Additionally, DDB2 undergoes degradation by the proteasome in response to UV irradiation (see below) (Fitch et al., 2003; Rapić-Otrin et al., 2002). Since this degradation is quite fast – particularly at relatively high UV doses – this situation further overshadows possible effects of UV-DDB on the repair of 6-4PPs.

Recently, the local UV irradiation technique has been applied to the quantification of 6-4PPs, which appear as fluorescent spots developed by an antibody specific for the photolesion (Moser et al., 2005). With this method, the total number of generated photolesions per cell was reduced substantially, and retardation of 6-4PP repair in the absence of UV-DDB became discernable. Similar conclusions were drawn from our experiments using fluorescence recovery after photobleaching (FRAP) (Nishi et al., 2009), which is a widely used method to assess the in vivo mobility of fluorescence-labeled proteins. With cells expressing NER factors fused to green fluorescent protein (GFP), global UV irradiation before photobleaching resulted in the significant retardation of fluorescence recovery within the bleached subnuclear region. This result indicated that the proteins concerned are sequestered at the sites of UV photolesions and engaged in NER (Houtsmuller et al., 1999).

The reduction in the mobility of GFP-XPC showed a unique biphasic relationship with the pre-UV dose. The immobilization of GFP-XPC was saturated at relatively low UV doses (5~10 J/m$^2$): higher UV doses resulted in further dose-dependent retardation of fluorescence recovery, which eventually became saturated again at extremely high doses (around 80~100 J/m²). Notably, the reduction in XPC mobility seemed to depend on the remaining 6-4PPs rather than on CPDs. Overexpression and siRNA knockdown of DDB2 revealed that the first immobilization of GFP-XPC (observed with low UV doses) was due to entrapment by UV-DDB bound to 6-4PPs (Nishi et al., 2009). These results indicate that UV-DDB likely contributes to the efficient detection of both of the major photolesions, particularly when the density of the induced lesions is low enough (in terms of physiologically relevant levels), and thereby recruits XPC and other NER factors.

Although the precise molecular mechanism underlying XPC recruitment by UV-DDB remains unclear, we have shown the presence of a direct physical interaction between these 2 damage-recognition factors by coimmunoprecipitation experiments (Sugasawa et al., 2005). Among the components of each complex, XPC and DDB2 appeared to be responsible for the interaction. More recently, researchers have solved the crystal structure for UV-DDB bound to a DNA duplex containing a 6-4PP (Scrima et al., 2008). DDB1 shows a unique structure containing 3 β-propeller domains (designated BPA, BPB, and BPC), whereas DDB2 has a β-propeller that is exclusively involved in its interaction with DNA. The N-terminal extension of DDB2 contains a helix-loop-helix motif, which mediates its interaction with DDB1. In this structure, UV-DDB approaches the lesion and inserts its evolutionarily conserved β-hairpin on the surface of the DDB2 β-propeller into the minor groove of the DNA, thereby causing a ~40° kink in the DNA. This β-hairpin seems to push the 2 affected bases out of the DNA duplex: these bases interact extensively with the amino acids that form a binding pocket on the surface of DDB2 (Fig. 2). The size of the binding pocket seems fit to accommodate 2 nucleotides, which suggests that DDB2 has evolved to recognize dinucleotide lesions, such as UV-induced photodimers. Considering that XPC interacts with the undamaged strand, XPC may gain access to the lesion from the side opposite to UV-DDB, sandwiching the DNA in between. However, the formation of such a ternary complex has not been demonstrated by EMSA or other methods.

## 2.3 Roles of ubiquitylation in GG-NER damage recognition

As mentioned above, UV-DDB is thought to be part of the ubiquitin ligase complex. Expression of the epitope-tagged DDB2 in cells and isolation of the protein complexes under relatively mild conditions have revealed that DDB2 associates in vivo with not only DDB1, but also with CUL4A-ROC1 and the COP9 signalosome (CSN) (Groisman et al., 2003). CSN, which is an 8-subunit complex possessing deneddylation and deubiquitylation activities, is believed to function as a negative regulator of the cullin-based ubiquitin ligase family (Lyapina et al., 2001; Yang et al., 2002). Upon UV irradiation of cells, UV-DDB relocates onto chromatin, where the associating ubiquitin ligase seems to be activated, judging from dissociation of CSN and neddylation of CUL4A (Groisman et al., 2003).

We have demonstrated that XPC is one of the substrates for this ubiquitin ligase (Sugasawa et al., 2005). After UV irradiation, slowly migrating, ubiquitylated forms of XPC became apparent. The appearance of these forms peaked around 1 h postirradiation, at which time the repair of 6-4PPs was rapidly ongoing. This transient ubiquitylation of XPC was detected even in NER-deficient XP and CS cells, with the only exception being XP-E cells. Notably, treatment of cells with a protein synthesis inhibitor, cycloheximide, revealed that ubiquitylated XPC had mostly reverted to its unmodified form, instead of being degraded. Subsequently, the recombinant DDB1-DDB2-CUL4A-ROC1 ubiquitin ligase complex was purified and successfully used for in vitro reconstitution of the XPC ubiquitylation. In this reaction, not only XPC but also DDB2 and CUL4A were found to be polyubiquitylated. It was previously reported that DDB2 undergoes degradation by the proteasome in response to UV irradiation (Fitch et al., 2003; Rapić-Otrin et al., 2002). These results suggest that the fates of the modified XPC and DDB2 are different, even though they seem to be ubiquitylated by the same ligase.

To elucidate the roles of ubiquitylation in the mechanism of GG-NER, we performed DNA-binding assays using paramagnetic beads immobilized with DNA containing the UV photolesions, CPD or 6-4PP (Sugasawa et al., 2005). In vitro ubiquitylation reactions in the presence of these DNA beads revealed that polyubiquitylation of DDB2 completely abolished the strong damaged DNA-binding activity of UV-DDB. In contrast, polyubiquitylated XPC in the same reaction continued to bind to DNA, with a slightly higher affinity than the unmodified form. Considering the remarkable difference in their affinities for UV-damaged DNA, it is conceivable that XPC cannot simply displace UV-DDB that is already bound to the site containing a photolesion.

When UV-DDB was added to cell-free NER reactions involving 6-4PP as a defined DNA substrate, only inhibition (and not stimulation) of dual incision was observed (Sugasawa et al., 2005). This finding suggested that UV-DDB tightly bound to the lesion may adversely block access to XPC and other NER factors, at least in vitro. Since this inhibition was partially alleviated by the addition of all of the components required for ubiquitylation, we proposed that damage handover from UV-DDB (strong binder) to XPC (weak binder) may be promoted by polyubiquitylation (Sugasawa et al., 2005; Sugasawa, 2006). Apart from these insights into the damage-recognition mechanism, the precise biological meanings of the UV-induced proteasomal degradation of DDB2 and the reversible polyubiquitylation of XPC remain to be understood.

*Ddb2*-deficient mice are characterized by a defect in UV-induced cellular apoptosis, in addition to a predisposition to skin cancer that was predicted from the phenotypes of human patients with XP-E (Itoh et al., 2004; Yoon et al., 2005). Although there have been some contradictory reports (Stubbert et al., 2007; Stubbert et al., 2009), the disappearance of DDB2 and/or the modification of XPC may be involved in a signal transduction pathway that regulates cellular responses to UV (Stoyanova et al., 2009). Among the known NER proteins, the expression of DDB2 and XPC is under the control of the p53 tumor suppressor (Adimoolam & Ford, 2002; Amundson et al., 2002; Hwang et al., 1999), whereas DDB2 conversely regulates p53 expression, thereby forming a regulatory circuit (Itoh et al., 2003).

Structural studies have suggested that the N-terminus of the rod-shaped CUL4 molecule anchors to the BPB domain of DDB1 (Angers et al., 2006; Scrima et al., 2008). In contrast to DDB2 and the other 2 β-propellers of DDB1 that seem to be fixed on the lesion, the BPB domain is supposed to exhibit considerable conformational flexibility. As a result, the ubiquitin ligase catalytic center assembled on the other tip of CUL4 is expected to move around within a certain spatial range (like a crane arm), and potentially ubiquitylate various targets around the lesion. Other substrates for the UV-DDB ubiquitin ligase include histones H2A (Kapetanaki et al., 2006), H3, and H4 (Wang et al., 2006). H3 and H4 ubiquitylation by the ligase reportedly leads to the dissociation of histone octamers from DNA. In this regard, it should be noted that the nucleosome assembly of DNA containing 6-4PPs interferes in vitro with lesion access to XPC, as well as the subsequent dual incision (Hara et al., 2000; Yasuda et al., 2005). On the other hand, owing to the substantial nonspecific DNA-binding activity of XPC, its specific binding to 6-4PP (observed with EMSAs) was easily competed out by the addition of undamaged DNA: this inhibition was dramatically attenuated by the organization of the competitor DNA into nucleosomes. Taken together, these studies indicate that nucleosome assembly may contribute to the masking of the undamaged part of the genomic DNA from useless surveillance by XPC, so that specific remodeling of the chromatin structures at relevant lesion sites can enhance damage discrimination tremendously.

In addition to ubiquitin ligase, the histone acetyltransferases CBP/p300 reportedly interact with UV-DDB (Datta et al., 2001; Rapić-Otrin et al., 2002), which suggests that multiple histone modifications may be involved in the reorganization of chromatin environments to allow the initiation of GG-NER. In the reconstituted cell-free system, UV-DDB is dispensable for and could even inhibit the repair of 6-4PPs, as described above. Moreover, its influence on CPD repair has been somewhat elusive, despite the obvious stimulatory effect observed in vivo. In some studies, significant stimulation of dual incision was obtained with the CPD substrate (Aboussekhra et al., 1995; Wakasugi et al., 2001; Wakasugi et al., 2002). However, other systems (including ours) showed no or only a minimal effect of UV-DDB on CPD repair, even in the presence of the components required for ubiquitylation (Reardon & Sancar, 2003; Sugasawa et al., 2005). As suggested by others, the involvement of chromatin structures may be important to reproduce the role for UV-DDB in the efficient recognition and repair of CPDs (Rapić Otrin et al., 1998).

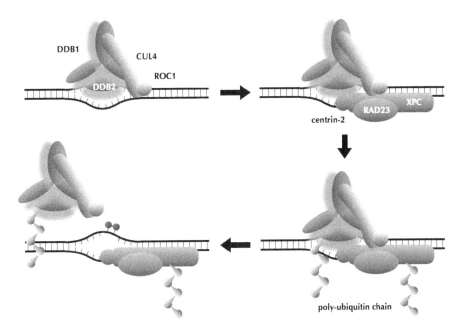

Fig. 3. Ubiquitylation-mediated damage handover model. Once UV-DDB binds to a UV photolesion, it recruits the XPC complex. The associating CUL4-ROC1 ubiquitin ligase is subsequently activated, thereby polyubiquitylating both XPC and DDB2. With the polyubiquitylation of DDB2, UV-DDB loses its affinity for damaged DNA, which results in the successful transfer of the lesion from UV-DDB to XPC.

## 3. The contribution of damage verification to the accuracy of GG-NER

As discussed in the previous section, there are at least 2 branches of damage-recognition pathways in GG-NER: sensing by XPC of unpaired bases associated with a wide variety of highly distorting lesions, and UV-DDB-dependent deployment of XPC that works specifically for UV-induced photolesions. However, particularly in the former pathway, XPC may bind to sites devoid of damage (e.g., bubble-like structures). The reason for this binding is that XPC can detect certain secondary structure of DNA, but not any feature of DNA chemistry. To avoid incision by NER at damage-free sites that could adversely challenge genomic stability, the verification of damage after XPC binding is fundamental.

### 3.1 Bipartite substrate discrimination model

Important clues to understand the structural determinants of NER substrate specificity were obtained from a series of biochemical studies. Among the key substrates were artificial DNA backbone lesions at the C4' position of the deoxyribose moiety (Buschta-Hedayat et al., 1999; Hess et al., 1997). Although these lesions were associated with little helix distortion and, thus, were hardly excised in human cell-free extracts, they were excised efficiently when combined with a small bubble structure. On the other hand, bubble structures devoid of lesions were never incised by NER. Based on these findings, the *bipartite substrate discrimination theory* was proposed, which states that efficient NER substrates must simultaneously contain 2 structural elements: disruption of canonical Watson-Crick base pairing (i.e., the presence of unpaired bases), and some aberrant modification of DNA chemistry. It should be noted that XPC senses the former, but not the latter, as described above. We later tested other DNA substrates containing a bubble structure and a dG-AAF adduct in various combinations (Sugasawa et al., 2001). XPC could bind to the bubble regardless of the presence or absence of the lesion, whereas in vitro NER incision occurred only when the AAF adduct existed at the bubble site. These results clearly indicate that DNA binding by XPC does not lead to dual incision in a straightforward manner. Instead, the presence of an alteration of DNA chemistry must be verified thereafter: in the case of no lesion, the repair process is aborted at a certain step.

One of difficulties with biochemical studies of the NER mechanism has been that its early process includes only assembly/disassembly of protein factors and unwinding of the DNA duplex: no chemical change in DNA occurs before dual incision, which is quite a late step in the repair reaction. However, mechanistic dissection of the early NER process was advanced recently by the finding that the 2 structural elements comprising NER substrates (i.e., unpaired base and chemical modification) are spatially separable (Sugasawa et al., 2009). With the C4' backbone lesions, it was already shown that those abnormal structures could be recognized and incised by NER in vitro, even if they resided a few bases apart from the end of a bubbled region (Buschta-Hedayat et al., 1999). Very recently, we showed that the distance between the 2 elements can be much longer (Sugasawa et al., 2009). Although CPDs are very poor substrates in our in vitro NER system because of the small helical distortion, enormous stimulation of dual incision was observed when a 3-base bubble was inserted about 60 bases on the 5' side of the lesion. Footprint analyses revealed that XPC was targeted to the bubble site, rather than to the CPD. This result indicated that the NER machinery was capable of searching around the XPC-bound site and finding the lesion at a distal position. More intriguingly, this stimulatory effect upon CPD recognition was abolished when the bubble was moved to the 3' side of the lesion.

The observed position specificity provided crucial insights into the molecular mechanism underlying the damage search. This mechanism was difficult to explain, if we assumed that the NER factors assembled at the XPC-bound site interacted in trans with the distal CPD. Instead, it seemed more likely that the damage search was accomplished by scanning the DNA strand in the 5' to 3' direction. This scanning mechanism was further supported by the observation that the stimulation of CPD removal was attenuated reciprocally by increasing the distance between the bubble and CPD. The damage search seemed to reach at least 160 bases from the bubble, but the efficiency declined if the distance was 400 bases or more.

### 3.2 Roles for TFIIH helicases in damage verification

Given the existence of a 5' to 3' scanning mechanism in damage verification, the transcription factor IIH (TFIIH) is thought to be the most likely candidate for performing the scan. TFIIH was originally identified as a basal transcription factor that is essential for the initiation of transcription by RNA polymerase II. TFIIH consists of 10 subunits, including 3 disease-related gene products, XPB, XPD, and TTDA (Giglia-Mari et al., 2004). Electron microscopic analyses of the purified TFIIH complex have revealed a ring-shaped structure, in which the spatial arrangement of individual subunits has been proposed (Chang & Kornberg, 2000; Schultz et al., 2000). Notably, the XPB and XPD subunits possess DNA-dependent ATPase and helicase activities: the XPD helicase translocates on a DNA strand in the 5' to 3' direction (Schaeffer et al., 1994; Sung et al., 1993), whereas the contribution of XPB helicase activity with the opposite (3' to 5') polarity seems only marginal (Coin et al., 1998; Schaeffer et al., 1994). These activities have been implicated in the local unwinding of the DNA duplex at promoter sites (for transcriptional initiation) (Holstege et al., 1996) and at sites containing DNA damage (for NER) (Evans et al., 1997; Mu et al., 1997).

The XPB ATPase activity is necessary for both transcription and NER (Hwang et al., 1996; Tirode et al., 1999). In contrast, ATP-hydrolysis by XPD seems dispensable for transcription, but not for NER (Winkler et al., 2000). TTDA (also known as p8) is a very small protein that recently was identified as a subunit of TFIIH (Giglia-Mari et al., 2004). TTDA stimulates the ATPase activity of XPB in the NER reaction, but it is not directly involved in transcription, which suggests that it performs NER-dedicated roles (Coin et al., 2006). However, TTDA appears to affect the stability of the gross TFIIH complex, because cells from patients with TTD-A show substantially reduced levels of TFIIH and transcriptional activity (Giglia-Mari et al., 2004).

The observed polarity of the XPD helicase coincides with the 5' to 3' scanning model of damage verification. In this regards, there have been notable reports that the helicase activity of Rad3p, the S. cerevisiae XPD homolog, is inhibited in the presence of DNA damage (Naegeli et al., 1992). This finding evokes the notion that damage verification may depend on obstruction of the TFIIH helicase translocation at sites where the DNA structure is chemically altered (Dip et al., 2004; Gillet & Schärer, 2006; Wood, 1999). Similar results were obtained recently with an archaeal XPD homologue (Mathieu et al., 2010), although some contradictory data have been also documented (Rudolf et al., 2010), which might be explained by differences in the DNA substrates used. Using paramagnetic beads immobilized with DNA containing a CPD and a 5'-loop, we showed that a certain NER protein complex assembled at the loop site indeed moves to the CPD in an ATP-dependent manner (Sugasawa et al., 2009). In addition to XPC, both XPB and XPD ATPase activities as well as XPA seemed to be involved in this process. Considering that XPB, another TFIIH-related helicase, exhibits the opposite (3' to 5') polarity, it has been proposed that XPB and XPD may be loaded onto different DNA strands and may move toward the same direction (Dip et al., 2004). This process would enable the simultaneous inspection of both strands, so that discrimination between damaged and undamaged strands can be made depending on which helicase is blocked. However, our results strongly suggest that only 1 strand is subjected to scanning, such that lesions on the other strand, if any, are ignored. Recent mutational analyses have revealed that the ATPase, but not the helicase, activity of XPB is required for NER (Coin et al., 2007): this finding implies that XPB may not mediate the opening of the DNA duplex or movement along a DNA strand.

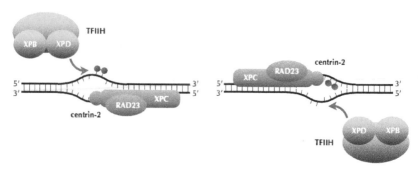

Fig. 4. Polarity of the XPC binding regulates which DNA strand is scanned by the XPD helicase. For successful loading of XPD onto the damaged strand, XPC must interact with the undamaged strand.

Another point made by this study was the importance of the XPC binding polarity. As demonstrated by the aforementioned biochemical and structural studies (Min & Pavletich, 2007; Sugasawa et al., 2002), XPC binds to a site containing unpaired bases in an asymmetric fashion. This binding polarity can be controlled intentionally by using a loop structure, in which only 1 DNA strand has unpaired bases. When a loop with either polarity was substituted for a bubble positioned on the 5' side of the CPD, incision at the lesion site was stimulated only by a looped-out sequence in the "undamaged" (CPD-free) DNA strand. In the case where both the loop and CPD were present in the same strand, incision was completely blocked (Sugasawa et al., 2009). These findings strongly suggest that, after XPC interacts with unpaired bases in 1 DNA strand, the XPD helicase in subsequently recruited TFIIH may be loaded onto the other strand and may start scanning in the 5' to 3' direction. According to this model, XPD would be forced to bind the undamaged strand erroneously, if the damage-containing strand is looped out.

Although this model was deduced from the results of in vitro experiments using rather artificial DNA substrates, it might also apply to normal NER reactions, in which unpaired bases and chemical modifications coexist in close proximity. To induce productive NER, XPC must interact with unpaired bases opposite the lesion, so that the XPD helicase can be loaded successfully onto the damaged strand immediately on the 5' side of the lesion. Intriguingly, with DNA containing a bulky lesion (such as the dG-AAF adduct), XPC exhibits a propensity to bind in a correct orientation in the absence of other factors, most likely because of steric effects preventing interactions between XPC and the modified base (Sugasawa et al., 2009). On the other hand, footprints of XPC on a 6-4PP appear rather symmetric (Sugasawa et al., 1998), which suggests that a substantial fraction of 6-4PP repair events that are initiated directly by XPC may be abortive. In the UV-DDB-mediated damage recognition pathway, however, XPC may be properly guided to interact with the undamaged strand, because the UV photolesions are already occupied by UV-DDB.

### 3.3 Possible roles for XPA and RPA

XPA, which was the first cloned XP gene (Tanaka et al., 1990), complemented UV sensitivity of fibroblasts from patients with XP-A. Cultured cells lacking expression of functional XPA are defective in both GG-NER and TC-NER, and show extreme sensitivity to killing by UV. The XPA gene product is a relatively small protein that is essential for in vitro NER. It shows a DNA-binding activity with a significant preference for various types of damaged DNA (Asahina et al., 1994; Jones & Wood, 1993).

Replication protein A (RPA) is a heterotrimeric protein complex exhibiting remarkable single-stranded DNA-binding activity. RPA is supposed to promote the unwinding of the DNA duplex, stabilize the single-stranded conformation, and stimulate various enzymatic activities, such as DNA polymerases. As the eukaryotic counterpart of bacterial SSB, RPA has been implicated in various DNA metabolisms, including replication, repair, and recombination (Wold, 1997). Its involvement in NER was demonstrated by fractionation and reconstitution of human cell-free extracts used for in vitro NER (Coverley et al., 1991). RPA also binds damaged DNA with significant specificity (Burns et al., 1996; Clugston et al., 1992; He et al., 1995), and the reported interaction between XPA and RPA seems to enhance their damage-specific DNA-binding activities (Buschta-Hedayat et al., 1999; He et al., 1995; Li et al., 1995; Wakasugi & Sancar, 1999).

Although the above findings suggest that the XPA-RPA complex could be responsible for initial damage recognition, the observed specificity and affinity of this complex for damaged DNA seem less pronounced than those of XPC or UV-DDB. In addition, accumulating evidence from biochemical and cell biological studies has supported the conclusion that these factors are more likely to be involved in later stages of the NER process. Both XPA and RPA are essential for the assembly of the NER preincision intermediate complex that contains the fully opened DNA duplex. RPA likely stabilizes the single-stranded conformation of DNA and protects the undamaged strand specifically, while XPA binds around the end of the unwound region on the 5' side of the lesion (Krasikova et al., 2010). Considering the reported physical interactions with a number of NER factors, one of the roles for these factors may be orchestrating the assembly of the preincision complex and correctly arranging other factors, including the 2 incision endonucleases, ERCC1-XPF and XPG.

The XPA protein possesses a zinc-finger domain, which NMR studies revealed is involved in the interaction with RPA. In contrast, its DNA-binding functionality was assigned to a different domain in the protein (Buchko et al., 1998; Buchko et al., 1999; Ikegami et al., 1998). Intriguingly, the DNA-binding domain in XPA shows structural resemblance to DNA binding β-hairpins (particularly BHD2) in XPC/Rad4p (Min & Pavletich, 2007), which suggests their evolutionary and functional relationship.

So far, the precise roles for the (rather weak) damage-specific DNA-binding activity of XPA remain unclear. XPA reportedly exhibits remarkable binding affinities for DNA containing highly kinked conformations, such as 3-way junctions and the Holliday junction-like structure (Camenisch et al., 2006; Missura et al., 2001). From mutational analyses, it has been proposed that XPA may be suitable for sensing abnormal electrostatic potentials of DNA, which could be caused by certain distorted DNA conformations in the damage-containing DNA duplex that are unwound by the helicase activities of TFIIH (Camenisch et al., 2007). In addition to such "proofreading" functions, our recent DNA-binding assays have raised the possibility that XPA may be required for launching the DNA scanning complex from the XPC-bound sites (Sugasawa et al., 2009). We also have shown that XPA may stimulate the TFIIH helicase activity under certain conditions, presumably through their reported physical interaction (Li et al., 1998; Park et al., 1995). Based on these findings, it is conceivable that a ternary complex involving XPC, XPA, and TFIIH scans DNA strands to search for damage: this model is reminiscent of the damage-recognition mechanism in the bacterial NER system. As for *E. coli*, 2 damage recognition pathways have been proposed (Van Houten et al., 2005): the UvrA homodimer directly recognizes and binds to distorted sites and then recruits UvrB, or preassembled complexes involving 2 UvrA and 1 or 2 UvrB molecules bind DNA in a nonspecific manner and then search for damage by scanning the DNA strands. In this analogy, UvrB seems to correspond to TFIIH as the driving subunit with ATPase/helicase activities, whereas UvrA may have evolved into XPC and/or XPA. Although little amino acid sequence homology exists between these bacterial and mammalian counterparts, the fundamental principles underlying NER damage recognition may have been conserved throughout evolution.

### 3.4 Implications in the damage surveillance mechanism

Although the specific DNA binding of UV-DDB and XPC has been observed in vitro, it still remains to be understood how these factors survey DNA and eventually reach relevant sites. For many DNA-binding proteins with sequence- and/or structure-specificity, it has been supposed that the proteins first bind DNA in a nonspecific manner and then "slide" or "hop" to search for their target sites (Gorman & Greene, 2008). A recent report has suggested that BHD1 and BHD2 in XPC may serve as dynamic damage sensors by binding to DNA and rapidly scanning for the integrity of base pairing (Camenisch et al., 2009). Once it encounters a distorted site, BHD3 may be inserted into the duplex to form a stabilized damage-recognition complex.

Apart from these models, our findings that the NER protein complex driven by the XPD helicase can scan DNA strands provides interesting insights into the molecular mechanism underlying in vivo damage surveillance: for instance, the association of XPC even with inappropriate (damage-free) sites could help the NER machinery to survey the local genomic region and find damage at rather distal positions. Possible candidates for such XPC anchoring sites include base mismatches (caused by errors of replication/repair and deamination of bases), thermodynamic "breathing" of the DNA duplex, and other sequences that are intrinsically prone to melting (e.g., transcriptional promoters and replication origins), especially in the presence of topological stresses imposed by chromatin structure. In addition, some endogenous DNA damage, such as abasic sites and single-strand breaks, also may target XPC and thereby launch the "patrolling" system. It would be of great interest to examine how the timing and efficiency of GG-NER are regulated at different genomic loci.

## 4. Conclusion

Multiple protein factors are involved in the detection and verification of DNA damage, which, in conjunction with the GG-NER system, determine whether to incise DNA or not.

These factors sample all different structural aspects of DNA damage. XPC senses the presence of oscillating unpaired bases, which allows GG-NER to target an extremely broad spectrum of DNA insults. UV-DDB seems more customized for the detection and repair of UV-induced photolesions through direct interaction with the affected bases. As for CPDs (which are refractory to detection by XPC), UV-DDB further extends the substrate specificity of GG-NER. The XPD helicase in TFIIH scans DNA strands as a fine sensor of chemical changes in DNA structure. By integrating these different strategies, GG-NER as a whole can work as a highly versatile, efficient, and accurate system. Numerous biochemical and cell biological studies have confirmed that checks for different structural abnormalities in DNA are conducted in a sequential manner. Additionally, possible stochastic mathematical models have been also discussed (Kesseler et al., 2007; Luijsterburg et al., 2010; Politi et al., 2005). Considering the in vivo situations, decondensation and some remodeling of the chromatin structure would also be expected to precede damage recognition by UV-DDB and XPC, although the underlying mechanism involved in this process remains unclear. These key issues need to be addressed at the molecular level in the near future.

## 5. Acknowledgement

This work was supported by Grants-in-Aid from the Ministry of Education, Culture, Sports, Science and Technology of Japan, by a Health and Labour Sciences Research Grant (Research on Intractable Diseases) from the Ministry of Health, Labour and Welfare of Japan, and by the Takeda Science Foundation.

## 6. References

Aboussekhra, A., Biggerstaff, M., Shivji, M.K., Vilpo, J.A., Moncollin, V., Podust, V.N., Protić, M., Hübscher, U., Egly, J.M., & Wood, R.D. (1995). Mammalian DNA nucleotide excision repair reconstituted with purified protein components. *Cell*, Vol.80, No.6, pp. 859-868, ISSN 0092-8674.

Adimoolam, S. & Ford, J.M. (2002). p53 and DNA damage-inducible expression of the xeroderma pigmentosum group C gene. *Proceedings of the National Academy of Sciences of the United States of America*, Vol.99, No.20, pp. 12985-12990, ISSN 0027-8424.

Amundson, S.A., Patterson, A., Do, K.T., & Fornace, A.J. (2002). A nucleotide excision repair master-switch: p53 regulated coordinate induction of global genomic repair genes. *Cancer Biology & Therapy*, Vol.1, No.2, pp. 145-149, ISSN 1538-4047.

Angers, S., Li, T., Yi, X., MacCoss, M.J., Moon, R.T., & Zheng, N. (2006). Molecular architecture and assembly of the DDB1-CUL4A ubiquitin ligase machinery. *Nature*, Vol.443, No.7111, pp. 590-593, ISSN 1476-4687.

Araki, M., Masutani, C., Takemura, M., Uchida, A., Sugasawa, K., Kondoh, J., Ohkuma, Y., & Hanaoka, F. (2001). Centrosome protein centrin 2/caltractin 1 is part of the xeroderma pigmentosum group C complex that initiates global genome nucleotide excision repair. *The Journal of Biological Chemistry*, Vol.276, No.22, pp. 18665-18672, ISSN 0021-9258.

Asahina, H., Kuraoka, I., Shirakawa, M., Morita, E.H., Miura, N., Miyamoto, I., Ohtsuka, E., Okada, Y., & Tanaka, K. (1994). The XPA protein is a zinc metalloprotein with an ability to recognize various kinds of DNA damage. *Mutation Research,* Vol.315, No.3, pp. 229-237, ISSN 0027-5107.

Batty, D., Rapic'-Otrin, V., Levine, A.S., & Wood, R.D. (2000). Stable binding of human XPC complex to irradiated DNA confers strong discrimination for damaged sites. *Journal of Molecular Biology,* Vol.300, No.2, pp. 275-290, ISSN 0022-2836.

Bootsma, D., Kraemer, K.H., Cleaver, J.E., & Hoeijmakers, J.H. (2001). Nucleotide excision repair syndromes: xeroderma pigmentosum, Cockayne syndrome, and trichothiodystrophy, In: *Metabolic and Molecular Bases of Inherited Disease,* Scriver, C. et al. (Eds.) Chap. 28, McGraw-Hill, New York.

Buchko, G.W., Daughdrill, G.W., de Lorimier, R., Rao B, K., Isern, N.G., Lingbeck, J.M., Taylor, J.S., Wold, M.S., Gochin, M., Spicer, L.D., Lowry, D.F., & Kennedy, M.A. (1999). Interactions of human nucleotide excision repair protein XPA with DNA and RPA70 Delta C327: chemical shift mapping and 15N NMR relaxation studies. *Biochemistry,* Vol.38, No.46, pp. 15116-15128, ISSN 0006-2960.

Buchko, G.W., Ni, S., Thrall, B.D., & Kennedy, M.A. (1998). Structural features of the minimal DNA binding domain (M98-F219) of human nucleotide excision repair protein XPA. *Nucleic Acids Research,* Vol.26, No.11, pp. 2779-2788, ISSN 0305-1048.

Bunick, C.G., Miller, M.R., Fuller, B.E., Fanning, E., & Chazin, W.J. (2006). Biochemical and structural domain analysis of xeroderma pigmentosum complementation group C protein. *Biochemistry,* Vol.45, No.50, pp. 14965-14979, ISSN 0006-2960.

Burns, J.L., Guzder, S.N., Sung, P., Prakash, S., & Prakash, L. (1996). An affinity of human replication protein A for ultraviolet-damaged DNA. *The Journal of Biological Chemistry,* Vol.271, No.20, pp. 11607-11610, ISSN 0021-9258.

Buschta-Hedayat, N., Buterin, T., Hess, M.T., Missura, M., & Naegeli, H. (1999). Recognition of nonhybridizing base pairs during nucleotide excision repair of DNA. *Proceedings of the National Academy of Sciences of the United States of America,* Vol.96, No.11, pp. 6090-6095, ISSN 0027-8424.

Camenisch, U., Dip, R., Schumacher, S.B., Schuler, B., & Naegeli, H. (2006). Recognition of helical kinks by xeroderma pigmentosum group A protein triggers DNA excision repair. *Nature Structural & Molecular Biology,* Vol.13, No.3, pp. 278-284, ISSN 1545-9993.

Camenisch, U., Dip, R., Vitanescu, M., & Naegeli, H. (2007). Xeroderma pigmentosum complementation group A protein is driven to nucleotide excision repair sites by the electrostatic potential of distorted DNA. *DNA Repair,* Vol.6, No.12, pp. 1819-1828, ISSN 1568-7864.

Camenisch, U., Träutlein, D., Clement, F.C., Fei, J., Leitenstorfer, A., Ferrando-May, E., & Naegeli, H. (2009). Two-stage dynamic DNA quality check by xeroderma pigmentosum group C protein. *The EMBO Journal,* Vol.28, No.16, pp. 2387-2399, ISSN 1460-2075.

Chang, W.H. & Kornberg, R.D. (2000). Electron crystal structure of the transcription factor and DNA repair complex, core TFIIH. *Cell,* Vol.102, No.5, pp. 609-613, ISSN 0092-8674.

Clugston, C.K., McLaughlin, K., Kenny, M.K., & Brown, R. (1992). Binding of human single-stranded DNA binding protein to DNA damaged by the anticancer drug cis-diamminedichloroplatinum (II). *Cancer Research,* Vol.52, No.22, pp. 6375-6379, ISSN 0008-5472.

Coin, F., Marinoni, J.C., Rodolfo, C., Fribourg, S., Pedrini, A.M., & Egly, J.M. (1998). Mutations in the XPD helicase gene result in XP and TTD phenotypes, preventing interaction between XPD and the p44 subunit of TFIIH. *Nature Genetics*, Vol.20, No.2, pp. 184-188, ISSN 1061-4036.

Coin, F., Oksenych, V., & Egly, J.M. (2007). Distinct roles for the XPB/p52 and XPD/p44 subcomplexes of TFIIH in damaged DNA opening during nucleotide excision repair. *Molecular Cell*, Vol.26, No.2, pp. 245-256, ISSN 1097-2765.

Coin, F., Proietti De Santis, L., Nardo, T., Zlobinskaya, O., Stefanini, M., & Egly, J.M. (2006). p8/TTD-A as a repair-specific TFIIH subunit. *Molecular Cell*, Vol.21, No.2, pp. 215-226, ISSN 1097-2765.

Coverley, D., Kenny, M.K., Munn, M., Rupp, W.D., Lane, D.P., & Wood, R.D. (1991). Requirement for the replication protein SSB in human DNA excision repair. *Nature*, Vol.349, No.6309, pp. 538-541, ISSN 0028-0836.

Datta, A., Bagchi, S., Nag, A., Shiyanov, P., Adami, G.R., Yoon, T., & Raychaudhuri, P. (2001). The p48 subunit of the damaged-DNA binding protein DDB associates with the CBP/p300 family of histone acetyltransferase. *Mutation Research*, Vol.486, No.2, pp. 89-97, ISSN 0027-5107.

Dip, R., Camenisch, U., & Naegeli, H. (2004). Mechanisms of DNA damage recognition and strand discrimination in human nucleotide excision repair. *DNA Repair*, Vol.3, No.11, pp. 1409-1423, ISSN 1568-7864.

Evans, E., Moggs, J.G., Hwang, J.R., Egly, J.M., & Wood, R.D. (1997). Mechanism of open complex and dual incision formation by human nucleotide excision repair factors. *The EMBO Journal*, Vol.16, No.21, pp. 6559-6573, ISSN 0261-4189.

Fitch, M.E., Cross, I.V., Turner, S.J., Adimoolam, S., Lin, C.X., Williams, K.G., & Ford, J.M. (2003). The DDB2 nucleotide excision repair gene product p48 enhances global genomic repair in p53 deficient human fibroblasts. *DNA Repair*, Vol.2, No.7, pp. 819-826, ISSN 1568-7864.

Fitch, M.E., Nakajima, S., Yasui, A., & Ford, J.M. (2003). In vivo recruitment of XPC to UV-induced cyclobutane pyrimidine dimers by the DDB2 gene product. *The Journal of Biological Chemistry*, Vol.278, No.47, pp. 46906-46910, ISSN 0021-9258.

Friedberg, E.C., Lehmann, A.R., & Fuchs, R.P. (2005). Trading places: how do DNA polymerases switch during translesion DNA synthesis? *Molecular Cell*, Vol.18, No.5, pp. 499-505, ISSN 1097-2765.

Friedberg, E.C., Walker, G.C., Siede, W., Wood, R.D., Schultz, R.A., & Ellenberger, T. (Eds.) (2006). *DNA Repair and Mutagenesis, Second Edition*, ASM Press, ISBN 1-55581-319-4, Washington, DC.

Fujiwara, Y., Masutani, C., Mizukoshi, T., Kondo, J., Hanaoka, F., & Iwai, S. (1999). Characterization of DNA recognition by the human UV-damaged DNA-binding protein. *The Journal of Biological Chemistry*, Vol.274, No.28, pp. 20027-20033, ISSN 0021-9258.

Giglia-Mari, G., Coin, F., Ranish, J.A., Hoogstraten, D., Theil, A., Wijgers, N., Jaspers, N.G., Raams, A., Argentini, M., van der Spek, P.J., Botta, E., Stefanini, M., Egly, J.M., Aebersold, R., Hoeijmakers, J.H., & Vermeulen, W. (2004). A new, tenth subunit of TFIIH is responsible for the DNA repair syndrome trichothiodystrophy group A. *Nature Genetics*, Vol.36, No.7, pp. 714-719, ISSN 1061-4036.

Gillet, L.C. & Schärer, O.D. (2006). Molecular mechanisms of mammalian global genome nucleotide excision repair. *Chemical Reviews,* Vol.106, No.2, pp. 253-276, ISSN 0009-2665.

Gorman, J. & Greene, E.C. (2008). Visualizing one-dimensional diffusion of proteins along DNA. *Nature Structural & Molecular Biology,* Vol.15, No.8, pp. 768-774, ISSN 1545-9985.

Groisman, R., Polanowska, J., Kuraoka, I., Sawada, J., Saijo, M., Drapkin, R., Kisselev, A.F., Tanaka, K., & Nakatani, Y. (2003). The ubiquitin ligase activity in the DDB2 and CSA complexes is differentially regulated by the COP9 signalosome in response to DNA damage. *Cell,* Vol.113, No.3, pp. 357-367, ISSN 0092-8674.

Hanawalt, P.C. & Spivak, G. (2008). Transcription-coupled DNA repair: two decades of progress and surprises. *Nature Reviews. Molecular Cell Biology,* Vol.9, No.12, pp. 958-970, ISSN 1471-0080.

Hara, R., Mo, J., & Sancar, A. (2000). DNA damage in the nucleosome core is refractory to repair by human excision nuclease. *Molecular and Cellular Biology,* Vol.20, No.24, pp. 9173-9181, ISSN 0270-7306.

He, Z., Henricksen, L.A., Wold, M.S., & Ingles, C.J. (1995). RPA involvement in the damage-recognition and incision steps of nucleotide excision repair. *Nature,* Vol.374, No.6522, pp. 566-569, ISSN 0028-0836.

Hess, M.T., Schwitter, U., Petretta, M., Giese, B., & Naegeli, H. (1997). Bipartite substrate discrimination by human nucleotide excision repair. *Proceedings of the National Academy of Sciences of the United States of America,* Vol.94, No.13, pp. 6664-6669, ISSN 0027-8424.

Hey, T., Lipps, G., Sugasawa, K., Iwai, S., Hanaoka, F., & Krauss, G. (2002). The XPC-HR23B complex displays high affinity and specificity for damaged DNA in a true-equilibrium fluorescence assay. *Biochemistry,* Vol.41, No.21, pp. 6583-6587, ISSN 0006-2960.

Holstege, F.C., van der Vliet, P.C., & Timmers, H.T. (1996). Opening of an RNA polymerase II promoter occurs in two distinct steps and requires the basal transcription factors IIE and IIH. *The EMBO Journal,* Vol.15, No.7, pp. 1666-1677, ISSN 0261-4189.

Houtsmuller, A.B., Rademakers, S., Nigg, A.L., Hoogstraten, D., Hoeijmakers, J.H., & Vermeulen, W. (1999). Action of DNA repair endonuclease ERCC1/XPF in living cells. *Science,* Vol.284, No.5416, pp. 958-961, ISSN 0036-8075.

Hwang, B.J., Ford, J.M., Hanawalt, P.C., & Chu, G. (1999). Expression of the p48 xeroderma pigmentosum gene is p53-dependent and is involved in global genomic repair. *Proceedings of the National Academy of Sciences of the United States of America,* Vol.96, No.2, pp. 424-428, ISSN 0027-8424.

Hwang, J.R., Moncollin, V., Vermeulen, W., Seroz, T., van Vuuren, H., Hoeijmakers, J.H., & Egly, J.M. (1996). A 3' --> 5' XPB helicase defect in repair/transcription factor TFIIH of xeroderma pigmentosum group B affects both DNA repair and transcription. *The Journal of Biological Chemistry,* Vol.271, No.27, pp. 15898-15904, ISSN 0021-9258.

Ikegami, T., Kuraoka, I., Saijo, M., Kodo, N., Kyogoku, Y., Morikawa, K., Tanaka, K., & Shirakawa, M. (1998). Solution structure of the DNA- and RPA-binding domain of the human repair factor XPA. *Nature Structural Biology,* Vol.5, No.8, pp. 701-706, ISSN 1072-8368.

Itoh, T., Cado, D., Kamide, R., & Linn, S. (2004). DDB2 gene disruption leads to skin tumors and resistance to apoptosis after exposure to ultraviolet light but not a chemical carcinogen. *Proceedings of the National Academy of Sciences of the United States of America*, Vol.101, No.7, pp. 2052-2057, ISSN 0027-8424.

Itoh, T., O'Shea, C., & Linn, S. (2003). Impaired regulation of tumor suppressor p53 caused by mutations in the xeroderma pigmentosum DDB2 gene: mutual regulatory interactions between p48(DDB2) and p53. *Molecular and Cellular Biology*, Vol.23, No.21, pp. 7540-7553, ISSN 0270-7306.

Janićijević, A., Sugasawa, K., Shimizu, Y., Hanaoka, F., Wijgers, N., Djurica, M., Hoeijmakers, J.H., & Wyman, C. (2003). DNA bending by the human damage recognition complex XPC-HR23B. *DNA Repair*, Vol.2, No.3, pp. 325-336, ISSN 1568-7864.

Jones, C.J. & Wood, R.D. (1993). Preferential binding of the xeroderma pigmentosum group A complementing protein to damaged DNA. *Biochemistry*, Vol.32, No.45, pp. 12096-12104, ISSN 0006-2960.

Kapetanaki, M.G., Guerrero-Santoro, J., Bisi, D.C., Hsieh, C.L., Rapić-Otrin, V., & Levine, A.S. (2006). The DDB1-CUL4ADDB2 ubiquitin ligase is deficient in xeroderma pigmentosum group E and targets histone H2A at UV-damaged DNA sites. *Proceedings of the National Academy of Sciences of the United States of America*, Vol.103, No.8, pp. 2588-2593, ISSN 0027-8424.

Kesseler, K.J., Kaufmann, W.K., Reardon, J.T., Elston, T.C., & Sancar, A. (2007). A mathematical model for human nucleotide excision repair: damage recognition by random order assembly and kinetic proofreading. *Journal of Theoretical Biology*, Vol.249, No.2, pp. 361-375, ISSN 0022-5193.

Krasikova, Y.S., Rechkunova, N.I., Maltseva, E.A., Petruseva, I.O., & Lavrik, O.I. (2010). Localization of xeroderma pigmentosum group A protein and replication protein A on damaged DNA in nucleotide excision repair. *Nucleic Acids Research*, Vol.38, No.22, pp. 8083-8094, ISSN 1362-4962.

Lee, J. & Zhou, P. (2007). DCAFs, the missing link of the CUL4-DDB1 ubiquitin ligase. *Molecular Cell*, Vol.26, No.6, pp. 775-780, ISSN 1097-2765.

Legerski, R. & Peterson, C. (1992). Expression cloning of a human DNA repair gene involved in xeroderma pigmentosum group C. *Nature*, Vol.359, No.6390, pp. 70-73, ISSN 0028-0836.

Li, L., Lu, X., Peterson, C.A., & Legerski, R.J. (1995). An interaction between the DNA repair factor XPA and replication protein A appears essential for nucleotide excision repair. *Molecular and Cellular Biology*, Vol.15, No.10, pp. 5396-5402, ISSN 0270-7306.

Li, R.Y., Calsou, P., Jones, C.J., & Salles, B. (1998). Interactions of the transcription/DNA repair factor TFIIH and XP repair proteins with DNA lesions in a cell-free repair assay. *Journal of Molecular Biology*, Vol.281, No.2, pp. 211-218, ISSN 0022-2836.

Luijsterburg, M.S., von Bornstaedt, G., Gourdin, A.M., Politi, A.Z., Moné, M.J., Warmerdam, D.O., Goedhart, J., Vermeulen, W., van Driel, R., & Höfer, T. (2010). Stochastic and reversible assembly of a multiprotein DNA repair complex ensures accurate target site recognition and efficient repair. *The Journal of Cell Biology*, Vol.189, No.3, pp. 445-463, ISSN 1540-8140.

Lutz, W., Lingle, W.L., McCormick, D., Greenwood, T.M., & Salisbury, J.L. (2001). Phosphorylation of centrin during the cell cycle and its role in centriole separation preceding centrosome duplication. *The Journal of Biological Chemistry,* Vol.276, No.23, pp. 20774-20780, ISSN 0021-9258.

Lyapina, S., Cope, G., Shevchenko, A., Serino, G., Tsuge, T., Zhou, C., Wolf, D.A., Wei, N., Shevchenko, A., & Deshaies, R.J. (2001). Promotion of NEDD-CUL1 conjugate cleavage by COP9 signalosome. *Science,* Vol.292, No.5520, pp. 1382-1385, ISSN 0036-8075.

Masutani, C., Sugasawa, K., Yanagisawa, J., Sonoyama, T., Ui, M., Enomoto, T., Takio, K., Tanaka, K., van der Spek, P.J., Bootsma, D., Hoeijmakers, J.H.J., & Hanaoka, F. (1994). Purification and cloning of a nucleotide excision repair complex involving the xeroderma pigmentosum group C protein and a human homologue of yeast RAD23. *The EMBO Journal,* Vol.13, No.8, pp. 1831-1843, ISSN 0261-4189.

Mathieu, N., Kaczmarek, N., & Naegeli, H. (2010). Strand- and site-specific DNA lesion demarcation by the xeroderma pigmentosum group D helicase. *Proceedings of the National Academy of Sciences of the United States of America,* Vol.107, No.41, pp. 17545-17550, ISSN 1091-6490.

Min, J.H. & Pavletich, N.P. (2007). Recognition of DNA damage by the Rad4 nucleotide excision repair protein. *Nature,* Vol.449, No.7162, pp. 570-575, ISSN 1476-4687.

Missura, M., Buterin, T., Hindges, R., Hübscher, U., Kaspárková, J., Brabec, V., & Naegeli, H. (2001). Double-check probing of DNA bending and unwinding by XPA-RPA: an architectural function in DNA repair. *The EMBO Journal,* Vol.20, No.13, pp. 3554-3564, ISSN 0261-4189.

Moser, J., Volker, M., Kool, H., Alekseev, S., Vrieling, H., Yasui, A., van Zeeland, A.A., & Mullenders, L.H. (2005). The UV-damaged DNA binding protein mediates efficient targeting of the nucleotide excision repair complex to UV-induced photo lesions. *DNA Repair,* Vol.4, No.5, pp. 571-582, ISSN 1568-7864.

Mu, D., Wakasugi, M., Hsu, D.S., & Sancar, A. (1997). Characterization of reaction intermediates of human excision repair nuclease. *The Journal of Biological Chemistry,* Vol.272, No.46, pp. 28971-28979, ISSN 0021-9258.

Naegeli, H., Bardwell, L., & Friedberg, E.C. (1992). The DNA helicase and adenosine triphosphatase activities of yeast Rad3 protein are inhibited by DNA damage. A potential mechanism for damage-specific recognition. *The Journal of Biological Chemistry,* Vol.267, No.1, pp. 392-398, ISSN 0021-9258.

Ng, J.M., Vermeulen, W., van der Horst, G.T., Bergink, S., Sugasawa, K., Vrieling, H., & Hoeijmakers, J.H. (2003). A novel regulation mechanism of DNA repair by damage-induced and RAD23-dependent stabilization of xeroderma pigmentosum group C protein. *Genes & Development,* Vol.17, No.13, pp. 1630-1645, ISSN 0890-9369.

Nishi, R., Alekseev, S., Dinant, C., Hoogstraten, D., Houtsmuller, A.B., Hoeijmakers, J.H., Vermeulen, W., Hanaoka, F., & Sugasawa, K. (2009). UV-DDB-dependent regulation of nucleotide excision repair kinetics in living cells. *DNA Repair,* Vol.8, No.6, pp. 767-776, ISSN 1568-7864.

Nishi, R., Okuda, Y., Watanabe, E., Mori, T., Iwai, S., Masutani, C., Sugasawa, K., & Hanaoka, F. (2005). Centrin 2 stimulates nucleotide excision repair by interacting with xeroderma pigmentosum group C protein. *Molecular and Cellular Biology,* Vol.25, No.13, pp. 5664-5674, ISSN 0270-7306.

Okuda, Y., Nishi, R., Ng, J.M., Vermeulen, W., van der Horst, G.T., Mori, T., Hoeijmakers, J.H., Hanaoka, F., & Sugasawa, K. (2004). Relative levels of the two mammalian

Rad23 homologs determine composition and stability of the xeroderma pigmentosum group C protein complex. *DNA Repair,* Vol.3, No.10, pp. 1285-1295, ISSN 1568-7864.

Park, C.H., Mu, D., Reardon, J.T., & Sancar, A. (1995). The general transcription-repair factor TFIIH is recruited to the excision repair complex by the XPA protein independent of the TFIIE transcription factor. *The Journal of Biological Chemistry,* Vol.270, No.9, pp. 4896-4902, ISSN 0021-9258.

Payne, A. & Chu, G. (1994). Xeroderma pigmentosum group E binding factor recognizes a broad spectrum of DNA damage. *Mutation Research,* Vol.310, No.1, pp. 89-102, ISSN 0027-5107.

Politi, A., Moné, M.J., Houtsmuller, A.B., Hoogstraten, D., Vermeulen, W., Heinrich, R., & van Driel, R. (2005). Mathematical modeling of nucleotide excision repair reveals efficiency of sequential assembly strategies. *Molecular Cell,* Vol.19, No.5, pp. 679-690, ISSN 1097-2765.

Popescu, A., Miron, S., Blouquit, Y., Duchambon, P., Christova, P., & Craescu, C.T. (2003). Xeroderma pigmentosum group C protein possesses a high affinity binding site to human centrin 2 and calmodulin. *The Journal of Biological Chemistry,* Vol.278, No.41, pp. 40252-40261, ISSN 0021-9258.

Rapić Otrin, V., Kuraoka, I., Nardo, T., McLenigan, M., Eker, A.P., Stefanini, M., Levine, A.S., & Wood, R.D. (1998). Relationship of the xeroderma pigmentosum group E DNA repair defect to the chromatin and DNA binding proteins UV-DDB and replication protein A. *Molecular and Cellular Biology,* Vol.18, No.6, pp. 3182-3190, ISSN 0270-7306.

Rapić-Otrin, V., McLenigan, M.P., Bisi, D.C., Gonzalez, M., & Levine, A.S. (2002). Sequential binding of UV DNA damage binding factor and degradation of the p48 subunit as early events after UV irradiation. *Nucleic Acids Research,* Vol.30, No.11, pp. 2588-2598, ISSN 1362-4962.

Rapić-Otrin, V., Navazza, V., Nardo, T., Botta, E., McLenigan, M., Bisi, D.C., Levine, A.S., & Stefanini, M. (2003). True XP group E patients have a defective UV-damaged DNA binding protein complex and mutations in DDB2 which reveal the functional domains of its p48 product. *Human Molecular Genetics,* Vol.12, No.13, pp. 1507-1522, ISSN 0964-6906.

Reardon, J.T. & Sancar, A. (2003). Recognition and repair of the cyclobutane thymine dimer, a major cause of skin cancers, by the human excision nuclease. *Genes & Development,* Vol.17, No.20, pp. 2539-2551, ISSN 0890-9369.

Reardon, J.T., Nichols, A.F., Keeney, S., Smith, C.A., Taylor, J.S., Linn, S., & Sancar, A. (1993). Comparative analysis of binding of human damaged DNA-binding protein (XPE) and Escherichia coli damage recognition protein (UvrA) to the major ultraviolet photoproducts: T[c,s]T, T[t,s]T, T[6-4]T, and T[Dewar]T. *The Journal of Biological Chemistry,* Vol.268, No.28, pp. 21301-21308, ISSN 0021-9258.

Riedl, T., Hanaoka, F., & Egly, J.M. (2003). The comings and goings of nucleotide excision repair factors on damaged DNA. *The EMBO Journal,* Vol.22, No.19, pp. 5293-5303, ISSN 0261-4189.

Roche, Y., Zhang, D., Segers-Nolten, G.M., Vermeulen, W., Wyman, C., Sugasawa, K., Hoeijmakers, J., & Otto, C. (2008). Fluorescence correlation spectroscopy of the binding of nucleotide excision repair protein XPC-hHr23B with DNA substrates. *Journal of Fluorescence*, Vol.18, No.5, pp. 987-995, ISSN 1053-0509.

Rudolf, J., Rouillon, C., Schwarz-Linek, U., & White, M.F. (2010). The helicase XPD unwinds bubble structures and is not stalled by DNA lesions removed by the nucleotide excision repair pathway. *Nucleic Acids Research*, Vol.38, No.3, pp. 931-941, ISSN 1362-4962.

Salisbury, J.L., Suino, K.M., Busby, R., & Springett, M. (2002). Centrin-2 is required for centriole duplication in mammalian cells. *Current Biology : CB*, Vol.12, No.15, pp. 1287-1292, ISSN 0960-9822.

Schaeffer, L., Moncollin, V., Roy, R., Staub, A., Mezzina, M., Sarasin, A., Weeda, G., Hoeijmakers, J.H., & Egly, J.M. (1994). The ERCC2/DNA repair protein is associated with the class II BTF2/TFIIH transcription factor. *The EMBO Journal*, Vol.13, No.10, pp. 2388-2392, ISSN 0261-4189.

Schultz, P., Fribourg, S., Poterszman, A., Mallouh, V., Moras, D., & Egly, J.M. (2000). Molecular structure of human TFIIH. *Cell*, Vol.102, No.5, pp. 599-607, ISSN 0092-8674.

Scrima, A., Koníčková, R., Czyzewski, B.K., Kawasaki, Y., Jeffrey, P.D., Groisman, R., Nakatani, Y., Iwai, S., Pavletich, N.P., & Thomä, N.H. (2008). Structural basis of UV DNA-damage recognition by the DDB1-DDB2 complex. *Cell*, Vol.135, No.7, pp. 1213-1223, ISSN 1097-4172.

Shivji, M.K., Eker, A.P., & Wood, R.D. (1994). DNA repair defect in xeroderma pigmentosum group C and complementing factor from HeLa cells. *The Journal of Biological Chemistry*, Vol.269, No.36, pp. 22749-22757, ISSN 0021-9258.

Stoyanova, T., Roy, N., Kopanja, D., Raychaudhuri, P., & Bagchi, S. (2009). DDB2 (damaged-DNA binding protein 2) in nucleotide excision repair and DNA damage response. *Cell Cycle (Georgetown, Tex.)*, Vol.8, No.24, pp. 4067-4071, ISSN 1551-4005.

Stubbert, L.J., Hamill, J.D., Spronck, J.C., Smith, J.M., Becerril, C., & McKay, B.C. (2007). DDB2-independent role for p53 in the recovery from ultraviolet light-induced replication arrest. *Cell Cycle (Georgetown, Tex.)*, Vol.6, No.14, pp. 1730-1740, ISSN 1551-4005.

Stubbert, L.J., Smith, J.M., Hamill, J.D., Arcand, T.L., & McKay, B.C. (2009). The anti-apoptotic role for p53 following exposure to ultraviolet light does not involve DDB2. *Mutation Research*, Vol.663, No.1-2, pp. 69-76, ISSN 0027-5107.

Sugasawa, K. (2006). UV-induced ubiquitylation of XPC complex, the UV-DDB-ubiquitin ligase complex, and DNA repair. *Journal of Molecular Histology*, Vol.37, No.5-7, pp. 189-202, ISSN 1567-2379.

Sugasawa, K., Akagi, J., Nishi, R., Iwai, S., & Hanaoka, F. (2009). Two-step recognition of DNA damage for mammalian nucleotide excision repair: Directional binding of the XPC complex and DNA strand scanning. *Molecular Cell*, Vol.36, No.4, pp. 642-653, ISSN 1097-4164.

Sugasawa, K., Ng, J.M., Masutani, C., Iwai, S., van der Spek, P.J., Eker, A.P., Hanaoka, F., Bootsma, D., & Hoeijmakers, J.H. (1998). Xeroderma pigmentosum group C protein complex is the initiator of global genome nucleotide excision repair. *Molecular Cell*, Vol.2, No.2, pp. 223-232, ISSN 1097-2765.

Sugasawa, K., Okamoto, T., Shimizu, Y., Masutani, C., Iwai, S., & Hanaoka, F. (2001). A multistep damage recognition mechanism for global genomic nucleotide excision repair. *Genes & Development*, Vol.15, No.5, pp. 507-521, ISSN 0890-9369.

Sugasawa, K., Okuda, Y., Saijo, M., Nishi, R., Matsuda, N., Chu, G., Mori, T., Iwai, S., Tanaka, K., Tanaka, K., & Hanaoka, F. (2005). UV-induced ubiquitylation of XPC protein mediated by UV-DDB-ubiquitin ligase complex. *Cell*, Vol.121, No.3, pp. 387-400, ISSN 0092-8674.

Sugasawa, K., Shimizu, Y., Iwai, S., & Hanaoka, F. (2002). A molecular mechanism for DNA damage recognition by the xeroderma pigmentosum group C protein complex. *DNA Repair*, Vol.1, No.1, pp. 95-107, ISSN 1568-7864.

Sung, P., Bailly, V., Weber, C., Thompson, L.H., Prakash, L., & Prakash, S. (1993). Human xeroderma pigmentosum group D gene encodes a DNA helicase. *Nature*, Vol.365, No.6449, pp. 852-855, ISSN 0028-0836.

Tanaka, K., Miura, N., Satokata, I., Miyamoto, I., Yoshida, M.C., Satoh, Y., Kondo, S., Yasui, A., Okayama, H., & Okada, Y. (1990). Analysis of a human DNA excision repair gene involved in group A xeroderma pigmentosum and containing a zinc-finger domain. *Nature*, Vol.348, No.6296, pp. 73-76, ISSN 0028-0836.

Tang, J. & Chu, G. (2002). Xeroderma pigmentosum complementation group E and UV-damaged DNA-binding protein. *DNA Repair*, Vol.1, No.8, pp. 601-616, ISSN 1568-7864.

Tang, J.Y., Hwang, B.J., Ford, J.M., Hanawalt, P.C., & Chu, G. (2000). Xeroderma pigmentosum p48 gene enhances global genomic repair and suppresses UV-induced mutagenesis. *Molecular Cell*, Vol.5, No.4, pp. 737-744, ISSN 1097-2765.

Thompson, J.R., Ryan, Z.C., Salisbury, J.L., & Kumar, R. (2006). The structure of the human centrin 2-xeroderma pigmentosum group C protein complex. *The Journal of Biological Chemistry*, Vol.281, No.27, pp. 18746-18752, ISSN 0021-9258.

Tirode, F., Busso, D., Coin, F., & Egly, J.M. (1999). Reconstitution of the transcription factor TFIIH: assignment of functions for the three enzymatic subunits, XPB, XPD, and cdk7. *Molecular Cell*, Vol.3, No.1, pp. 87-95, ISSN 1097-2765.

Trego, K.S. & Turchi, J.J. (2006). Pre-steady-state binding of damaged DNA by XPC-hHR23B reveals a kinetic mechanism for damage discrimination. *Biochemistry*, Vol.45, No.6, pp. 1961-1969, ISSN 0006-2960.

Treiber, D.K., Chen, Z., & Essigmann, J.M. (1992). An ultraviolet light-damaged DNA recognition protein absent in xeroderma pigmentosum group E cells binds selectively to pyrimidine (6-4) pyrimidone photoproducts. *Nucleic Acids Research*, Vol.20, No.21, pp. 5805-5810, ISSN 0305-1048.

Van Houten, B., Croteau, D.L., DellaVecchia, M.J., Wang, H., & Kisker, C. (2005). 'Close-fitting sleeves': DNA damage recognition by the UvrABC nuclease system. *Mutation Research*, Vol.577, No.1-2, pp. 92-117, ISSN 0027-5107.

Venema, J., van Hoffen, A., Natarajan, A.T., van Zeeland, A.A., & Mullenders, L.H. (1990). The residual repair capacity of xeroderma pigmentosum complementation group C fibroblasts is highly specific for transcriptionally active DNA. *Nucleic Acids Research*, Vol.18, No.3, pp. 443-448, ISSN 0305-1048.

Volker, M., Moné, M.J., Karmakar, P., van Hoffen, A., Schul, W., Vermeulen, W., Hoeijmakers, J.H., van Driel, R., van Zeeland, A.A., & Mullenders, L.H. (2001).

Sequential assembly of the nucleotide excision repair factors in vivo. *Molecular Cell*, Vol.8, No.1, pp. 213-224, ISSN 1097-2765.

Wakasugi, M. & Sancar, A. (1999). Order of assembly of human DNA repair excision nuclease. *The Journal of Biological Chemistry*, Vol.274, No.26, pp. 18759-18768, ISSN 0021-9258.

Wakasugi, M., Kawashima, A., Morioka, H., Linn, S., Sancar, A., Mori, T., Nikaido, O., & Matsunaga, T. (2002). DDB accumulates at DNA damage sites immediately after UV irradiation and directly stimulates nucleotide excision repair. *The Journal of Biological Chemistry*, Vol.277, No.3, pp. 1637-1640, ISSN 0021-9258.

Wakasugi, M., Shimizu, M., Morioka, H., Linn, S., Nikaido, O., & Matsunaga, T. (2001). Damaged DNA-binding protein DDB stimulates the excision of cyclobutane pyrimidine dimers in vitro in concert with XPA and replication protein A. *The Journal of Biological Chemistry*, Vol.276, No.18, pp. 15434-15440, ISSN 0021-9258.

Wang, H., Zhai, L., Xu, J., Joo, H.Y., Jackson, S., Erdjument-Bromage, H., Tempst, P., Xiong, Y., & Zhang, Y. (2006). Histone H3 and H4 ubiquitylation by the CUL4-DDB-ROC1 ubiquitin ligase facilitates cellular response to DNA damage. *Molecular Cell*, Vol.22, No.3, pp. 383-394, ISSN 1097-2765.

Winkler, G.S., Araújo, S.J., Fiedler, U., Vermeulen, W., Coin, F., Egly, J.M., Hoeijmakers, J.H., Wood, R.D., Timmers, H.T., & Weeda, G. (2000). TFIIH with inactive XPD helicase functions in transcription initiation but is defective in DNA repair. *The Journal of Biological Chemistry*, Vol.275, No.6, pp. 4258-4266, ISSN 0021-9258.

Wittschieben, B.Ø., Iwai, S., & Wood, R.D. (2005). DDB1-DDB2 (xeroderma pigmentosum group E) protein complex recognizes a cyclobutane pyrimidine dimer, mismatches, apurinic/apyrimidinic sites, and compound lesions in DNA. *The Journal of Biological Chemistry*, Vol.280, No.48, pp. 39982-39989, ISSN 0021-9258.

Wold, M.S. (1997). Replication protein A: a heterotrimeric, single-stranded DNA-binding protein required for eukaryotic DNA metabolism. *Annual Review of Biochemistry*, Vol.66, pp. 61-92, ISSN 0066-4154.

Wood, R.D. (1999). DNA damage recognition during nucleotide excision repair in mammalian cells. *Biochimie*, Vol.81, No.1-2, pp. 39-44, ISSN 0300-9084.

Yang, X., Menon, S., Lykke-Andersen, K., Tsuge, T., Di Xiao, Wang, X., Rodriguez-Suarez, R.J., Zhang, H., & Wei, N. (2002). The COP9 signalosome inhibits p27(kip1) degradation and impedes G1-S phase progression via deneddylation of SCF Cul1. *Current Biology : CB*, Vol.12, No.8, pp. 667-672, ISSN 0960-9822.

Yasuda, T., Sugasawa, K., Shimizu, Y., Iwai, S., Shiomi, T., & Hanaoka, F. (2005). Nucleosomal structure of undamaged DNA regions suppresses the non-specific DNA binding of the XPC complex. *DNA Repair*, Vol.4, No.3, pp. 389-395, ISSN 1568-7864.

Yoon, T., Chakrabortty, A., Franks, R., Valli, T., Kiyokawa, H., & Raychaudhuri, P. (2005). Tumor-prone phenotype of the DDB2-deficient mice. *Oncogene*, Vol.24, No.3, pp. 469-478, ISSN 0950-9232.

# Part 2

# Polymorphism of DNA Repair Genes

# 5

# DNA Repair Capacity-Related to Genetic Polymorphisms of DNA Repair Genes and Aflatoxin B1-Related Hepatocellular Carcinoma Among Chinese Population

Xi-Dai Long[1,2] et al.[*]
*[1]Department of Pathology, Youjiang Medical College for Nationalities*
*[2]Department of Pathology, Shanghai Jiao Tong University School of Medicine*
*China*

## 1. Introduction

Primary liver cancer (PLC) is the sixth most commonly occurring cancer and the third most common cause of cancer deaths in the world (1). This tumor has two main pathological types: hepatocellular carcinoma (HCC) and cholangiocellular carcinoma. HCC, the most common pathological form of PLC, occurs more often in specific regions which include eastern and southeastern Asia, Melanesia, and sub-Saharan Africa (1, 2). Once diagnosed, survival rates for HCC are poor: 75% of patients die within 1 year, and 5-year survival rate is only 3 - 5% (3, 4). Therefore, insight into the tumorigenesis mechanisms of HCC will broaden and deepen implications in understanding and preventing occurrence of the cancer.

It has been known that chronic infection with hepatitis virus [including hepatitis virus B (HBV) and hepatitis virus C (HCV)] is the most common cause of HCC worldwide (3). In sub-Saharan Africa and Southern China, chronic exposure of aflatoxin B1 (AFB1) may present a special environmental hazard, especially in individuals chronically infected with HBV (1, 2, 5-8). However, increasing epidemiological evidence has exhibited that although many people are exposed to these risk factors, only a relatively small proportion of chronic infectors or exposure person develop HCC (3, 9, 10). This indicates an individual susceptibility related to genetic factors such as DNA repair capacity might be associated with HCC carcinogenesis (3, 11). In recent years, evidence has been accumulated to support the hypothesis that common genetic polymorphisms in genes involved in long process of

[*] Jin-Guang Yao[3], Zhi Zeng[2], Cen-Han Huang[3], Pinhu Liao[3], Zan-Song Huang[3], Yong-Zhi Huang[1], Fu-Zhi Ban[4], Xiao-Yin Huang[1], Li-Min Yao[5], Lu-Dan Fan[6], and Guo-Hui Fu[2]
[3]*Department of Medicine, Youjiang Medical College for Nationalities, China.*
[4]*Department of Medicine, the Southwestern Affiliated Hospital of Youjiang Medical College for Nationalities, China.*
[5]*Department of Imaging Medicine (Grade 2008), Youjiang Medical College for Nationalities, China.*
[6]*Department of Clinic Medicine (Grade 2009), Youjiang Medical College for Nationalities, China.*

carcinogenesis may be of importance in determining individual susceptibility to HCC (3, 9, 12). Therefore, the existence of low penetrate genetic polymorphisms may explain the reason why only a small portion of individuals, even in high-risk areas, develop HCC in their life span. This study reviews recent efforts in identifying genetic variants which may have impact on risk of HCC.

## 2. Epidemiology of AFB1-related HCC in China

In China, HCC is the third or fourth most common malignant tumors and accounts for about 55% of the world's HCC cases, more than 340,000 each year (1, 13). This tumor occurs more often in eastern and southeastern China, including Jiangsu, Shanghai, Zhejiang, Fujiang, Guangdong, and Guangxi, mainly because of high AFB1 exposure and/or chronic infection of HBV and HCV (13). In the high AFB1-exposure areas such as Guangxi Zhuang Autonomous Region, this tumor is the most common occuring cancer (13, 14). Moreover, the incidence rate gradually increases with age increasing in above-mentioned AFB1-exposure areas (15). Males are always more frequently affected than females but high male to female ratios of > 3 in the high AFB1-exposure areas (15). Although the incidence rates of this tumor in low AFB1-exposure areas in China have markedly decreased (because of the control of hepatitis virus infection), they have changed little in high AFB1-exposure areas (13, 15). For example, during May 2007 to April 2008, incidence rates were 117.8/100,000 and 103.1/100,000 for Xiangzhou and Fusui (two main high AFB1-exposure areas of China), respectively (13, 16). This was similar to the results before ten years (17).

Because of the very poor prognosis, HCC is the second most common cause of death from cancer in China (18). In the past thirty years, total mortality rate of HCC gradually increased from 12.5/100,000 to 26.26/100,000 (Fig 1A), regardless of countryside areas or urban areas (Fig 1B). This trend was more noticeable in male population than female population (Fig 1C), possibly because male individuals featured more high AFB1 exposure. Supporting aforementioned hypothesis, a recent study from high AFB1-exposure areas has demonstrated these having longer exposure years or higher exposure levels of AFB1 would face lower 5-years survival rate (4).

## 3. AFB1 exposure and DNA damage and repair

AFB1 is an important mycotoxin produced by the moulds Aspergillus parasiticus and Aspergillus flavus (19). This toxic agent has been found as contaminants of human and animal food, particularly ground nuts (peanuts) and core, in tropical areas such as the Southeastern China as a result of fungal contamination during growth and after harvest which under hot and humid conditions (8, 14, 19, 20). Epidemiological evidence has shown dietary ingestion of high levels of AFB1 presents a significant environmental hazard of HCC (16, 17, 21). Experimental animal models have also shown that AFB1 can induce HCC; whereas DNA damage should play an important role during hepatocellular carcinogenesis (19, 22, 23). Therefore, AFB1 has been classified as a category I known human carcinogen by the International Agency for Research on Cancer (24).

AFB1 is metabolized by cytochrome P450 enzymes to its reactive form, AFB1-8,9-epoxide (AFB1-epoxide), which covalently binds to DNA and induces DNA damage (19, 25-28). DNA damage induced by AFB1 includes AFB1-DNA adducts, oxidative DNA damage,

DNA Repair Capacity-Related to Genetic Polymorphisms of DNA Repair Genes and Aflatoxin B1-Related
Hepatocellular Carcinoma Among Chinese Population

109

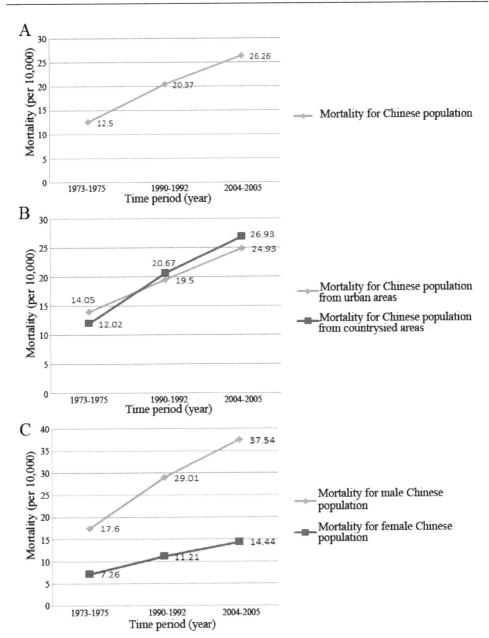

Fig. 1. The mortality rates of HCC in China during 1973 and 2005. Total mortality rates (A),
regardless of in urban areas or countryside areas (B), were significantly increasing from
during 1973 and 1975 to during 1990 and 1992 or to during 2004 and 2005. This increasing
mortality rates were more remarkable among male population (C).

and gene mutation (Fig. 2). Among these AFB1-DNA adducts, 8,9-di-hydro-8-(N⁷-guanyl)-9-hydroxy–AFB1 (AFB1-N⁷-Gua) adduct is the most common type identified and confirmed in vivo researches (19, 25-27, 29, 30). The formation of this adduct proceeds by a precovalent intercalation complex between double-stranded DNA and the highly electrophilic, unstable AFB1-epoxide isomer (31, 32). After that, the induction of a positive charge on the imidazole portion of the formed AFB1-N⁷-Gua adduct gives rise to another important a DNA adduct, a ring-opened formamidopyridine AFB1 (AFB1-FAPy) adduct (33, 34). Accumulation of AFB1-FAPy adduct is characterized by time-dependence, non-enzyme, and may be of biological basis of genes mutation because of its apparent persistence in DNA (19, 33, 34). Furthermore, above adducts are capable of forming subsequent repair-resistant adducts, depurination, or lead to error-prone DNA repair resulting in single-strand breaks (SSBs), double-strand breaks (DSBs), base pair substitution, or frame shift mutations (35, 36). Additionally, AFB1 exposure also induces the formation of such oxidation DNA damage as 8-oxodeoxyguanosine (8-$_{oxod}$G), a common endogenous DNA adduct (36-38). Although these DNA adducts are mainly produced in liver cells, they are also found in the peripheral blood white cells (39, 40). Recent studies have shown that the levels of AFB1-DNA adduct of the peripheral blood white cells are positively and lineally correlated with that of liver cells, implying analysis of AFB1-DNA adducts in the peripheral blood white cells may substitute for the elucidation of tissular levels of adducts (39, 41).

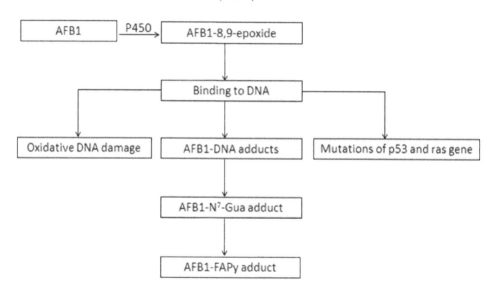

Fig. 2. The DNA damage induced by AFB1.

For genes mutations induced by AFB1 exposure, the experimental and theoretical researches are briefly on the p53 gene (42-49). Reaction with DNA at the N⁷ position of guanine preferentially causes a G:C > T:A mutation in codon 249 of this gene, leading to an amino acid substitution of arginine to serine (44-50). In high AFB1-exposure areas, this mutation is present in more than 40% of HCC and can be detected in serum DNA of patients with preneoplastic lesions and HCC (41). While codon 249 transversion mutations are either very rare or absent in low or no AFB1-exposure areas (49, 51, 52). Using the human p53 gene in an in vitro assay, codon 249 has been exhibited to be a preferential site for formation of AFB1-N$^7$-Gua adducts, evidence consistent with a role for AFB1 in the mutations observed in HCC (50, 53). Therefore, the codon 249 mutation of p53 gene has been defined as the hot-spot mutation of p53 gene resulting from AFB1 and has become the molecular symbol of HCC induced by AFB1 exposure (54-56).

A wide diversity of DNA damage produced by AFB1 exposure, if not repaired, may cause chromosomal aberrations, micronuclei, sister chromatid exchange, unscheduled DNA synthesis, and chromosomal strand breaks, and can be converted into gene mutations and genomic instability, which in turn results in cellular malignant transformation (19). Nevertheless, human cells have evolved surveillance mechanisms that monitor the integrity of genome to minimize the consequences of detrimental mutations (54). AFB1-induced DNA damage can be repaired through the following pathways: nucleotide excision repair (NER), base excision repair (BER), single-strand break repair (SSBR), and double-strand break repair (DSBR) (12, 28, 57). During the process of damage removed by aforementioned repair pathways, DNA repair genes play a central role, because their function determines DNA repair capacity (12). It has been shown that reduction in DNA repair capacity related to DNA repair genes is associated with increased risk of cancers (4, 39-41, 58-62). Thus, genetic polymorphisms in DNA repair genes which contribute to the variation in DNA repair capacity may be correlated with risk of developing cancers, including AFB1-related HCC.

## 4. Genetic polymorphisms in genes involved in NER pathway and risk of HCC

NER pathway, a major DNA repair pathways in human cells featuring genomic DNA damage, can remove structurally such diverse lesions as pyrimidine dimers, irradiative damage, and bulky chemical adducts, and DNA damage from carcinogens and some chemotherapeutic drugs (63, 64). To date, the mechanism of this pathway is well understood and has been reconstituted in vitro. It consists of several sequential steps: lesion sensing, opening of a denaturation bubble, incision of the damaged strand, displacement of the lesion-containing oligonucleotide, gap filling, and ligation (63, 64). In the fibroblast cells with the deficiency of xeroderma pigmentosum A (XPA) gene, conversion of the initial AFB1-N$^7$-Gua adduct to the AFB1-FAPy adduct has been found to be more extensive (53). This suggests that NER should be a major mechanism for enzymatic repair of AFB1 adducts (12). It's defects lead to severe diseases related AFB1 exposure, including liver injury and HCC. Accumulating evidence has implied that genetic polymorphisms in NER genes are associated with DNA repair capacity and modulate the risk of cancers (65-69). Molecular epidemiology studies of AFB1-related HCC in China have investigated the associations with several genes involved in NER pathway such as xeroderma pigmentosum C (XPC) and xeroderma pigmentosum D (XPD)(4, 39, 70, 71).

*XPC.* XPC gene spans 33kb on chromosome 3p25 and contains 16 exons and 15 introns (Genbank accession no. AC090645). This gene encodes a 940-amino acid protein, an important DNA damage recognition molecule which plays an important role in NER pathway (72). It binds tightly with HR23B to form a stable XPC-HR23B complex, the first protein component that recognizes and binds to the DNA damage sites. XPC-HR23B complex can recognize a variety of DNA adducts formed by exogenous carcinogens such as AFB1 and binds to the DNA damage sites (72). Thus, it may play a role in the pathogenesis of HCC-related AFB1. Some recent studies have showed that defects in XPC have been related to many types of malignant tumors (73-82). Transgenic mice studies also revealed predisposition to many types of tumors in XPC gene knockout mice (83). Furthermore, pathological and cellular researches have exhibited that the abnormal expression of this gene is related to hepatocarcinogenesis (84). These studies suggests the polymorphisms localizing at conserved sites of XPC gene might modify the risk of HCC induced by AFB1 exposure. Recently, four studies from high AFB1-exposure areas of China have approved aforementioned hypothesis (4, 70, 71, 85).

The first study conducted by Cai et al.(85) is from Shunde area, Guangdong Province. In this 1-1 case-control study (including 78 HCC cases and 78 age- and sex-matching controls), researchers analyzed between two common polymorphisms—Ala499Val and Lys939Gln—of XPC gene and risk of HCC and found these two polymorphisms modified HCC risk [adjusted odds ratios (ORs) were 3.77 with 95% confidence interval (CI)1.34-12.89 for Ala499Lys and 6.78 with 95% CI 2.03-22.69], especially under HBV and HCV infection condition. Although they evaluated the effects of XPC-hepatitis viruses interaction on HCC risk, they did not elucidate the possible interaction of AFB1 exposure.

The other three studies are from Guangxi Zhuang Autonomous Region (4, 70). Li et al.(71), Wu et al. (70), and Long et al. (4) investigated the modifying effects of genetic polymorphisms XPC on HCC based hospitals. The results showed XPC codon 939 Gln alleles increased about 2-times risk of HCC. Furthermore, Wu, et al.(70), and Long, et al. (4) quantitatively elucidated AFB1 exposure years and levels and their interactive effects with XPC Lys939Gln polymorphism. They found some evidence of AFB1 exposure-risk genotypes of XPC codon 939 on HCC risk (22.33 > 1.88 × 8.69 for the interaction of AFB1-exposure years and XPC risk genotypes and 18.38 > 1.11 × 4.62 for the interaction of AFB1-exposure levels and XPC risk genotypes). Additionally, Gln alleles at codon 939 of XPC gene are observed to be correlated with the decrease of XPC expression levels in cancerous tissues ($r = -0.369$, $P < 0.001$) and with the overall survival of HCC patients (the median survival times are 30, 25, and 19 months for patients with XPC gene codon 939 Lys/Lys, Lys/Gln, and Gln/Gln respectively). This decreasing 5-years survival rates would be noticeable under high AFB1 exposure conditions (the median survival times are 15 months for the joint of XPC gene codon 939 Gln/Gln and long-term AFB1-exposure years and 17 month for the joint of XPC gene codon 939 Gln/Gln and high AFB1-exposure level) (4).

These results demonstrate that polymorphism at codon 939 of XPC gene is not only a genetic determinant in the development of HCC induced by AFB1 exposure in Chinese population, but also is an independent prognostic factor influencing the survival of HCC, like AFB1 exposure. However, Li et al. (71) reported that the proportional distribution of the Val/Val genotype at codon 499 of XPC gene did not differ between cases with HCC and controls in Guangxi Zhuang Autonomous Region, China ($P > 0.05$), dissimilar to the data from another area of China, Guangdong Province (85). Possible explanations for these inconsistent finding may be either due to unknown confounders or due to small sample size.

*XPD.* XPD gene-encoding protein, a DNA-dependent ATPase/helicase, is associated with the TFIIH transcription-factor complex and plays a role in NER pathway (86, 87). During NER, XPD participates in the opening of the DNA helix to allow the excision of the DNA fragment containing the damaged base. There are two described polymorphisms that induce amino acid changes in the protein: at codons 312 (Asp to Asn) and 751 (Lys to Gln) (87-89). To date, these two polymorphisms have been extensively studied (87, 88, 90-95).

DNA Repair Capacity-Related to Genetic Polymorphisms of DNA Repair Genes and Aflatoxin B1-Related
Hepatocellular Carcinoma Among Chinese Population

113

Several groups have done genotype-phenotype analyses with these two polymorphisms and have shown that the variant allele genotypes are associated with low DNA repair ability (96, 97). Recent studies have showed the polymorphisms at codon 312 and 751 of XPD are correlated with DNA-adducts levels, p53 gene mutation, and cancers risk (88, 94, 98-100).

In a hospital-based case-control study in Guangxi (39), we found that the variant XPD codon 751 genotypes (namely Lys/Gln and Gln/Gln) detected by TaqMan-MGB PCR was significantly different between controls (26.3% and 8.6% for Lys/Gln and Gln/Gln, respectively) and HCC cases (35.9% and 20.1% for Lys/Gln and Gln/Gln, respectively, $P < 0.001$). Individuals with variant alleles had about 1.5- to 2.5-fold risk of developing the cancer (adjusted OR 1.75 and 95% CI 1.30-2.37 for Lys/Gln; adjusted OR 2.47 and 95% CI 1.62-3.76 for Gln/Gln). Based on relative sample size (including 618 HCC cases and 712 controls), we stratified genotypes of XPD codon 751 according to matching factors and observed some evidence of interaction between XPD codon 751 Gln alleles and sex. These female having Gln alleles, compared to those without these alleles, featured increased HCC risk. Furthermore, the interactive effects of between variant genotypes of XPD gene codon 751 environment variant AFB1 or another NER gene XPC on HCC risk were also found, with interactive value 0.85, 1.04, and 1.71 for AFB1-exposure years, AFB1-exposure levels, and XPC gene codon 939 risk genotypes ($P_{interaction} < 0.05$). Therefore, the XPD gene codon 751 polymorphism may have potential effect on AFB1-related HCC susceptibility among Chinese population. However, the study from AFB1-exposure areas don't exhibit polymorphism at codon 312 of XPD gene significantly associates with the risk of HCC induced by AFB1.

## 5. Genetic polymorphisms in genes involved in SSBR pathway and risk of HCC

SSB is a common type of DNA damage produced by AFB1 exposure (36). If not repaired, it can disrupt transcription and replication and can be converted into potentially clastogenic and/or lethal DSBs. This DNA damage is repaired via SSBR pathway (101, 102). SSBR pathway includes four basic steps: *a*. SSB detection and signaling, through poly (ADP-ribose) polymerase (PARP); *b*. DNA break end processing, through the role of polynucleotide kinase (PNK), AP endonuclease-1 (APE1), DNA polymerase β (Pol β), tyrosyl phosphodiesterase 1 (TDP1), and flap endonuclease-1 (FEN-1); *c*. gap filling, involving in multiple DNA polymerases; *d*. DNA ligation, involving in multiple DNA ligases. Of the later three steps of SSBR pathway, x-ray repair cross complementary 1 (XRCC1) is indispensible, because it not only acts as the scaffolding protein of SSBR, but also stimulates the activity of PNK (103).

XRCC1 gene encoding protein (633 amino acids), consists of three functional domains — N-terminal domain (NTD), central breast cancer susceptibility protein-1 homology C-terminal (BRCT I), and C-terminal breast cancer susceptibility protein-1 homology C-terminal (BRCT II) (103-106). This protein is directly associated with Pol β, DNA ligase III, and PARP, via their three functional domains and is implicated in the core processes in SSBR and BER pathway (103). Mutant hamster ovary cell lines that lack XRCC1 genes are hypersensitive to DNA damage agents such as ionizing radiation, hydrogen peroxide, and alkylating agents (103). Furthermore, this kind of cells usually face increasing frequency of spontaneous chromosome aberrations and deletions. Three single nucleotide polymorphisms in the coding region of XRCC1 gene that lead to amino acid substitution have been described and investigated (12). Of these polymorphisms, the codon 399 polymorphism is of special concern, because this polymorphism resides in functionally significant regions (BECT II) and may be related to decreasing DNA repair capacity, increasing genes mutation, and running-up risk of cancers (12, 107-114).

In AFB1-exposure areas from China, a total of six molecular epidemiological studies were found in PubMed database, Wangfang Database, and Weipu database (61, 62, 115-118). However, associations between XRCC1 gene codon 399 polymorphism and individual susceptibility to HCC have been reported in these case-control studies with the results being contradictory. We analyzed the possible causes of contradictory using meta-analysis method (Comprehensive Meta Analysis Version 2, http://www.meta-analysis.com/). Fig. 3 showed the meta-analysis results of the modifying effects of XRCC1 gene codon 399 polymorphism on HCC risk. We found these subjects with Gln alleles had increasing risk of HCC (total crude adjusted OR = 1.34, $P < 0.01$), moreover, there were larger relative weight to assign to those studies with OR-value more than 1. Actually, although Yang *et al.* (116) and Ren *et al.* (118) did not observed significantly risk of XRCC1 gene codon 399 polymorphism in crude logistic regression, they found Gln alleles would increase HCC risk in stratified analysis with susceptive environment variants. A individually matching case-controls demonstrated that subjects having Gln alleles might feature remarkably increasing risk of HCC under longer-term AFB1-exposure years or higher AFB1-exposure levels conditions (adjusted OR > 10) (61). This suggests that the genotypes with codon 399 Gln alleles of XRCC1 should be a risk biomarker of Chinese HCC related to AFB1 exposure.

## 6. Genetic polymorphisms in genes involved in BER pathway and risk of HCC

Of the oxidative DNA damage resulting from AFB1 exposure, the formation of 8-$_{oxod}$G is thought to be important due to being abundant and highly mutagenic and hepatocarcinogenesis (21, 36-38). The 8-$_{oxod}$G lesions are repaired primarily through the BER pathway (119). The BER pathway facilitates DNA repair through two general pathways: *a.* the short-patch BER pathway, leading to a repair tract of a single nucleotide; *b.* the long-patch BER pathway, producing a repair tract of at least two nucleotides (120). In these two repair sub-pathways, DNA glycosylases play a central role because they can recognize and catalyze the removal of damaged bases (120). This suggests that the defect of DNA glycosylases should be related to the decreasing capacity of the BER pathway and might increase the risk of such cancers as HCC.

Human oxoguanine glycosylase 1(hOGG1) is a specific DNA glycosylase that catalyzes the release of 8-$_{oxod}$G and the cleavage of DNA at the AP site (121, 122). Genetic structure study has revealed the presence of several polymorphisms within hOGG1 locus (123). Among them, the polymorphism at position 1245 in exon 7 causes an amino acid substitution (Ser to Cys) at codon 326, suggesting this polymorphism may glycosylase function (123). A functional complementation activity assay showed that hOGG1 protein encoded by the 326 Cys allele had substantially lower DNA repair activity than that encoded by the 326 Ser allele (124). Similar results were observed in human cells in vivo (122, 125). Therefore, low capacity of 8-$_{oxod}$G repair resulting from hOGG1 326Cys polymorphism might contribute to

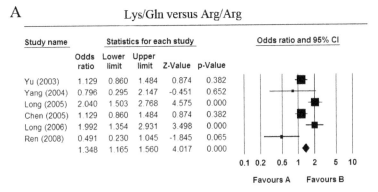

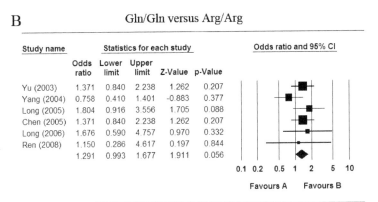

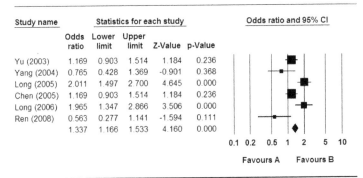

Fig. 3. The meta-analysis of the relationship between XRCC1 codon Lys399Gln
polymorphism and HCC risk among China population. Compared with Arg/Arg genotype,
Arg/Gln (A) and Gln/Gln (B) genotypes increased HCC risk. This risk effect was also
observed in the binding of Arg/Gln and Gln/Gln genotypes (C).

the persistence of 8-$_{oxod}$G in genomic DNA in vivo, which, in turn, could be associated with increased cancer risk (125, 126).

In 2003, Peng *et al.* (126) investigated the correlation among 8-$_{oxod}$G levels, hOGG1 expression, and hOGG1 Cys326Ser polymorphism in Guangxi Autonomous Region. They found that individuals with genotypes with hOGG1 codon 326 Cys alleles faced lower level of hOGG1 expression and higher 8-$_{oxod}$G levels. Supporting their results, Cheng *et al.* (21) reported that hOGG1 expression was significantly linear correlated with HCC. Recently, using the molecular epidemiological methods, Zhang *et al.*(127) found that the distribution of Cys alleles at codon 326 of hOGG1 in HCC cases (43.0%) significantly differed from in controls (33.1%). Logistic regression analysis showed that the genotypes with Cys alleles, compared to without this alleles, increased HCC risk of Chinese population, with adjusted OR-value (95% CI) 1.5 (0.79-2.93) for Cys/Ser and 1.9 (0.83-4.55) for Cys/Cys. These findings suggested pathogenic role of hOGG1 Cys326Ser polymorphism in the hepatocarcinogenesis.

## 7. Genetic polymorphisms in genes involved in DSBR pathway and risk of HCC

DSBs, although only make up a very small proportion of AFB1-induced DNA damage, are critical lesions that can result in cell death or a wide variety of genetic alterations including large- or small-scale deletions, loss of heterozygosity, translocations, and chromosome loss (19, 128, 129). This type damage is repaired DSBR consisting of non-homologous end-joining (NHEJ) and homologous recombination (HR) (130-133). There are several decades DNA repair genes involves in DSBR pathway and the defects in these genes cause genome instability and promote tumorigenesis (128, 134, 135). In published molecular epidemiological studies, only XRCC3 gene codon Thr241Met polymorphism effects the risk of AFB1-related HCC risk among Chinese population (58, 60).

The product of the XRCC3 gene is one of identified paralogs of the strand-exchange protein RAD51 in human beings (136). This protein correlates directly with DNA breaks and facilitates of the formation of the RAD51 nucleoprotein filament, which is crucial both for homologous recombination and HRR (136-138). Previous studies have shown that a common polymorphism at codon 241 of XRCC3 gene (Thr to Met) modifies the function of this gene ad increases cancers risk (139-143). Two reports from high AFB1-exposure areas of China supported above-mentioned conclusions (58, 60).

In the first frequent case-control study in Guangxiese (58), we observed that the genotypes with XRCC3 codon 241 Met alleles (namely Thr/Met and Met/Met) was significantly different between controls (33.01%) and HCC cases (61.48%, P < 0.001). Met alleles increases about 2- to 10-fold risk of HCC and this running-up risk is modulated by the number of Met alleles (adjusted OR 2.48 and 10.06 for one and two this alleles). Considering small sample size in this study, we recruited, in another independent frequent case-control study (60), a relatively larger sample size to compare the results. Subjects included in this study, 491 HCC cases and 862 age-, sex, race, hepatitis virus infection information-matching controls, were permanent residents of Guangxi areas. Similar to the results of the first report, the distribution of XRCC3 codon 241 Met allele frequency was found to be significantly different between cases (59.7%) and controls (32.1%). Individuals having the Thr/Met or Met/Met were at a 2.22-fold or 7.19 fold increased risk of developing HCC cancer. Above two studies showed this allele multiplicatively interacted with AFB1 exposure in the process

DNA Repair Capacity-Related to Genetic Polymorphisms of DNA Repair Genes and Aflatoxin B1-Related
Hepatocellular Carcinoma Among Chinese Population

117

of hepato-tumorigenesis. These results exhibits that the polymorphism at codon 241 of XRCC3 gene is a genetic determinant in the development of HCC induced by AFB1 exposure among Chinese population.

## 8. Summary

Like most other human malignant tumors, HCC is a complex disease attributed to environment variation and genetic susceptive factors. In high incidence areas of HCC in China, AFB1 is an important environment variation as well as chronic HBV and HCV infection. This toxic variation is characterized by: *a*. the attraction of specific organs, especially liver; *b*. genotoxicity, mainly inducing the formation of AFB1-DNA adducts and the hot-spot mutation of p53 gene; and *c*. carcinogenicity, primarily causing HCC. In the process of AFB1 hepatocarcinogenesis, AFB1-DNA adducts play a central role because of their genotoxicity and interactions with genetic susceptive factors. Numerous studies reviewed in this paper have demonstrated that the hereditary variations in DNA repair genes are associated with susceptibility to AFB1-related HCC among Chinese population. These molecular epidemiological studies have significantly contributed to our knowledge of the importance of genetic polymorphisms in DNA repair genes in the etiology of HCC related to AFB1 exposure. It would be expected that genetic susceptibility factors involved in DNA repair genes for HCC could serve as useful biomarkers for identifying at-risk individuals and, therefore, targeting prevention of this malignant tumor.

However, there are several issues to be noted. Firstly, the conclusions should be drawn carefully, because of conflicting data existing in the same ethnic population in view of between some genotypes of DNA repair genes and the risk of HCC. Secondly, caution should be taken particularly in extrapolating these data to other ethnic populations, because of the difference of population frequencies corresponding to genetic polymorphisms that depends on ethnicity. Thirdly, when risk of a specific polymorphism is considered, AFB1 exposure should be stressed because AFB1 exposure may differ from areas to areas and from individuals to individuals. Lastly, because of the fact that AFB1-related hepatocarcinogenesis is polygenic, no single genetic marker may sufficiently predict HCC risk. Therefore, a panel of susceptive biomarkers is warranted to define individuals at high-risk for this cancer.

## 9. Acknowledgments

We are grateful to Yuan-Feng Zhou for the collection and management of data. This study is supported by the Youth Science Foundation of Guangxi (NO. 0832097), the Science Foundation of Youjiang Medical College for Nationalities (NO. 2005 and 2008), and Guangxi Key Construction Project of Laboratory Room (NO. 2009).

## 10. Abbreviations

AFB1, Aflatoxin B1; AFB1-epoxide, AFB1-8,9-epoxide; AFB1-N[7]-Gua, 8,9-di-hydro-8-(N[7]-guanyl)-9-hydroxy–AFB1; AFB1-FAPy, ring-opened formamidopyridine AFB1; APE1, AP endonuclease-1; BER, base excision repair; CI, confidence interval; DSB, double-strand break; DSBR, double-strand break repair; HBV, hepatitis virus B; HCV, hepatitis virus C; HCC, hepatocellular carcinoma; hOGG1, Human oxoguanine glycosylase 1; NER, nucleotide excision repair; OR, odds ratio; 8-$_{oxod}$G, 8-oxodeoxyguanosine; PARP, poly (ADP-ribose) polymerase; PLC, Primary liver cancer; PNK, polynucleotide kinase; Pol β, DNA polymerase β; SSB, single-strand break; SSBR, single-strand break repair; XPA, xeroderma pigmentosum A; XPC, xeroderma pigmentosum C; XPD, xeroderma pigmentosum D; XRCC1, x-ray repair cross complementary 1; XRCC3, x-ray repair cross complementary 3; XRCC4, x-ray repair cross complementary 4.

## 11. References

[1] Parkin DM, Bray F, Ferlay J, Pisani P. Global cancer statistics, 2002. CA Cancer J Clin. 2005;55:74-108.

[2] Henry SH, Bosch FX, Troxell TC, Bolger PM. Reducing Liver Cancer--Global Control of Aflatoxin. Science. 1999;286:2453-4.

[3] Dominguez-Malagon H, Gaytan-Graham S. Hepatocellular carcinoma: an update. Ultrastruct Pathol. 2001;25:497-516.

[4] Long XD, Ma Y, Zhou YF, Ma AM, Fu GH. Polymorphism in xeroderma pigmentosum complementation group C codon 939 and aflatoxin B1-related hepatocellular carcinoma in the Guangxi population. Hepatology. 2010;52:1301-9.

[5] de Oliveira CA, Germano PM. [Aflatoxins: current concepts on mechanisms of toxicity and their involvement in the etiology of hepatocellular carcinoma]. Rev Saude Publica. 1997;31:417-24.

[6] Makarananda K, Pengpan U, Srisakulthong M, Yoovathaworn K, Sriwatanakul K. Monitoring of aflatoxin exposure by biomarkers. J Toxicol Sci. 1998;23 Suppl 2:155-9.

[7] Sell S. Mouse models to study the interaction of risk factors for human liver cancer. Cancer research. 2003;63:7553-62.

[8] Wang JS, Huang T, Su J, Liang F, Wei Z, Liang Y, et al. Hepatocellular carcinoma and aflatoxin exposure in Zhuqing Village, Fusui County, People's Republic of China. Cancer Epidemiol Biomarkers Prev. 2001;10:143-6.

[9] Clifford RJ, Zhang J, Meerzaman DM, Lyu MS, Hu Y, Cultraro CM, et al. Genetic variations at loci involved in the immune response are risk factors for hepatocellular carcinoma. Hepatology. 2010;52:2034-43.

[10] Pang RW, Joh JW, Johnson PJ, Monden M, Pawlik TM, Poon RT. Biology of hepatocellular carcinoma. Annals of surgical oncology. 2008;15:962-71.

[11] Wilson DM, 3rd, Thompson LH. Life without DNA repair. Proceedings of the National Academy of Sciences of the United States of America. 1997;94:12754-7.

[12] Long XD, Tang YH, Qu DY, Ma Y. The toxicity and role of aflatoxin B1 and DNA repair (corresponding DNA repair enzymes) Youjiang Medical College for Nationalities Xue Bao. 2006;28:278-80.

[13] J.G. C, sONG XM. An evaluation on incidence cases of liver cancer in China. Zhongguo Zhongliu. 2005;14:28-31.

[14] Wang YW, Lan LZ, Ye BF, Xu YC, Liu YY, Li WG. The association among geographic distribution, climate, and aflatoxin B1 exposure in China. Zhonggou Kexue. 1983;B:431-7.

[15] Zhang WS, Chen WQ, Kong LZ, Li GL, Zhao P. An Annual Repor t: Cancer Incidence in 35 Cancer Registr ies in China, 2003. Zhongguo Zhongliu. 2007;16:497-507.

DNA Repair Capacity-Related to Genetic Polymorphisms of DNA Repair Genes and Aflatoxin B1-Related
Hepatocellular Carcinoma Among Chinese Population

119

[16] Ma AM, Luo YX, Wan JL, Long XD, Huang YZ. An investigative analysis of incidence, prevention, and treatment of hepatocellular carcinoma in Xiangzhou, Baise, Guangxi Autonomous Region, China. Youjiang Minzu Yixueyuan Bao. 2010;32:165-6.

[17] Youjiang Medical College for Nationalities and Public Health Bureau of Tiandong. General investigation of hepatocellular carcinoma in Tiandong, China. 1994—2008 China Academic Journal electronic Publishing Househttp://wwwenkinet.

[18] the Ministry of Health of People's Republic of China. Annals of health statistics in China in 2010. ttp://www.moh.gov.cn/publicfiles/business/htmlfiles/zwgkzt/ptjnj/year2010/index2010.html.

[19] Wang JS, Groopman JD. DNA damage by mycotoxins. Mutation research. 1999;424:167-81.

[20] Yeh FS, Mo CC, Yen RC. Risk factors for hepatocellular carcinoma in Guangxi, People's Republic of China. National Cancer Institute monograph. 1985;69:47-8.

[21] Cheng B, Jungst C, Lin J, Caselmann WH. [Potential role of human DNA-repair enzymes hMTH1, hOGG1 and hMYHalpha in the hepatocarcinogenesis]. Journal of Huazhong University of Science and Technology Medical sciences. 2002;22:206-11, 15.

[22] Scholl PF, McCoy L, Kensler TW, Groopman JD. Quantitative analysis and chronic dosimetry of the aflatoxin B1 plasma albumin adduct Lys-AFB1 in rats by isotope dilution mass spectrometry. Chem Res Toxicol. 2006;19:44-9.

[23] Sudhakar P, Latha P, Sreenivasulu Y, Reddy BV, Hemalatha TM, Balakrishna M, et al. Inhibition of Aspergillus flavus colonization and aflatoxin (AfB1) in peanut by methyleugenol. Indian J Exp Biol. 2009;47:63-7.

[24] International Agency for Research on Cancer, Some naturally occurring substances: food items and constituents, heterocyclic aromatic amines and mycotoxins, IARC Monogr. Eval Carcinog Risks Hum 1993;56:245-540.

[25] Ben Rejeb I, Arduini F, Arvinte A, Amine A, Gargouri M, Micheli L, et al. Development of a bio-electrochemical assay for AFB1 detection in olive oil. Biosens Bioelectron. 2009;24:1962-8.

[26] Giri I, Stone MP. Wobble dC.dA pairing 5' to the cationic guanine N7 8,9-dihydro-8-(N7-guanyl)-9-hydroxyaflatoxin B1 adduct: implications for nontargeted AFB1 mutagenesis. Biochemistry. 2003;42:7023-34.

[27] Habib SL, Said B, Awad AT, Mostafa MH, Shank RC. Novel adenine adducts, N7-guanine-AFB1 adducts, and p53 mutations in patients with schistosomiasis and aflatoxin exposure. Cancer Detect Prev. 2006;30:491-8.

[28] Guengerich FP, Johnson WW, Shimada T, Ueng YF, Yamazaki H, Langouet S. Activation and detoxication of aflatoxin B1. Mutation research. 1998;402:121-8.

[29] Essigmann JM, Croy RG, Nadzan AM, Busby WF, Jr., Reinhold VN, Buchi G, et al. Structural identification of the major DNA adduct formed by aflatoxin B1 in vitro. Proceedings of the National Academy of Sciences of the United States of America. 1977;74:1870-4.

[30] Croy RG, Essigmann JM, Reinhold VN, Wogan GN. Identification of the principal aflatoxin B1-DNA adduct formed in vivo in rat liver. Proceedings of the National Academy of Sciences of the United States of America. 1978;75:1745-9.

[31] Gopalakrishnan S, Harris TM, Stone MP. Intercalation of aflatoxin B1 in two oligodeoxynucleotide adducts: comparative 1H NMR analysis of d(ATCAFBGAT).d(ATCGAT) and d(ATAFBGCAT)2. Biochemistry. 1990;29:10438-48.

[32] Johnson WW, Guengerich FP. Reaction of aflatoxin B1 exo-8,9-epoxide with DNA: kinetic analysis of covalent binding and DNA-induced hydrolysis. Proceedings of the National Academy of Sciences of the United States of America. 1997;94:6121-5.

[33] Croy RG, Wogan GN. Temporal patterns of covalent DNA adducts in rat liver after single and multiple doses of aflatoxin B1. Cancer research. 1981;41:197-203.

[34] Groopman JD, Croy RG, Wogan GN. In vitro reactions of aflatoxin B1-adducted DNA. Proceedings of the National Academy of Sciences of the United States of America. 1981;78:5445-9.

[35] Chen JD, Liu CJ, Lee PH, Chen PJ, Lai MY, Kao JH, et al. Hepatitis B genotypes correlate with tumor recurrence after curative resection of hepatocellular carcinoma. Clin Gastroenterol Hepatol. 2004;2:64-71.

[36] Gradelet S, Le Bon AM, Berges R, Suschetet M, Astorg P. Dietary carotenoids inhibit aflatoxin B1-induced liver preneoplastic foci and DNA damage in the rat: role of the modulation of aflatoxin B1 metabolism. Carcinogenesis. 1998;19:403-11.

[37] Farombi EO, Adepoju BF, Ola-Davies OE, Emerole GO. Chemoprevention of aflatoxin B1-induced genotoxicity and hepatic oxidative damage in rats by kolaviron, a natural bioflavonoid of Garcinia kola seeds. European journal of cancer prevention : the official journal of the European Cancer Prevention Organisation. 2005;14:207-14.

[38] Liu J, Yang CF, Lee BL, Shen HM, Ang SG, Ong CN. Effect of Salvia miltiorrhiza on aflatoxin B1-induced oxidative stress in cultured rat hepatocytes. Free radical research. 1999;31:559-68.

[39] Long XD, Ma Y, Zhou YF, Yao JG, Ban FZ, Huang YZ, et al. XPD Codon 312 and 751 Polymorphisms, and AFB1 Exposure, and Hepatocellular Carcinoma Risk. BMC cancer. 2009;9:400.

[40] Long XD, Ma Y, Deng ZL. GSTM1 and XRCC3 polymorphisms: Effects on levels of aflatoxin B1-DNA adducts. Chinese J Cancer Res. 2009;21:177-84.

[41] Long XD, Ma Y, Huang HD, Yao JG, Qu de Y, Lu YL. Polymorphism of XRCC1 and the frequency of mutation in codon 249 of the p53 gene in hepatocellular carcinoma among Guangxi population, China. Mol Carcinog. 2008;47:295-300.

[42] Wild CP, Montesano R. A model of interaction: aflatoxins and hepatitis viruses in liver cancer aetiology and prevention. Cancer letters. 2009;286:22-8.

[43] Shen HM, Ong CN. Mutations of the p53 tumor suppressor gene and ras oncogenes in aflatoxin hepatocarcinogenesis. Mutation research. 1996;366:23-44.

[44] Golli-Bennour EE, Kouidhi B, Bouslimi A, Abid-Essefi S, Hassen W, Bacha H. Cytotoxicity and genotoxicity induced by aflatoxin B1, ochratoxin A, and their combination in cultured Vero cells. J Biochem Mol Toxicol. 2010;24:42-50.

[45] Paget V, Sichel F, Garon D, Lechevrel M. Aflatoxin B1-induced TP53 mutational pattern in normal human cells using the FASAY (Functional Analysis of Separated Alleles in Yeast). Mutation research. 2008;656:55-61.

[46] Van Vleet TR, Watterson TL, Klein PJ, Coulombe RA, Jr. Aflatoxin B1 alters the expression of p53 in cytochrome P450-expressing human lung cells. Toxicol Sci. 2006;89:399-407.

DNA Repair Capacity-Related to Genetic Polymorphisms of DNA Repair Genes and Aflatoxin B1-Related
Hepatocellular Carcinoma Among Chinese Population

121

[47] Vahakangas K. TP53 mutations in workers exposed to occupational carcinogens. Hum Mutat. 2003;21:240-51.

[48] Staib F, Hussain SP, Hofseth LJ, Wang XW, Harris CC. TP53 and liver carcinogenesis. Hum Mutat. 2003;21:201-16.

[49] Chan KT, Hsieh DP, Lung ML. In vitro aflatoxin B1-induced p53 mutations. Cancer letters. 2003;199:1-7.

[50] Bressac B, Kew M, Wands J, Ozturk M. Selective G to T mutations of p53 gene in hepatocellular carcinoma from southern Africa. Nature. 1991;350:429-31.

[51] Li D, Cao Y, He L, Wang NJ, Gu JR. Aberrations of p53 gene in human hepatocellular carcinoma from China. Carcinogenesis. 1993;14:169-73.

[52] Kirk GD, Camus-Randon AM, Mendy M, Goedert JJ, Merle P, Trepo C, et al. Ser-249 p53 mutations in plasma DNA of patients with hepatocellular carcinoma from The Gambia. Journal of the National Cancer Institute. 2000;92:148-53.

[53] Loechler, E.L. Mechanisms by which aflatoxins and other bulky carcinogens induce mutations, in: D.L. Eaton, J.D. Groopman (Eds.), The Toxicology of Aflatoxins: Human Health, Veterinary, and Agricultural Significance, Academic Press, San Diego, 1994, pp. 149–178.

[54] Aguilar F, Hussain SP, Cerutti P. Aflatoxin B1 induces the transversion of G-->T in codon 249 of the p53 tumor suppressor gene in human hepatocytes. Proceedings of the National Academy of Sciences of the United States of America. 1993;90:8586-90.

[55] Denissenko MF, Koudriakova TB, Smith L, O'Connor TR, Riggs AD, Pfeifer GP. The p53 codon 249 mutational hotspot in hepatocellular carcinoma is not related to selective formation or persistence of aflatoxin B1 adducts. Oncogene. 1998;17:3007-14.

[56] Duflot A, Hollstein M, Mehrotra R, Trepo C, Montesano R, Cova L. Absence of p53 mutation at codon 249 in duck hepatocellular carcinomas from the high incidence area of Qidong (China). Carcinogenesis. 1994;15:1353-7.

[57] Waters R, Jones CJ, Martin EA, Yang AL, Jones NJ. The repair of large DNA adducts in mammalian cells. Mutation research. 1992;273:145-55.

[58] Long XD, Ma Y, Deng ZL, Huang YZ, Wei NB. [Association of the Thr241Met polymorphism of DNA repair gene XRCC3 with genetic susceptibility to AFB1-related hepatocellular carcinoma in Guangxi population]. Zhonghua yi xue yi chuan xue za zhi = Zhonghua yixue yichuanxue zazhi = Chinese journal of medical genetics. 2008;25:268-71.

[59] Long XD, Ma Y, Huang YZ, Yi Y, Liang QX, Ma AM, et al. Genetic polymorphisms in DNA repair genes XPC, XPD, and XRCC4, and susceptibility to Helicobacter pylori infection-related gastric antrum adenocarcinoma in Guangxi population, China. Mol Carcinog. 2010;49:611-8.

[60] Long XD, Ma Y, Qu de Y, Liu YG, Huang ZQ, Huang YZ, et al. The polymorphism of XRCC3 codon 241 and AFB1-related hepatocellular carcinoma in Guangxi population, China. Ann Epidemiol. 2008;18:572-8.

[61] Long XD, Ma Y, Wei YP, Deng ZL. The polymorphisms of GSTM1, GSTT1, HYL1*2, and XRCC1, and aflatoxin B1-related hepatocellular carcinoma in Guangxi population, China. Hepatol Res. 2006;36:48-55.

[62] Long XD, Ma Y, Wei YP, Deng ZL. X-ray repair cross-complementing group 1 (XRCC1) Arg 399 Gln polymorphism and aflatoxin B1 (AFB1)-related hepatocellular carcinoma (HCC) in Guangxi population. Chinese J Cancer Res. 2005;17:17-21.

[63]  Nouspikel T. DNA repair in mammalian cells : Nucleotide excision repair: variations on versatility. Cellular and molecular life sciences : CMLS. 2009;66:994-1009.

[64]  Rechkunova NI, Lavrik OI. Nucleotide excision repair in higher eukaryotes: mechanism of primary damage recognition in global genome repair. Sub-cellular biochemistry. 2010;50:251-77.

[65]  Rajaraman P, Hutchinson A, Wichner S, Black PM, Fine HA, Loeffler JS, et al. DNA repair gene polymorphisms and risk of adult meningioma, glioma, and acoustic neuroma. Neuro Oncol. 2010;12:37-48.

[66]  Matakidou A, Eisen T, Fleischmann C, Bridle H, Houlston RS. Evaluation of xeroderma pigmentosum XPA, XPC, XPD, XPF, XPB, XPG and DDB2 genes in familial early-onset lung cancer predisposition. International journal of cancer. 2006;119:964-7.

[67]  Wang M, Yuan L, Wu D, Zhang Z, Yin C, Fu G, et al. A novel XPF -357A>C polymorphism predicts risk and recurrence of bladder cancer. Oncogene. 2010;29:1920-8.

[68]  Dogru-Abbasoglu S, Inceoglu M, Parildar-Karpuzoglu H, Hanagasi HA, Karadag B, Gurvit H, et al. Polymorphisms in the DNA repair genes XPD (ERCC2) and XPF (ERCC4) are not associated with sporadic late-onset Alzheimer's disease. Neurosci Lett. 2006;404:258-61.

[69]  Romanowicz-Makowska H, Smolarz B, Kulig A. [Polymorphisms in XRCC1 and ERCC4/XPF DNA repair genes and associations with breast cancer risk in women]. Pol Merkur Lekarski. 2007;22:200-3.

[70]  X.M. W, Ma Y, Deng ZL, Long XD. The polymorphism at codon 939 of xeroderma pigmentosum C gene and hepatocellular carcinoma among Guangxi population. Zhonghua Xiaohua Zazhi. 2010;30:846-8.

[71]  Li LM, Zeng XY, Ji L, Fan XJ, Li YQ, Hu XH, et al. [Association of XPC and XPG polymorphisms with the risk of hepatocellular carcinoma]. Zhonghua gan zang bing za zhi = Zhonghua ganzangbing zazhi = Chinese journal of hepatology. 2010;18:271-5.

[72]  Sugasawa K. XPC: its product and biological roles. Advances in experimental medicine and biology. 2008;637:47-56.

[73]  Strom SS, Estey E, Outschoorn UM, Garcia-Manero G. Acute myeloid leukemia outcome: role of nucleotide excision repair polymorphisms in intermediate risk patients. Leuk Lymphoma. 2010;51:598-605.

[74]  Sakano S, Matsumoto H, Yamamoto Y, Kawai Y, Eguchi S, Ohmi C, et al. Association between DNA repair gene polymorphisms and p53 alterations in Japanese patients with muscle-invasive bladder cancer. Pathobiology. 2006;73:295-303.

[75]  Qiu L, Wang Z, Shi X, Wang Z. Associations between XPC polymorphisms and risk of cancers: A meta-analysis. Eur J Cancer. 2008;44:2241-53.

[76]  Zhou RM, Li Y, Wang N, Zhang XJ, Dong XJ, Guo W. [Correlation of XPC Ala499Val and Lys939Gln polymorphisms to risks of esophageal squamous cell carcinoma and gastric cardiac adenocarcinoma]. Ai zheng = Aizheng = Chinese journal of cancer. 2006;25:1113-9.

[77]  Zhang D, Chen C, Fu X, Gu S, Mao Y, Xie Y, et al. A meta-analysis of DNA repair gene XPC polymorphisms and cancer risk. Journal of human genetics. 2008;53:18-33.

[78]  Shore RE, Zeleniuch-Jacquotte A, Currie D, Mohrenweiser H, Afanasyeva Y, Koenig KL, et al. Polymorphisms in XPC and ERCC2 genes, smoking and breast cancer risk. International journal of cancer. 2008;122:2101-5.

DNA Repair Capacity-Related to Genetic Polymorphisms of DNA Repair Genes and Aflatoxin B1-Related
Hepatocellular Carcinoma Among Chinese Population

123

[79]  Hansen RD, Sorensen M, Tjonneland A, Overvad K, Wallin H, Raaschou-Nielsen O, et al. XPA A23G, XPC Lys939Gln, XPD Lys751Gln and XPD Asp312Asn polymorphisms, interactions with smoking, alcohol and dietary factors, and risk of colorectal cancer. Mutation research. 2007;619:68-80.

[80]  Liang J, Gu A, Xia Y, Wu B, Lu N, Wang W, et al. XPC gene polymorphisms and risk of idiopathic azoospermia or oligozoospermia in a Chinese population. International journal of andrology. 2009;32:235-41.

[81]  Gangwar R, Mandhani A, Mittal RD. XPC gene variants: a risk factor for recurrence of urothelial bladder carcinoma in patients on BCG immunotherapy. Journal of cancer research and clinical oncology. 2010;136:779-86.

[82]  Francisco G, Menezes PR, Eluf-Neto J, Chammas R. XPC polymorphisms play a role in tissue-specific carcinogenesis: a meta-analysis. Eur J Hum Genet. 2008;16:724-34.

[83]  Cheo DL, Burns DK, Meira LB, Houle JF, Friedberg EC. Mutational inactivation of the xeroderma pigmentosum group C gene confers predisposition to 2-acetylaminofluorene-induced liver and lung cancer and to spontaneous testicular cancer in Trp53-/- mice. Cancer research. 1999;59:771-5.

[84]  Fautrel A, Andrieux L, Musso O, Boudjema K, Guillouzo A, Langouet S. Overexpression of the two nucleotide excision repair genes ERCC1 and XPC in human hepatocellular carcinoma. J Hepatol. 2005;43:288-93.

[85]  Cai XL, Gao YH, Yu ZW, Wu ZQ, Zhou WP, Yang Y, et al. [A 1:1 matched case-control study on the interaction between HBV, HCV infection and DNA repair gene XPC Ala499Val, Lys939Gln for primary hepatocellular carcinoma]. Zhonghua liu xing bing xue za zhi = Zhonghua liuxingbingxue zazhi. 2009;30:942-5.

[86]  Aloyz R, Xu ZY, Bello V, Bergeron J, Han FY, Yan Y, et al. Regulation of cisplatin resistance and homologous recombinational repair by the TFIIH subunit XPD. Cancer research. 2002;62:5457-62.

[87]  Manuguerra M, Saletta F, Karagas MR, Berwick M, Veglia F, Vineis P, et al. XRCC3 and XPD/ERCC2 single nucleotide polymorphisms and the risk of cancer: a HuGE review. American journal of epidemiology. 2006;164:297-302.

[88]  Benhamou S, Sarasin A. ERCC2/XPD gene polymorphisms and cancer risk. Mutagenesis. 2002;17:463-9.

[89]  Shen MR, Jones IM, Mohrenweiser H. Nonconservative amino acid substitution variants exist at polymorphic frequency in DNA repair genes in healthy humans. Cancer research. 1998;58:604-8.

[90]  Sturgis EM, Dahlstrom KR, Spitz MR, Wei Q. DNA repair gene ERCC1 and ERCC2/XPD polymorphisms and risk of squamous cell carcinoma of the head and neck. Arch Otolaryngol Head Neck Surg. 2002;128:1084-8.

[91]  Chen S, Tang D, Xue K, Xu L, Ma G, Hsu Y, et al. DNA repair gene XRCC1 and XPD polymorphisms and risk of lung cancer in a Chinese population. Carcinogenesis. 2002;23:1321-5.

[92]  Park JY, Lee SY, Jeon HS, Park SH, Bae NC, Lee EB, et al. Lys751Gln polymorphism in the DNA repair gene XPD and risk of primary lung cancer. Lung cancer (Amsterdam, Netherlands). 2002;36:15-6.

[93]  Xing D, Qi J, Miao X, Lu W, Tan W, Lin D. Polymorphisms of DNA repair genes XRCC1 and XPD and their associations with risk of esophageal squamous cell carcinoma in a Chinese population. International journal of cancer. 2002;100:600-5.

[94]   Stern MC, Johnson LR, Bell DA, Taylor JA. XPD codon 751 polymorphism, metabolism
       genes, smoking, and bladder cancer risk. Cancer Epidemiol Biomarkers Prev.
       2002;11:1004-11.

[95]   Sturgis EM, Castillo EJ, Li L, Eicher SA, Strom SS, Spitz MR, et al. XPD/ERCC2 EXON
       8 Polymorphisms: rarity and lack of significance in risk of squamous cell carcinoma
       of the head and neck. Oral Oncol. 2002;38:475-7.

[96]   Spitz MR, Wu X, Wang Y, Wang LE, Shete S, Amos CI, et al. Modulation of nucleotide
       excision repair capacity by XPD polymorphisms in lung cancer patients. Cancer
       research. 2001;61:1354-7.

[97]   Rzeszowska-Wolny J, Polanska J, Pietrowska M, Palyvoda O, Jaworska J, Butkiewicz
       D, et al. Influence of polymorphisms in DNA repair genes XPD, XRCC1 and
       MGMT on DNA damage induced by gamma radiation and its repair in
       lymphocytes in vitro. Radiation research. 2005;164:132-40.

[98]   Tang D, Cho S, Rundle A, Chen S, Phillips D, Zhou J, et al. Polymorphisms in the DNA
       repair enzyme XPD are associated with increased levels of PAH-DNA adducts in a
       case-control study of breast cancer. Breast cancer research and treatment.
       2002;75:159-66.

[99]   Terry MB, Gammon MD, Zhang FF, Eng SM, Sagiv SK, Paykin AB, et al.
       Polymorphism in the DNA repair gene XPD, polycyclic aromatic hydrocarbon-
       DNA adducts, cigarette smoking, and breast cancer risk. Cancer epidemiology,
       biomarkers & prevention : a publication of the American Association for Cancer
       Research, cosponsored by the American Society of Preventive Oncology.
       2004;13:2053-8.

[100]  Mechanic LE, Marrogi AJ, Welsh JA, Bowman ED, Khan MA, Enewold L, et al.
       Polymorphisms in XPD and TP53 and mutation in human lung cancer.
       Carcinogenesis. 2005;26:597-604.

[101]  Caldecott KW. Mammalian single-strand break repair: mechanisms and links with
       chromatin. DNA repair. 2007;6:443-53.

[102]  Fortini P, Dogliotti E. Base damage and single-strand break repair: mechanisms and
       functional significance of short- and long-patch repair subpathways. DNA repair.
       2007;6:398-409.

[103]  Thompson LH, West MG. XRCC1 keeps DNA from getting stranded. Mutation
       research. 2000;459:1-18.

[104]  Yamane K, Katayama E, Tsuruo T. The BRCT regions of tumor suppressor BRCA1
       and of XRCC1 show DNA end binding activity with a multimerizing feature.
       Biochem Biophys Res Commun. 2000;279:678-84.

[105]  Bhattacharyya N, Banerjee S. A novel role of XRCC1 in the functions of a DNA
       polymerase beta variant. Biochemistry. 2001;40:9005-13.

[106]  Dulic A, Bates PA, Zhang X, Martin SR, Freemont PS, Lindahl T, et al. BRCT domain
       interactions in the heterodimeric DNA repair protein XRCC1-DNA ligase III.
       Biochemistry. 2001;40:5906-13.

[107]  Divine KK, Gilliland FD, Crowell RE, Stidley CA, Bocklage TJ, Cook DL, et al. The
       XRCC1 399 glutamine allele is a risk factor for adenocarcinoma of the lung.
       Mutation research. 2001;461:273-8.

[108]  Lee JM, Lee YC, Yang SY, Yang PW, Luh SP, Lee CJ, et al. Genetic polymorphisms of
       XRCC1 and risk of the esophageal cancer. International journal of cancer.
       2001;95:240-6.

[109]  Ratnasinghe D, Yao SX, Tangrea JA, Qiao YL, Andersen MR, Barrett MJ, et al. Polymorphisms of the DNA repair gene XRCC1 and lung cancer risk. Cancer Epidemiol Biomarkers Prev. 2001;10:119-23.

[110]  Kim SU, Park SK, Yoo KY, Yoon KS, Choi JY, Seo JS, et al. XRCC1 genetic polymorphism and breast cancer risk. Pharmacogenetics. 2002;12:335-8.

[111]  Schneider J, Classen V, Helmig S. XRCC1 polymorphism and lung cancer risk. Expert Rev Mol Diagn. 2008;8:761-80.

[112]  Kiran M, Saxena R, Chawla YK, Kaur J. Polymorphism of DNA repair gene XRCC1 and hepatitis-related hepatocellular carcinoma risk in Indian population. Mol Cell Biochem. 2009;327:7-13.

[113]  Li Y, Long C, Lin G, Marion MJ, Freyer G, Santella RM, et al. Effect of the XRCC1 codon 399 polymorphism on the repair of vinyl chloride metabolite-induced DNA damage. J Carcinog. 2009;8:14.

[114]  Saadat M, Ansari-Lari M. Polymorphism of XRCC1 (at codon 399) and susceptibility to breast cancer, a meta-analysis of the literatures. Breast cancer research and treatment. 2009;115:137-44.

[115]  Yu MW, Yang SY, Pan IJ, Lin CL, Liu CJ, Liaw YF, et al. Polymorphisms in XRCC1 and glutathione S-transferase genes and hepatitis B-related hepatocellular carcinoma. Journal of the National Cancer Institute. 2003;95:1485-8.

[116]  Yang JL, Han YN, Zheng SG. Influence of human XRCC1-399 single nucleotide polymorphism on primary hepatocytic carcinoma. Zhongliu. 2004;24:322-4.

[117]  Chen CC, Yang SY, Liu CJ, Lin CL, Liaw YF, Lin SM, et al. Association of cytokine and DNA repair gene polymorphisms with hepatitis B-related hepatocellular carcinoma. International journal of epidemiology. 2005;34:1310-8.

[118]  Ren y, Wang DS, Li Z, Xin YM, Yin JM, Zahgn B, et al. Study on the Relationship between Gene XRCC1 Codon 399 Single Nucleotide Polymorphisms and Primary Hepatic Carcinoma in Han Nationality. Linchuang Gandang Bing Zazhi. 2008;24:361-4.

[119]  Donigan KA, Sweasy JB. Sequence context-specific mutagenesis and base excision repair. Mol Carcinogen. 2009;48:362-8.

[120]  Robertson AB, Klungland A, Rognes T, Leiros I. DNA repair in mammalian cells: Base excision repair: the long and short of it. Cellular and molecular life sciences : CMLS. 2009;66:981-93.

[121]  Boiteux S, Radicella JP. The human OGG1 gene: structure, functions, and its implication in the process of carcinogenesis. Archives of biochemistry and biophysics. 2000;377:1-8.

[122]  Nishimura S. Involvement of mammalian OGG1(MMH) in excision of the 8-hydroxyguanine residue in DNA. Free radical biology & medicine. 2002;32:813-21.

[123]  Weiss JM, Goode EL, Ladiges WC, Ulrich CM. Polymorphic variation in hOGG1 and risk of cancer: a review of the functional and epidemiologic literature. Mol Carcinogen. 2005;42:127-41.

[124]  Kohno T, Shinmura K, Tosaka M, Tani M, Kim SR, Sugimura H, et al. Genetic polymorphisms and alternative splicing of the hOGG1 gene, that is involved in the repair of 8-hydroxyguanine in damaged DNA. Oncogene. 1998;16:3219-25.

[125]  Sakamoto T, Higaki Y, Hara M, Ichiba M, Horita M, Mizuta T, et al. hOGG1 Ser326Cys polymorphism and risk of hepatocellular carcinoma among Japanese. Journal of epidemiology / Japan Epidemiological Association. 2006;16:233-9.

[126]   Peng T, Shen HM, Liu ZM, Yan LN, Peng MH, Li LQ, et al. Oxidative DNA damage in peripheral leukocytes and its association with expression and polymorphisms of hOGG1: a study of adolescents in a high risk region for hepatocellular carcinoma in China. World J Gastroenterol. 2003;9:2186-93.

[127]   Zhang H, He BC, He FC. [Impact of DNA repair gene hOGG1 Ser326Cys polymorphism on the risk of hepatocellular carcinoma]. Shijie Huaren xiaohua zazhi. 2006;14:2311-4.

[128]   Mills KD, Ferguson DO, Alt FW. The role of DNA breaks in genomic instability and tumorigenesis. Immunol Rev. 2003;194:77-95.

[129]   Morgan WF, Corcoran J, Hartmann A, Kaplan MI, Limoli CL, Ponnaiya B. DNA double-strand breaks, chromosomal rearrangements, and genomic instability. Mutation research. 1998;404:125-8.

[130]   Kanaar R, Hoeijmakers JH, van Gent DC. Molecular mechanisms of DNA double strand break repair. Trends Cell Biol. 1998;8:483-9.

[131]   Pardo B, Gomez-Gonzalez B, Aguilera A. DNA repair in mammalian cells: DNA double-strand break repair: how to fix a broken relationship. Cell Mol Life Sci. 2009;66:1039-56.

[132]   Shrivastav M, De Haro LP, Nickoloff JA. Regulation of DNA double-strand break repair pathway choice. Cell Res. 2008;18:134-47.

[133]   Sonoda E, Hochegger H, Saberi A, Taniguchi Y, Takeda S. Differential usage of non-homologous end-joining and homologous recombination in double strand break repair. DNA repair. 2006;5:1021-9.

[134]   Cahill D, Connor B, Carney JP. Mechanisms of eukaryotic DNA double strand break repair. Front Biosci. 2006;11:1958-76.

[135]   Valerie K, Povirk LF. Regulation and mechanisms of mammalian double-strand break repair. Oncogene. 2003;22:5792-812.

[136]   Bishop DK, Ear U, Bhattacharyya A, Calderone C, Beckett M, Weichselbaum RR, et al. Xrcc3 is required for assembly of Rad51 complexes in vivo. J Biol Chem. 1998;273:21482-8.

[137]   Brenneman MA, Wagener BM, Miller CA, Allen C, Nickoloff JA. XRCC3 controls the fidelity of homologous recombination: roles for XRCC3 in late stages of recombination. Mol Cell. 2002;10:387-95.

[138]   Brenneman MA, Weiss AE, Nickoloff JA, Chen DJ. XRCC3 is required for efficient repair of chromosome breaks by homologous recombination. Mutation research. 2000;459:89-97.

[139]   Sun H, Qiao Y, Zhang X, Xu L, Jia X, Sun D, et al. XRCC3 Thr241Met polymorphism with lung cancer and bladder cancer: a meta-analysis. Cancer science. 2010;101:1777-82.

[140]   Jiang Z, Li C, Xu Y, Cai S. A meta-analysis on XRCC1 and XRCC3 polymorphisms and colorectal cancer risk. International journal of colorectal disease. 2010;25:169-80.

[141]   Fang F, Wang J, Yao L, Yu XJ, Yu L. Relationship between XRCC3 T241M polymorphism and gastric cancer risk: a meta-analysis. Med Oncol. 2010.

[142]   Economopoulos KP, Sergentanis TN. XRCC3 Thr241Met polymorphism and breast cancer risk: a meta-analysis. Breast cancer research and treatment. 2010;121:439-43.

[143]   Zhou K, Liu Y, Zhang H, Liu H, Fan W, Zhong Y, et al. XRCC3 haplotypes and risk of gliomas in a Chinese population: a hospital-based case-control study. International journal of cancer. 2009;124:2948-53.

# Polymorphisms in Nucleotide Excision Repair Genes and Risk of Colorectal Cancer: A Systematic Review

Rikke Dalgaard Hansen and Ulla Vogel

*Danish Cancer Society & National Research Centre for the Working Environment*
*Denmark*

## 1. Introduction

Various DNA alterations can be caused by exposure to environmental and endogenous carcinogens through direct binding of metabolites (adduct formation). If not repaired the DNA lesions may lead to genetic instability, mutagenesis and oncogenesis. Thus, DNA repair constitutes a first line of defence against cancer.

Environmental factors are likely to cause damage to DNA through direct binding of metabolites (adduct formation). The nucleotide excision repair (NER) pathway is the primary mechanism for removal of large and bulky adducts from DNA.

### 1.1 Single nucleotide polymorphisms

Common occurring single nucleotide polymorphisms (SNPs) in genes involved in DNA repair may possibly contribute to the variation in the capacity of repair of bulky DNA adducts. Hence, these SNPs may be important biomarkers of susceptibility to cancer.

The present book chapter includes a systematic review of the available scientific literature on associations between SNPs in genes involved in NER and risk of colorectal adenomas and colorectal cancer. The present review of colorectal cancer studies includes 19 studies on 22 different SNPs. The review is focused on SNPs in four genes: *XPD, XPC, XPA* and *ERCC1* encoding the essential components of NER: xeroderma pigmentosum complementation group A, C, and D and excision repair cross complementary group 1 and risk of colorectal adenomas and colorectal cancer, and on interaction between the polymorphisms and various life style factors in relation to colorectal cancer risk.

The NER polymorphisms studied in the work underlying this book chapter include the polymorphisms: *XPD* Lys751Gln, *XPD* Asp312Asn, *XPA* G23A, *XPC* Lys939Gln, and *ERCC1* Asn118Asn.

### 1.2 Colorectal cancer

Colorectal cancer is the third most common cancer and the leading cause of cancer deaths in Western industrialised countries. Thus, every year nearly one million people worldwide develop colorectal cancer. Lifetime risk of colorectal cancer may reach 6% of the population in the Western industrialised countries (Jemal et al., 2006). The age-specific incidence of colorectal cancer increases sharply after 35 years of age, with approximately 90% of cancers occurring in persons older than 50 years (Schottenfeld & Winawer, 1996) . The mean age at time for diagnosis in Danish colorectal cancer patients is approximately 70 years for men and 72 years for women (Iversen et al., 2005) . The disease develops either sporadically, as a part of a hereditary cancer syndrome, or induced by inflammatory bowel disease. Ten to fifteen percent of colorectal cancer cases are caused by hereditary syndromes (Schottenfeld & Winawer, 1996) .

Migrant studies and large international variation in incidence rates indicate that life style factors, including dietary, are associated with risk of colorectal cancer, but traditional epidemiological studies based on life style questionnaires and outcome have mostly failed in identifying the exact risk and beneficial factors. Our current knowledge of colorectal carcinogenesis indicates a multi-factorial and multi-step process that involves various genetic alterations and several biological pathways. An understanding of differences in individual susceptibility and better exposure assessment may be crucial in identifying life style risk factors and possible interactions between susceptibility and exposures in relation to risk of colorectal cancer.

## 2. DNA adducts

Several life style factors and dietary components are suggested to be associated with risk of colorectal cancer, listed in Table 1. The associations may possibly be caused by increased formation of DNA adducts.

### 2.1 NOC, HCA and PAH

N-nitroso compounds (NOCs) are present in tobacco smoke and in nitrate- or nitrite-treated meats (Hotchkiss, 1989; Hecht & Hoffmann, 1988). NOCs are alkylating agents able to react with DNA and form adducts. More than 85% of 300 NOCs tested for carcinogenicity in experimental animals were observed to be carcinogenic (Mirvish, 1995), but epidemiologic studies have been inconclusive in finding association between the exposure of NOCs and risk of various cancer forms in humans (Burch et al., 1987; Preston-Martin & Mack, 1991; Carozza et al., 1995), although an increased endogenous production of NOCs, suggested primarily by bacterial catalysis, are proposed associated to the etiology of colorectal cancer (Bingham et al., 1996).

Polycyclic aromatic hydrocarbons (PAHs) and heterocyclic aromatic amines (HCAs) constitute a major class of chemical carcinogens present in the environment. When metabolically activated, these compounds act as mutagens and carcinogens in animal models (Culp et al., 1998; Moller et al., 2002; Dingley et al., 2003) and are able to form bulky DNA adducts in humans (Hecht, 2003), (Phillips, 2002) . Many PAHs and HCAs are found to be tumourigenic in humans or experimental animals (International Agency for Research on Cancer (IARC), 1983). Cooking meat at high temperatures and certain preservation and processing procedures leads to the formation of PAHs and HCAs (Sinha et al., 2005; Guillen et al., 1997) . PAHs are ubiquitous environmental contaminants formed by incomplete combustion of organic matter. They are one of several classes of carcinogenic chemicals present in tobacco smoke (Benhamou et al., 2003; Melikian et al., 1999). PAH compounds may not only be formed by high cooking temperatures but are also found in uncooked food, like sea food and plants, due to contamination of the aquatic environment (Meador et al., 1995) or via atmospheric exposure (Guillen et al., 1997).

## 2.2 Life style factors and DNA adduct formation

Air pollution is not an established risk factor for colorectal cancer in humans, although several studies have shown higher risk among workers exposed to diesel exhaust (Goldberg et al., 2001). Some studies have found an association between ambient air pollution and DNA adduct levels (Poirier et al., 1998; Hemminki et al., 1990b; Binkova et al., 1995; Palli et al., 2001; Nielsen et al., 1996a; Nielsen et al., 1996c), whereas others failed to find such an association (Kyrtopoulos et al., 2001; Peluso et al., 1998). DNA adduct levels are increased following occupational exposure among foundry and coke oven workers and among workers exposed to diesel exhaust (Hemminki et al., 1997; Hemminki et al., 1990a; Hemminki et al., 1994; Perera et al., 1988; Perera et al., 1994; Lewtas et al., 1997; Nielsen et al., 1996a; Nielsen et al., 1996b), while among fire-fighters (Rothman et al., 1993), traffic exposed policemen (Peluso et al., 1998) and aluminium workers (Yang et al., 1998), no associations between occupational exposures and DNA adducts have been found.

Tobacco smoking is an established risk factor for development of adenomas (Ji et al., 2006), and recently an association between tobacco smoking and risk of colorectal cancer has been recognized by IARC. Following tobacco smoking, adducts formed by metabolites of NOCs and PAHs are not only located in airway tissue, but are also found in bladder and cervical tissue from smokers (Benhamou et al., 2003; Melikian et al., 1999).

| Life style factor | Risk of CRC | DNA adduct formation |
|---|---|---|
| Air pollution | ↑ | PAH |
| Tobacco smoking | ↑ | PAH, NOC |
| Alcohol | ↑ | Acetaldehyde |
| Red meat | ↑ | PAH, NOC, HCA |
| Processed meat | ↑ | PAH, NOC, HCA |
| Vegetables | ↓ | - |
| Fruit | ↓ | - |

Table 1. Possible environmental risk and beneficial factors of colorectal cancer and their association with DNA adduct formation. Arrows indicate adverse (↑) or preventive (↓) association with risk of colorectal cancer.

A growing body of evidence supports that avoidance of alcohol is recommended to prevent colorectal cancer (Correa Lima & Gomes-da-Silva, 2005). Acetaldehyde is the primary oxidative metabolite of ethanol. Acetaldehyde and malondialdehyde, the end-product of lipid peroxidation by reactive oxygen species, can combine to form the malondialdehyde-acetaldehyde adduct, which is very reactive and avidly binds to DNA (Brooks & Theruvathu, 2005). The level of acetaldehyde DNA adducts in white blood cell DNA in alcohol abusers have been measured up to 13-fold higher than in subjects from the non-drinking control group (Fang & Vaca, 1997).

There is some evidence for adverse associations between intake of red and processed meat and risk of colorectal cancer (Johnson & Lund, 2007; Doyle, 2007; Norat et al., 2005). The elevated risk may be due to an increased endogenous production of NOC, which may enhance the colonic formation of the DNA adduct O6-carboxymethyl guanine (Bingham et al., 1996; Lewin et al., 2006). Cooking meat at high temperatures leads to the formation of polycyclic aromatic hydrocarbons (PAHs) and heterocyclic amines (HCAs) (Sinha et al., 2005). Additionally, intake of charbroiled or smoked meat may be associated with increased levels of DNA adducts (Rothman et al., 1990; van Maanen et al., 1994; Georgiadis et al., 2001; Rothman et al., 1993), due to HCAs and PAHs (Bruemmer et al., 1996; Balbi et al., 2001; Peters et al., 2004; Skog et al., 1995). The levels of some HCAs and PAHs are comparable for red meat, fish and poultry smoked or cooked at high temperatures (Sinha et al., 1995; Gomaa et al., 1993). Intake of red meat, but not of fish and poultry, increases the luminal contents of N-nitrosocompounds (NOCs) in colon (Bingham et al., 1996; Lewin et al., 2006). The increase in endogenous N-nitrosation can be attributed to heme iron (Cross et al., 2003), which is 10-fold higher in red meat than in white meat (Pierre et al., 2003).

There is limited evidence for a preventive effect of intake of fruit and vegetables for cancer in colon and rectum (International Agency for Research on Cancer (IARC), 2003). Intake of fruit, vegetables or antioxidant vitamins have been shown to be negatively associated with DNA adduct levels (Palli et al., 2000; Mooney et al., 1997; Palli et al., 2003; Palli et al., 2004), although some studies found no effect (Georgiadis et al., 2001; Nielsen et al., 1996b) and one study found an effect of increased vitamin intake only in females (Mooney et al., 2005).

## 3. Nucleotide excision repair

The nucleotide excision repair (NER) pathway is the primary mechanism for removal of helix-distorting damages from DNA, including bulky adducts and UV-induced photolesions. The mechanism of NER includes five steps: 1. Damage recognition, 2. Assembly of the repair factors at the site of damage, 3. Dual incisions and excision of the damage-containing oligomers, 4. Resynthesis to fill in the gap, and 5. Ligation of the strands. All these steps involve more than 20 proteins, like recognition factors, replication protein, transcription factor, helicases, endonucleases and polymerases. Steps 1 and 2 are illustrated in Figure 1.

### 3.1 The NER pathway

There are two sub-pathways of NER, termed the global genome NER (GG-NER), which corrects lesions in the entire genome including the non-transcribed strands of active genes, and transcription-coupled NER (TC-NER), that only repairs lesions in transcribed strands in active genes. The major differences of the two pathways are the damage recognition step: In GG-NER the proteins Xeroderma Pigmentosum complementation group A and C (XPA/XPC) make the recognition complex (Hanawalt, 2002; Reardon & Sancar, 2002; You et al., 2003; Volker et al., 2001), while in TC-NER a stalled RNA polymerase II (blocked by a lesion) and Cockayne syndrome proteins have this function to act as a signal to recruit NER proteins (Kobayashi et al., 2005; Hanawalt, 2002).

In global genomic NER the XPA and XPC enzymes are involved in the damage recognition-complex of NER. Several studies have shown the XPC-hHR23B complex to function at a very early stage of DNA damage recognition (Reardon & Sancar, 2002; You et al., 2003;

Hanawalt, 2002; Volker et al., 2001). The hHR23B (also called Rad23) NER factor co-purifies with XPC (Masutani et al., 1994) and is essential for high XPC activity in NER (Batty et al., 2000; Guzder et al., 1998). XPC-hHR23B complex exhibit a very strong affinity for damaged DNA (Reardon et al., 1996; Batty et al., 2000; Sugasawa et al., 1998), why it is thought to be the initiator in GG-NER. By interaction with the XPC complex XPA and the transcription factor II H (TFIIH) may be recruited to the damaged DNA site (You et al., 2003; Volker et al., 2001). TFIIH is a nine sub-unit protein complex required for opening the DNA helix at the vicinity of the lesion (Schaeffer et al., 1993; Feaver et al., 1993; Drapkin et al., 1994). Biochemical studies have generated conflicting results with regard to association between the XPC-hHR23B complex, XPA and TFIIH. Some have found recruitment of TFIIH to the site of DNA damage to be dependent on XPC (Volker et al., 2001; Yokoi et al., 2000), while others have found XPA to be interacting with TFIIH (Park et al., 1995). Undoubtedly, both XPC and XPA are vital factors in the very early steps of GG-NER, but exactly when

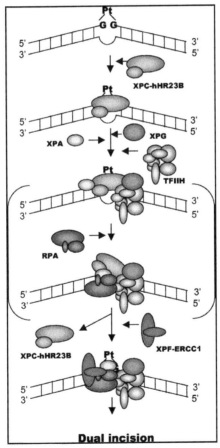

Fig. 1. A proposed molecular mechanism of damage recognition process in the early stage of global genome nucleotide excision repair. Transient steps are indicated with brackets. Adapted from (You et al., 2003) .

XPA enters the site of damage is not clear. XPA physically interacts with replication factor A (RPA) and is essential to efficient NER (Stigger et al., 1998) by stabilizing the interaction between XPA and the damaged DNA. XPA is capable of binding to the XPF-ERCC1 complex with very high affinity (Park & Sancar, 1994). The XPF-ERCC1 is a specific 5′ endonuclease complex, and thus must be located near the site of 5′ incision (Niedernhofer et al., 2001). XPG, a 3′ endonuclease, seems to be the next factor recruited to the site, and is probably positioned at the 3′ incision site (Reardon & Sancar, 2002). Previous studies have observed XPG to co-purify with TFIIH, like XPC, and that XPG exclude XPC when binding to TFIIH (Wakasugi & Sancar, 1999; Wakasugi & Sancar, 1998), which may suggest that the binding of XPG to the NER complex displaces XPC. Hence, XPA is thought to be crucial to the subsequent positioning of the involved NER enzymes by binding to XPF-ERCC1 complex and possibly recruit XPG to the site of DNA damage. XPD and XPB are helicases and parts of the large TFIIH complex. They participate in the unwinding of helix in opposite directions of the region of damaged DNA (Reardon & Sancar, 2002; Schaeffer et al., 1993). When the DNA around the DNA lesion is unwound, the endonucleases XPG and XPF-ERCC1 complex excises an oligonucleotide of 24-32 bases including the damaged site (Mu et al., 1996). The two endonucleases require an opening of approximately 5-8 bases (Evans et al., 1997; de Laat et al., 1998). The final steps of NER are re-synthesis of the strand to fill in the gap and ligation of the new strand with the remaining strand. In mammals the synthesis requires the DNA polymerases δ and/or ε (Hunting et al., 1991; Coverley et al., 1992), the replication protein A (RPA) and replication factor C (RFC) (Shivji et al., 1995) and proliferating cell nuclear antigen (PCNA) (Shivji et al., 1992). The XPF-ERCC1 5′ incision leaves a hydroxyl-group at the 3′ terminus of the gap. This terminus may act as a DNA primer for DNA polymerases (Sijbers et al., 1996). RPA is required for the gap-filling DNA synthesis (Shivji et al., 1995), possibly to protect the template strand against nucleases, and RFC and PCNA as a complex that facilitates the assembly of the polymerases (Shivji et al., 1992). The new fragment of DNA is synthesized and the final step is ligation of the new patch to the original sequence, which possibly may be performed by DNA ligase I (Tomkinson & Levin, 1997).

### 3.2 SNPs in NER genes and colorectal cancer risk

The variant alleles of *XPA* G23A (Wu et al., 2003), *XPD* Asp312Asn and *XPD* Lys751Gln (Spitz et al., 2001; Qiao et al., 2002) polymorphisms and a polymorphism in *XPC* (Qiao et al., 2002), in full linkage disequilibrium with the *XPC* Lys939Gln polymorphism (Khan et al., 2000), have been associated with a lowered DNA repair capacity compared to the wild type allele. *ERCC1* gene polymorphism is a predictor for clinical outcome in advanced colorectal cancer patients treated with platinum-based chemotherapy (Viguier et al., 2005). Furthermore, the variant alleles of the polymorphisms *XPD* Asp312Asn and *XPD* Lys751Gln have been associated with higher DNA adduct levels (Hou et al., 2002; Matullo et al., 2001; Palli et al., 2001) than the wild type alleles.

Mutations in the NER gene *XPD* are associated with the rare, autosomal-recessive inherited disorder Xeroderma Pigmentosum, where patients suffer from severe photosensitivity and actinic changes leading to early onset of skin cancers induced by sunlight (Cleaver, 2005). Recently the first case of human inherited ERCC1 deficiency was reported (Jaspers et al., 2007). Cells from the patient showed moderate hypersensitivity to ultraviolet rays, but the clinical features were very severe and compatible with a diagnosis of cerebro-oculo-facio-skeletal syndrome. This discovery represents a novel complementation group of patients with defective NER and suggests novel functions for ERCC1.

Overall, the above mentioned studies of the polymorphisms in the genes involved in NER, *XPD* Lys751Gln, *XPD* Asp312Asn, *XPA* G23A, *XPC* Lys939Gln, and *ERCC1* Asn118Asn, indicate that the polymorphisms may modulate DNA repair capacity and may thereby possibly be associated with development of cancer.

There are limited numbers of studies of NER genes in relation to risk of colorectal cancer. A search on the PubMed database of NCBI on January 26th 2011 on the MeSH terms "polymorphism, single nucleotide AND colorectal neoplasms" resulted in 148 hits of which seven studies included polymorphisms in *XPD*, *XPA*, *XPC*, and *ERCC1*. In combination with a new search on the PubMed database of NCBI by using different combinations of the words: "XPD XPA XPC ERCC1 polymorphism colorectal colon rectum cancer" 19 studies of SNPs in the four genes in relation to risk of colorectal cancer or prestages to colorectal cancer were identified. The studies are listed in Table 2.

### 3.2.1 *XPD* Lys751Gln and *XPD* Asp312Asn

The *XPD* Lys751Gln polymorphism is the most frequently studied of the NER polymorphisms in association with risk of cancer. In our Danish prospective study on the Diet, Cancer and Health cohort, we observed no association of the *XPD* Lys751Gln and *XPD* Asp312Asn polymorphisms with risk of colorectal cancer (Hansen et al., 2007). Previously, several studies had similar findings of no association between the *XPD* Lys751Gln (Moreno et al., 2006; Huang et al., 2006; Berndt et al., 2006; Mort et al., 2003; Starinsky et al., 2005; Skjelbred et al., 2006b; Engin et al., 2010; Stern et al., 2009; Stern et al., 2007; Yeh et al., 2005; Joshi et al., 2009; Wang et al., 2010) and the *XPD* Asp312Asn (Moreno et al., 2006; Huang et al., 2006; Berndt et al., 2006; Goodman et al., 2006; Stern et al., 2009; Stern et al., 2007; Joshi et al., 2009) polymorphisms and risk of colorectal cancer. Additionally, Bigler and colleagues found no association of the two polymorphisms with development of adenomas (Bigler et al., 2005). However, they detected a higher risk of colorectal adenomas among individuals with at least two variant alleles of the *XPD* polymorphisms, with an OR of 1.57 (CI: 1.04-2.38). When stratifying by age the association of the two polymorphisms with risk of adenomatous polyps was restricted to the individuals younger than 60 years when diagnosed (OR=3.77, CI: 1.94-7.35). The risk of adenomatous polyps was higher among smokers carrying the homozygous *XPD* variant alleles (OR=3.93, OR: 1.68-9.21) compared with non-smokers carrying the homozygous wild type. A similar finding could not be detected on risk of hyper-plastic polyps. In our Danish study (Hansen et al., 2007) and in a Singapore Chinese study (Stern et al., 2007) did neither of the two XPD polymorphisms, *XPD* Lys751Gln or *XPD* Asp312Asn, modify the effect of smoking on risk of colorectal cancer.

Goodman *et al.*, did not detect any SNP-SNP interaction between the *XPD* Asp312Asn polymorphism and other NER polymorphisms (Goodman et al., 2006). Skjelbred and colleagues detected an association between the *XPD* Lys751Gln polymorphism and development of colorectal adenomas, with an OR of 1.40 (CI: 1.08-1.81), among carriers of the variant allele compared to carriers of the homozygous wild type allele (Skjelbred et al., 2006b). The statistical significance was limited to the low-risk adenoma group (OR: 1.46, CI: 1.11-1.90). The results were contradicted by a large study by Stern *et al.*, including 740 cases with adenomas and 789 controls, where a lower risk of adenomas was observed (OR=0.7, CI:

0.4-1.0) among homozygous carriers of the *XPD* 751Gln allele (Stern et al., 2006). The result was not stratified for ethnicity (Caucasian, African-American, Latinos, Asian-Pacific Islander). When excluding the 1 case and the 17 controls of Latinos, the OR increased to 0.9 (confidence intervals were not reported). An interaction between the *XPD* Lys751Gln polymorphism and alcohol consumption was observed (P=0.04), with higher risk of adenomas among ever-drinkers carrying the *XPD* 751 Gln/Gln genotype (OR=2.5, CI: 1.2-5.2) compared with never-drinkers carrying the same genotype. There was no interaction between the polymorphisms *XPD* Lys751Gln or *XPD* Asp312Asn, respectively, and alcohol consumption on risk of colorectal cancer in our Danish study (Hansen et al., 2007) and in the Singapore Chinese study (Stern et al., 2007).

In a family-based case-control study using a case-only design, an interaction was observed between the two polymorphisms, *XPD* Lys751Gln and *XPD* Asp312Asn, and intake of heavily browned red meat on colorectal cancer risk (Joshi et al., 2009). Intake of red meat heavily brown on the outside or inside increased the risk for colorectal cancer only among carriers of the *XPD* codon 751 Lys/Lys genotype or the *XPD* codon 312 Asp/Asp genotype (case-only interaction P <0.006). There was no association between the meat intake and colorectal cancer risk when the individuals carried at least one copy of the Asn[321] or Gln[751] alleles. The results remained statistically significant after accounting for multiple testing. No interaction was observed in our Danish study between the two *XPD* polymorphisms and intake of red meat on risk of colorectal cancer (Hansen et al., 2007).

A higher risk of colorectal cancer has been observed among Ashkenazi Jews below 50 years of age when diagnosed (Starinsky et al., 2005). The risk was higher among carriers of the *XPD* 751Gln allele, but it may be a chance finding due to low number of cases (only 15 cases were diagnosed before their 50 years birthday). Furthermore, the Ashkenazi population is known to have particular genetic characteristics, why the result may not be generalized to other populations.

A large study from Taiwan observed a non-significant tendency for higher risk of colorectal cancer among men carrying the *XPD* 751Gln allele (OR=1.5, CI: 0.9-2.3), while no association was observed for women (OR=0.9, CI: 0.6-1.5) (Yeh et al., 2007). A similar tendency for a gender specific effect of the *XPD* Lys751Gln polymorphism was observed in our Danish study, with lower risk of colorectal cancer among women carrying the variant allele of *XPD* Lys751Gln with an IRR less than 0.62 among carriers of the *XPD* 751Gln allele, compared to women carrying the wild type allele (Hansen et al., 2007) . No association was found among men. The gender differences could hypothetically be caused by a hormonal interaction. However, we observed no interaction between the use of hormone replacement therapy among women and the polymorphism. Thus, we did not find the hypothesis plausible and conclude that our result in the Danish study may be a chance finding.

### 3.2.2 *XPC* Lys939Gln

In our Danish study and in a Turkish study by Engin *et al.* (Engin et al., 2010) , the *XPC* Lys939Gln polymorphism was not associated with risk of colorectal cancer (Hansen et al., 2007). However, we did observe an interaction between the polymorphism and intake of red meat, with an IRR of 3.70 (CI: 1.70-8.04) for colorectal cancer per 100g red meat intake per day among homozygous carriers of the *XPC* Lys939Gln variant allele (Hansen et al., 2007) . In the light of the sample size and the multiple comparisons being made, this result may be a chance finding. The association was not statistically significant after a Bonferroni correction.

In a large American study by Huang three polymorphisms in *XPC* was studied, including the *XPC* Lys939Gln polymorphism. No association was found between the *XPC* Lys939Gln polymorphism and risk of adenomas (Huang et al., 2006). However, higher risk for development of adenomas was observed among current or recent smokers carrying the *XPC* 939Gln allele (OR=2.0, CI: 1.3-3.0) or a *XPC* haplotype encompassing three linked SNPs in *XPC* (Arg492His, Ala499Val, Lys939Gln) compared with never-smokers carrying the homozygous wild type allele. A study by Joshi *et al.* observed no association between the *XPC* intron 11 polymorphism and risk of colorectal cancer (Joshi et al., 2009) .

In a small study by Berndt *et al.* a tendency for higher risk of proximal colon cancer was observed among homozygous carriers of the variant *XPC* Lys939Gln allele, with an OR of 1.74 (CI: 0.98-3.08) (Berndt et al., 2006). The result may possibly be a chance finding due to sample size and multiple testing. Three other SNPs in the *XPC* gene, see Table 2, were not associated with colorectal cancer risk.

### 3.2.3 *XPA* G23A

To our knowledge, only three studies have been published on the association of polymorphisms in the *XPA* gene with risk of colorectal cancer: The studies by Berndt *et al.*, Joshi *et al.*, and our study. For a polymorphism positioned in the *XPA* 5′ UTR region, a lower risk for colon cancer cancer was observed among carriers of the T-allele (OR=0.4, 95% CI: 0.2-0.8) compared with homozygous carriers of the C-allele (Joshi et al., 2009) . There was no association for risk of rectal cancer. No association was observed of the *XPA* G23A polymorphism (Hansen et al., 2007) or a polymorphism in the 3′ un-translated region of *XPA* (Berndt et al., 2006) with risk of colorectal cancer.

### 3.2.4 *ERCC1* Asn118Asn

The results from studies by Skjelbred *et al.* (Skjelbred et al., 2006a) , Joshi *et al.* (Joshi et al., 2009) , and our Danish study (Hansen et al., 2008) on the *ERCC1* Asn118Asn polymorphism suggest no association with risk of colorectal cancer.

Moreno *et al.* examined five polymorphisms in the *ERCC1* gene. A haplotype containing the minor allele of three of the *ERCC1* polymorphisms was associated with a higher risk of colorectal cancer (OR=2.3, 95% CI: 1.0-5.3) compared with carriers of the most frequent haplotype (Moreno et al., 2006). Two other SNPs in the *ERCC1* gene were not associated with risk of colorectal cancer (Mort et al., 2003; Berndt et al., 2006).

### 3.3 SNPs in NER and risk of other types of cancer than colorectal cancer

Numerous association studies of polymorphisms in genes involved in NER are reported on various types of cancer, with the majority of studies focused on the *XPD* Lys751Gln and *XPD* Asp312Asn polymorphisms. A meta-analysis of lung cancer by Kiyohara *et al.* (with 1913 cases and 1882 controls of different ethnicities) (Kiyohara & Yoshimasu, 2007) suggested among other studies (Xing et al., 2002; Hu et al., 2004; Yin et al., 2006), that carriers of the variant alleles of either of the two *XPD* polymorphisms were found to be at higher risk of lung cancer, while a number of other studies did not observe any association of the two polymorphisms with lung cancer risk (De et al., 2007; Vogel et al., 2005b; Popanda et al., 2004; Hu et al., 2006).

| Reference | Polymorphism | Endpoint | Study design | Cases | Controls | Ethnicity | DNA source | Associations (main results) | Interactions |
|---|---|---|---|---|---|---|---|---|---|
| Yeh et al., 2007 | XPD Lys751Gln | Colorectal cancer | Case-control | 727 with carcinomas | 736 negative colonoscopy screening | Asian (Taiwan) | Blood samples | Tendency of XPD 751Gln ↑ risk of CRC among men (69 cases/55 controls) | ↑risk for colorectal cancer with XPD variant in combinations with several genotypes |
| Yeh et al., 2005 | XPD Lys751Gln | Colorectal cancer | Case-control | 727 with carcinomas | 736 negative colonoscopy screening | Asian (Taiwan) | Blood samples | No association of single SNP | - |
| Wang et al., 2010 | XPD Lys751Gln | Colorectal cancer | Case-control | 302 with primary colorectal carcinoma | 291 free of cancer | Caucasian (India) | Blood samples | No association of single SNP | No GE-interactions with smoking or alcohol consumption |
| Hansen et al., 2007 | XPD Lys751Gln, XPD Asp312Asn, XPA A23G, XPC Lys939Gln | Colorectal cancer | Prospective case-cohort | 397 with primary colorectal cancer | 800 randomly selected from the cohort (10 with colorectal cancer) | Caucasian (Denmark) | Blood samples | No association of single SNPs | GE-interaction between XPC polymorphism and intake of red meat. No GE-interactions with smoking, alcohol consumption, intake of processed meat, fruit and vegetables |
| Skjelbred et al., 2006 | XPD Lys751Gln | Carcinomas and adenomas (high and low-risk) | Case-control | 157 with carcinomas, 983 with adenomas (227 high-risk/756 low-risk) | 399 negative flexible sigmoidoscopy screening | Caucasian (Norway) | Blood samples | ↑ risk for low-risk adenomas among carriers of the XPD 751Gln allele compared to homozygous carriers of the wild type allele | No GE-interactions with cigarette smoking |
| Mort et al., 2003 | XPD exon 6, XPD exon 22, XPD exon 23, ERCC1 exon 4 | Colorectal cancer | Case-control | 45 with carcinomas | 71 hospitalized, not cancer | Caucasian? (England) | Carcinomas/blood samples | No association of single SNPs | - |
| Engin et al., 2010 | XPD Lys751Gln, XPD Asp312Asn | Colorectal cancer | Case-control | 110 with carcinomas | 116 free of cancer | Caucasian (Turkey) | Blood samples | No association of single SNPs | - |
| Stern et al., 2009 | XPD Lys751Gln, XPD Asp312Asn | Colorectal cancer | Case-control | 311 with colorectal cancer | 1181 free of cancer | Chinese (Singapore) | Blood/buccal cell samples | No association of single SNPs | - |
| Stern et al., 2007 | XPD Lys751Gln, XPD Asp312Asn | Colorectal cancer | Case-control | 310 with colorectal cancer | 1176 free of cancer | Chinese (Singapore) | Blood/buccal cell samples | No association of single SNPs | No interaction between XPD polymorphisms and smoking or alcohol consumption, respectively, on colorectal cancer risk |
| Goodman et al., 2006 | XPD Asp312Asn | Colon cancer | Case-control | 216 men with carcinomas | 255 hospitalized men, not cancer, HBV, HIV or HCV | Caucasian, African American (USA) | Primarily blood samples, otherwise colon tissue (some cases) | No association of SNP | No SNP-SNP interaction between the XPD polymorphism and other NER polymorphisms |
| Bigler et al., 2005 | XPD Lys751Gln, XPD Asp312Asn | Adenomatous and hyperplastic polyps | Case-control | 694 with polyps (384 adenomatous/191 hyperplastic/119 both types) | 621 negative colonoscopy screening | Caucasian and Afroamerican (USA) | Blood samples | No association of single SNPs. ↑risk of adenomas for individuals with two XPD variant alleles | ↑risk of adenomatous polyps among heavy smokers carrying homozygous XPD variant compared with nonsmokers who were homozygous wild type. |
| Stern et al., 2006 | XPD Lys751Gln | Adenomas | Case-control | 740 with adenomas | 789 hospitalized, no current or past polyps | Caucasian, African American, Latinos, Asian-Pacific Islander (USA) | Blood samples | ↓risk of adenomas among homozygous carriers of the variant allele compared with carriers of the wild type allele | No GE-interactions with consumption of alcohol or meat intake; GE-interaction between XPD polymorphism and alcohol consumption |
| Starinsky et al., 2005 | XPD Lys751Gln | Colorectal cancer | Case-control | 456 diagnosed or treated for colorectal cancer | 87 hospitalized, not cancer | Jewish (64% Ashkenazi among cases) (Israel) | Blood samples | No association of SNP | No GE-interaction with smoking; ↑ risk of colorectal cancer among Ashkenazi Jews age below 50 years when diagnosed, carrying |

| Study | Polymorphisms | Design | Cases | Controls | Ethnicity | Samples | Results |
|---|---|---|---|---|---|---|---|
| **Moreno et al., 2006** | XPD Lys751Gln<br>XPD Asp312Asn<br>ERCC1 19716 G→C<br>ERCC1 19007 T→C<br>ERCC1 17677 A→C<br>ERCC1 15310 G→C<br>ERCC1 8092 C→A | Case-control | Colorectal adeno-carcinoma | 377 with carcinomas | 329 hospitalized, not cancer | Caucasian (Spain) | Not mentioned | ↑ risk of colorectal cancer among carriers of an ERCC1 haplotype (19716C, 19007C, 17677C)<br><br>No association of single SNPs |
| **Huang et al., 2006** | XPD Lys751Gln<br>XPD Asp312Asn<br>XPC Arg492His<br>XPC Ala499Val<br>XPC Lys939Gln | Case-control | Adenomas | 772 with advanced adenomas in the distal colon | 777 negative colonoscopy screening, no family history of CRC | Mixed (USA) | Blood samples | ↑ risk of advanced adenomas among current or recent smokers carrying XPC haplotype (492Arg, 499Ala, 939Gln)<br><br>No interactions with age, gender or ethnicity |
| **Joshi et al., 2009** | XPD Lys751Gln<br>XPD Asp312Asn<br>XPC intron 11<br>XPA 5'UTR<br>ERCC1 3'UTR | Case-control (case-only analyses) | Colorectal cancer | 307 with colorectal cancer | 307 unaffected, siblings to the cancer cases | Not mentioned (USA) | Blood samples | No association of SNPs in XPD, XPC and ERCC1.<br>↑ risk of CRC among homozygous carriers of the variant XPA 5'UTR allele (C/C) compared to carriers of the wild type allele. | ↑ risk of colorectal cancer from intake of red meat heavily browned on the outside or inside among carriers of the XPD codon 751 Lys/Lys or XPD codon 312 Asp/Asp genotypes only<br><br>No GE- interactions with total red meat intake or total intake of red meat cooked by high-temperature methods |
| **Berndt et al., 2006** | XPD Lys751Gln<br>XPD Asp312Asn<br>XPD IVS19-70<br>G→A<br>XPC Arg492His<br>XPC Ala499Val<br>XPC Arg687Arg<br>XPC Lys939Gln<br>XPA 3'UTR 32T<br>C→G<br>ERCC1 IVS74 G→C | Case-cohort | Colorectal cancer | 250 with carcinomas | 2224 (no colorectal cancer diagnosis) | Mixed (98% caucasian among sub-cohort and full cohort) (USA) | Blood samples | Borderline significant ↑ risk of proximal colon cancer among homozygous carriers of the variant XPC Lys99Gln allele compared to carriers of the wild type allele<br><br>↑ risk of colorectal cancer among carriers of 2 or more risk alleles for XPC Arg492His and ERCC6 R1213G (a NER gene) | No interactions with age, gender, smoking, or intake of red meat |
| **Skjelbred et al., 2006** | ERCC1 Asn118Asn | Case-control | Carcinomas and adenomas (high and low-risk) | 156 with carcinomas, 981 with adenomas (227 high-risk/ 754 low-risk) | 399 negative flexible sigmoidoscopy screening | Caucasian (Norway) | Blood samples | No association of single SNP | No interactions with smoking or alcohol consumption |
| **Hansen et al., 2007** | ERCC1 Asn118Asn | Prospective case-cohort | Colorectal cancer | 394 with colorectal cancer | 791 randomly selected from the cohort (10 with colorectal cancer) | Caucasian (Denmark) | Blood samples | No association of single SNP | No interactions with smoking or alcohol consumption |

Table 2. Studies of possible associations between polymorphisms in *XPD, XPA, XPC,* and *ERCC1* and risk of colorectal adenomas or colorectal cancer and gene-environment (G E) interaction on risk of colorectal adenomas or colorectal cancer. The results reviewed and included are solely on polymorphisms in the four genes and the environmental factors that are the topic of the present book chapter.

Two large meta-analyses (with 3725 cases and 4152 controls) included identical nine case-control studies but made two dissimilar conclusions: The *XPD* Lys751Gln and *XPD* Asp312Asn polymorphisms are associated with risk of lung cancer (Hu et al., 2004) or no clear association was found (Benhamou & Sarasin, 2005). Some studies suggest an interaction between the two *XPD* polymorphisms and smoking in relation to risk of lung cancer (De et al., 2007; Hu et al., 2006; Xing et al., 2002).

Combinations of the *XPD*, *XPC* and *XPA* genotypes, variant alleles, is suggested to be associated with higher risk of lung cancer (Vogel et al., 2005b). This may be plausible but in the light of multiple testing and the low number of cases this may be a chance finding. The largest breast cancer studies by the number of individuals, 1053 cases/1102 controls (Terry et al., 2004) and 1830 cases/1262 controls (Debniak et al., 2006) observed modest associations of the *XPD* polymorphisms with breast cancer risk. Carriers of the variant *XPD* Lys751Gln allele was associated with a 20% higher risk (OR=1.21, CI: 1.01-1.44) compared with homozygous carriers of the wild type allele. The risk seemed limited to those with a PAH-DNA adduct level above the median, with an OR of 1.61 (CI: 0.99-2.63) among homozygous carriers of the *XPD* 751Gln allele (Terry et al., 2004). Several other studies observed no association of the *XPD* Lys751Gln polymorphism (Debniak et al., 2006; Dufloth et al., 2005; Brewster et al., 2006; Costa et al., 2007; Mechanic et al., 2006; Jorgensen et al., 2007) or the *XPD* Asp312Asn polymorphism (Mechanic et al., 2006; Forsti et al., 2004)to risk of breast cancer. However, higher risk has been detected among ever smoking women carrying the *XPD* 751Gln allele (OR=2.52, CI: 1.27-5.03) compared to ever smoking women carrying the homozygous wild type allele (Metsola et al., 2005). Association with breast cancer risk has been detected when the homozygous variant *XPD* Lys751Gln allele and the homozygous variant *XPD* Asp312Asn allele segregated together, with OR=1.5 (p<0.05) and OR=3.69 (CI: 1.76-7.74), respectively (Debniak et al., 2006; Justenhoven et al., 2004). A large study including 2485 cases with single primary melanoma and 1238 cases with second or higher order primary melanomas detected higher melanoma risk among homozygous carriers of the variant *XPD* Lys751Gln allele (OR=1.4, CI: 1.1-1.7) or the variant *XPD* Asp312Asn allele (OR=1.5, CI: 1.2-1.9), respectively (Millikan et al., 2006). Similar results were obtained in a study by Li *et al.* (Li et al., 2006b), while another study observed the inverse association for both polymorphisms (Han et al., 2005). When stratifying by age Baccarelli *et al.* observed an association of the two *XPD* polymorphisms to risk of melanoma only among the individuals older than 50 years when diagnosed (Baccarelli et al., 2004). The *XPD* Lys751Gln (Andrew et al., 2006) and the *XPD* Asp312Asn polymorphism (Wu et al., 2006) have been associated with risk of bladder cancer. An interaction is suggested between the *XPD* Lys751Gln polymorphism and smoking in relation to bladder cancer risk (Andrew et al., 2006; Stern et al., 2002; Schabath et al., 2005). Individuals carrying both the variant *XPD* alleles were more susceptible to development of bladder cancer (Wu et al., 2006; Andrew et al., 2006) than carriers of wild type alleles. The *XPD* Lys751Gln and *XPD* Asp312Asn polymorphisms have not been associated to risk of basal cell carcinoma (Vogel et al., 2005a; Festa et al., 2005; Han et al., 2005; Lovatt et al., 2005), endometrial cancer (Weiss et al., 2006) prostate cancer (Ritchey et al., 2005) or gastric cancer (Huang et al., 2005).

A small study suggest that the variant allele of the polymorphism XPC Lys939Gln is associated with higher risk of bladder cancer (OR=1.49, CI:1.16-1.92) (Sanyal et al., 2004). No association is observed between the polymorphism and risk of lung cancer (Vogel et al., 2005b; Lee et al., 2005; Hu et al., 2006) but a haplotype encompassing more polymorphisms in *XPC* may contribute to a higher risk of lung cancer (Vogel et al., 2005b; Lee et al., 2005;

Hu et al., 2006): Individuals with both putative genotypes of *XPC* Lys939Gln and *XPC* Ala499Val polymorphisms are observed with a 2.4-fold (OR=2.37, CI: 1.33-4.21) higher risk of lung cancer compared with individuals with both wild type genotypes (Vogel et al., 2005b; Lee et al., 2005; Hu et al., 2006), with the higest risk observed among smokers. Polymorphisms in *XPC* have not been associated to risk of basal cell carcinoma (Festa et al., 2005; Nelson et al., 2005), cutaneous melanoma (Blankenburg et al., 2005; Li et al., 2006a) or breast cancer (Mechanic et al., 2006; Jorgensen et al., 2007; Forsti et al., 2004). A lower risk of endometrial cancer may be associated with carriage of at least one variant allele for both *XPC* Lys939Gln and *XPC* Ala499Val polymorphisms (Weiss et al., 2005).

In a Korean population carriers of the wild type allele (G/G or A/G) in the *XPA* G23A polymorphism were reported to have a lower risk of lung cancer compared to carriers of the A/A genotype, with an OR of 0.56 (CI:0.35-0.90) (Park et al., 2002). Similar results were obtained in studies on lung cancer risk in Caucasians and Mexican-Americans (Vogel et al., 2005b; Wu et al., 2003) (Popanda et al., 2004), while a Norwegian study observed the inverse effect with a 1.6-fold higher risk (OR=1.59, CI:1.12-2.27) of lung cancer among carriers of the G/G genotype compared with carriers of the A-allele (Zienolddiny et al., 2006). When stratifying by smoking status the protective effect for lung cancer was only observed among ever smokers (Wu et al., 2003) or current smokers (Park et al., 2002) carrying at least one G-allele or the G/G genotype, respectively. A tendency for lower risk of basal cell carcinoma has been observed among carriers of the variant G-allele, with an OR of 0.82 (CI: 0.66-1.01) and an OR of 0.74 (CI: 0.53-1.03) for homozygous and heterozygous carriers, respectively (Miller et al., 2006). The same tendency was observed for risk of squamous cell carcinoma (Miller et al., 2006). Carriage of at least one A-allele for *XPA* G23A was associated with decreased risk of endometrial cancer, OR=0.47 (CI:0.25-0.82) compared with carriers of the G/G genotype, but only among women with a history of using oral contraceptives (Weiss et al., 2006).

The *ERCC1* Asn118Asn polymorphism is not associated with testicular cancer (Laska et al., 2005). Furthermore, no association has been observed for the *ERCC1* Asn118Asn polymorphism to risk of endometrial cancer (Jo et al., 2007; Weiss et al., 2006), ovarian cancer (Jo et al., 2007) and adult glioma (Wrensch et al., 2005).

All in all the studies suggest that the two *XPD* polymorphisms at amino acid position 312 or 751, the *XPD* Lys751Gln in particular, are associated with risk of cancer in the lung, breast and bladder and seems to modify the effect of smoking on risk of the three cancer forms. The *XPC* Lys939Gln polymorphism may possibly be associated with risk of bladder cancer, and the *XPA* G23A polymorphism may be associated with risk of skin cancer (basal cell carcinoma), endometrial cancer and lung cancer. However, the studies are few and the results are inconsistent.

## 4. Discussion

In summary, this review, limited by the bias against publication of null findings, highlights the complexities inherent in epidemiological research and, particularly, in molecular epidemiological research on colorectal cancer. Studies on possible associations between SNPs in genes involved in defence of oxidative DNA damages and in nucleotide excision repair and risk of colorectal cancer have not obtained consistent results, why the issue of whether the SNPs are possible biomarkers of susceptibility for colorectal cancer is not satisfactorily clarified at present.

Sample size coupled with allele frequency may have influenced the validity of the results. Differences in the study design, like distribution of gender, age, topology, ethnicity and criteria for recruitment of comparison individuals may have contributed to the dissimilar findings. The application of large, well-designed association studies of the polymorphisms will make it statistically reasonable to make stratified analyses for obtaining information on risk factors in sub-groups and will generally decrease the risk of chance findings. Furthermore, studies including both cases with pre-stages of colorectal cancer and cancer cases will contribute with valuable information of the processes during colorectal carcinogenesis.

Most of the studies analyze individual polymorphisms in genes with modest effect in relation to risk of cancer. Cancer is a complex multigenic and multistage disease involving the interplay of many genetic and environmental factors. Hence, it is unlikely that a single genetic polymorphism in low-penetrance genes would have a dramatic effect on cancer risk. More information may be obtained from haplotyping multiple polymorphisms within genes or from combining multiple polymorphisms within pathways. The continued advances in SNP maps and in high-throughput genotyping methods will facilitate these analyses. Defining haplotypes and whole genome association studies may yield information on un-explored regions of the genome that has impact on colorectal cancer risk and development.

Colorectal cancer is probably caused by a complex interaction between many genetic and environmental factors over time. More and large studies with information on life style factors are required to assess these very possible gene-environment interactions. Identification of gene-environment interactions in cohorts with large relevant exposures has proven to be a useful approach.

Most environmental carcinogens require metabolic activation before they are able to form DNA damages. These activated forms may be detoxified or induce DNA repair or apoptosis. Thus, genetically determined susceptibility to colorectal cancer may depend on the balance among enzymes involved in metabolism and detoxification of carcinogens and on the balance between induction of DNA repair or apoptosis. Further investigations of the combined effects of polymorphisms between genes involved in these four mechanisms may help to clarify the influence of genetic variation in the carcinogenic process and may shed light on the complexities of the many pathways involved in colorectal cancer development, providing hypotheses for future functional studies.

## 5. Conclusion

In general, the studies suggest that the *XPD* Lys751Gln and *XPD* Asp312Asn polymorphisms may be associated with risk of colorectal adenomas with the possibility of interaction with smoking and alcohol consumption. The reported studies of polymorphisms in *XPC* and *XPA* in relation to risk of colorectal cancer are few, but the results are relatively consistent: In general, no association of the polymorphisms in the genes involved in NER (*XPD*, *XPC*, *XPA* and *ERCC1*) was observed with risk of colorectal cancer. A possible interpretation of the results may be that the polymorphisms in the genes *XPD*, *XPC*, *XPA* and *ERCC1* are not of major importance in colorectal cancer carcinogenesis, which points towards that lowered repair capacity of the NER pathway may not be a risk factor for development of colorectal cancer.

The results were generally inconsistent or too few to compare to highlight any trend and no strong associations were observed for risk of colorectal adenomas or colorectal cancer.

Overall, the role of genetic variants as SNPs in genes involved in NER is not satisfactorily clarified at present. It is possible that some of the SNPs may contribute to development of adenomas or colorectal cancer only in concomitance with certain dietary and life style factors. Furthermore, it may be only the joint effect of multiple polymorphisms that will provide us with information about genetic susceptibility for colorectal cancer. Larger carefully designed studies with stratified/adjusted analyses of gene-gene and gene-environment interactions may be required in the future to achieve convincing statistically significant results on factors involved in colorectal carcinogenesis.

## 6. Acknowledgement

Our special thanks go to Anne Tjønneland and Kim Overvad for giving us the opportunity to study the polymorphisms in the Danish prospective cohort Diet, Cancer and Health. We thank Anne-Karin Jensen and Lourdes M. Pedersen for excellent technical assistance for the genotyping in our studies on the Danish cohort. And a thank goes to our co-authors of the manuscripts on the Danish studies for the valuable feedback when writing the manuscripts. The work of the present book chapter was supported by a research grant from the Danish Cancer Society (Grant number R2-A84-09-S2).

## 7. References

Andrew, A.S., Nelson, H.H., Kelsey, K.T., Moore, J.H., Meng, A.C., Casella, D.P., Tosteson, T.D., Schned, A.R. & Karagas, M.R. (2006). Concordance of multiple analytical approaches demonstrates a complex relationship between DNA repair gene SNPs, smoking and bladder cancer susceptibility. *Carcinogenesis*, Vol. 27, No. 5, pp. (1030-1037)

Baccarelli, A., Calista, D., Minghetti, P., Marinelli, B., Albetti, B., Tseng, T., Hedayati, M., Grossman, L., Landi, G., Struewing, J.P. & Landi, M.T. (2004). XPD gene polymorphism and host characteristics in the association with cutaneous malignant melanoma risk. *Br.J.Cancer*, Vol. 90, No. 2, pp. (497-502)

Balbi, J.C., Larrinaga, M.T., De Stefani, E., Mendilaharsu, M., Ronco, A.L., Boffetta, P. & Brennan, P. (2001). Foods and risk of bladder cancer: a case-control study in Uruguay. *Eur J Cancer Prev*, Vol. 10, No. 5, pp. (453-458)

Batty, D., Rapic'-Otrin, V., Levine, A.S. & Wood, R.D. (2000). Stable binding of human XPC complex to irradiated DNA confers strong discrimination for damaged sites. *J.Mol.Biol.*, Vol. 300, No. 2, pp. (275-290)

Benhamou, S., Laplanche, A., Guillonneau, B., Mejean, A., Desgrandchamps, F., Schrameck, C., Degieux, V. & Perin, F. (2003). DNA adducts in normal bladder tissue and bladder cancer risk. *Mutagenesis*, Vol. 18, No. 5, pp. (445-448)

Benhamou, S. & Sarasin, A. (2005). ERCC2 /XPD gene polymorphisms and lung cancer: a HuGE review. *Am.J.Epidemiol.*, Vol. 161, No. 1, pp. (1-14)

Berndt, S.I., Platz, E.A., Fallin, M.D., Thuita, L.W., Hoffman, S.C. & Helzlsouer, K.J. (2006). Genetic variation in the nucleotide excision repair pathway and colorectal cancer risk. *Cancer Epidemiol.Biomarkers Prev.*, Vol. 15, No. 11, pp. (2263-2269)

Bigler, J., Ulrich, C.M., Kawashima, T., Whitton, J. & Potter, J.D. (2005). DNA repair polymorphisms and risk of colorectal adenomatous or hyperplastic polyps. *Cancer Epidemiol.Biomarkers Prev.*, Vol. 14, No. 11 Pt 1, pp. (2501-2508)

Bingham, S.A., Pignatelli, B., Pollock, J.R., Ellul, A., Malaveille, C., Gross, G., Runswick, S., Cummings, J.H. & O'Neill, I.K. (1996). Does increased endogenous formation of N-nitroso compounds in the human colon explain the association between red meat and colon cancer? *Carcinogenesis*, Vol. 17, No. 3, pp. (515-523)

Binkova, B., Lewtas, J., Miskova, I., Lenicek, J. & Sram, R. (1995). DNA adducts and personal air monitoring of carcinogenic polycyclic aromatic hydrocarbons in an environmentally exposed population. *Carcinogenesis*, Vol. 16, No. 5, pp. (1037-1046)

Blankenburg, S., Konig, I.R., Moessner, R., Laspe, P., Thoms, K.M., Krueger, U., Khan, S.G., Westphal, G., Berking, C., Volkenandt, M., Reich, K., Neumann, C., Ziegler, A., Kraemer, K.H. & Emmert, S. (2005). Assessment of 3 xeroderma pigmentosum group C gene polymorphisms and risk of cutaneous melanoma: a case-control study. *Carcinogenesis*, Vol. 26, No. 6, pp. (1085-1090)

Brewster, A.M., Jorgensen, T.J., Ruczinski, I., Huang, H.Y., Hoffman, S., Thuita, L., Newschaffer, C., Lunn, R.M., Bell, D. & Helzlsouer, K.J. (2006). Polymorphisms of the DNA repair genes XPD (Lys751Gln) and XRCC1 (Arg399Gln and Arg194Trp): relationship to breast cancer risk and familial predisposition to breast cancer. *Breast Cancer Res.Treat.*, Vol. 95, No. 1, pp. (73-80)

Brooks, P.J. & Theruvathu, J.A. (2005). DNA adducts from acetaldehyde: implications for alcohol-related carcinogenesis. *Alcohol*, Vol. 35, No. 3, pp. (187-193)

Bruemmer, B., White, E., Vaughan, T.L. & Cheney, C.L. (1996). Nutrient intake in relation to bladder cancer among middle-aged men and women. *Am J Epidemiol*, Vol. 144, No. 5, pp. (485-495)

Burch, J.D., Craib, K.J., Choi, B.C., Miller, A.B., Risch, H.A. & Howe, G.R. (1987). An exploratory case-control study of brain tumors in adults. *J.Natl.Cancer Inst.*, Vol. 78, No. 4, pp. (601-609)

Carozza, S.E., Olshan, A.F., Faustman, E.M., Gula, M.J., Kolonel, L.N., Austin, D.F., West, E.D., Weiss, N.S., Swanson, G.M., Lyon, J.L. & . (1995). Maternal exposure to N-nitrosatable drugs as a risk factor for childhood brain tumours. *Int.J.Epidemiol.*, Vol. 24, No. 2, pp. (308-312)

Cleaver, J.E. (2005). Cancer in xeroderma pigmentosum and related disorders of DNA repair. *Nat.Rev.Cancer*, Vol. 5, No. 7, pp. (564-573)

Correa Lima, M.P. & Gomes-da-Silva, M.H. (2005). Colorectal cancer: lifestyle and dietary factors. *Nutr.Hosp.*, Vol. 20, No. 4, pp. (235-241)

Costa, S., Pinto, D., Pereira, D., Rodrigues, H., Cameselle-Teijeiro, J., Medeiros, R. & Schmitt, F. (2007). DNA repair polymorphisms might contribute differentially on familial and sporadic breast cancer susceptibility: a study on a Portuguese population. *Breast Cancer Res.Treat.*, Vol. 103, No. 2, pp. (209-217)

Coverley, D., Kenny, M.K., Lane, D.P. & Wood, R.D. (1992). A role for the human single-stranded DNA binding protein HSSB/RPA in an early stage of nucleotide excision repair. *Nucleic Acids Res.*, Vol. 20, No. 15, pp. (3873-3880)

Cross, A.J., Pollock, J.R. & Bingham, S.A. (2003). Haem, not protein or inorganic iron, is responsible for endogenous intestinal N-nitrosation arising from red meat. *Cancer Res.*, Vol. 63, No. 10, pp. (2358-2360)

Culp, S.J., Gaylor, D.W., Sheldon, W.G., Goldstein, L.S. & Beland, F.A. (1998). A comparison of the tumors induced by coal tar and benzo[a]pyrene in a 2-year bioassay. *Carcinogenesis*, Vol. 19, No. 1, pp. (117-124) de Laat, W.L., Sijbers, A.M., Odijk, H., Jaspers, N.G. & Hoeijmakers, J.H. (1998). Mapping of interaction domains between human repair proteins ERCC1 and XPF. *Nucleic Acids Res.*, Vol. 26, No. 18, pp. (4146-4152)

De, R.K., Szaumkessel, M., De, R., I, Dehoorne, A., Vral, A., Claes, K., Velghe, A., Van, M.J. & Thierens, H. (2007). Polymorphisms in base-excision repair and nucleotide-excision repair genes in relation to lung cancer risk. *Mutat.Res.*, Vol. 631, No. 2, pp. (101-110)

Debniak, T., Scott, R.J., Huzarski, T., Byrski, T., Masojc, B., van de, W.T., Serrano-Fernandez, P., Gorski, B., Cybulski, C., Gronwald, J., Debniak, B., Maleszka, R., Kladny, J., Bieniek, A., Nagay, L., Haus, O., Grzybowska, E., Wandzel, P., Niepsuj, S., Narod, S.A. & Lubinski, J. (2006). XPD common variants and their association with melanoma and breast cancer risk. *Breast Cancer Res.Treat.*, Vol. 98, No. 2, pp. (209-215)

Dingley, K.H., Ubick, E.A., Chiarappa-Zucca, M.L., Nowell, S., Abel, S., Ebeler, S.E., Mitchell, A.E., Burns, S.A., Steinberg, F.M. & Clifford, A.J. (2003). Effect of dietary constituents with chemopreventive potential on adduct formation of a low dose of the heterocyclic amines PhIP and IQ and phase II hepatic enzymes. *Nutr.Cancer*, Vol. 46, No. 2, pp. (212-221)

Doyle, V.C. (2007). Nutrition and colorectal cancer risk: a literature review. *Gastroenterol.Nurs.*, Vol. 30, No. 3, pp. (178-182)

Drapkin, R., Reardon, J.T., Ansari, A., Huang, J.C., Zawel, L., Ahn, K., Sancar, A. & Reinberg, D. (1994). Dual role of TFIIH in DNA excision repair and in transcription by RNA polymerase II. *Nature*, Vol. 368, No. 6473, pp. (769-772)

Dufloth, R.M., Costa, S., Schmitt, F. & Zeferino, L.C. (2005). DNA repair gene polymorphisms and susceptibility to familial breast cancer in a group of patients from Campinas, Brazil. *Genet.Mol.Res.*, Vol. 4, No. 4, pp. (771-782)

Engin, A.B., Karahalil, B., Engin, A. & Karakaya, A.E. (2010). Oxidative stress, Helicobacter pylori, and OGG1 Ser326Cys, XPC Lys939Gln, and XPD Lys751Gln polymorphisms in a Turkish population with colorectal carcinoma. *Genet.Test.Mol.Biomarkers*, Vol. 14, No. 4, pp. (559-564)

Evans, E., Fellows, J., Coffer, A. & Wood, R.D. (1997). Open complex formation around a lesion during nucleotide excision repair provides a structure for cleavage by human XPG protein. *EMBO J.*, Vol. 16, No. 3, pp. (625-638)

Fang, J.L. & Vaca, C.E. (1997). Detection of DNA adducts of acetaldehyde in peripheral white blood cells of alcohol abusers. *Carcinogenesis*, Vol. 18, No. 4, pp. (627-632)

Feaver, W.J., Svejstrup, J.Q., Bardwell, L., Bardwell, A.J., Buratowski, S., Gulyas, K.D., Donahue, T.F., Friedberg, E.C. & Kornberg, R.D. (1993). Dual roles of a multiprotein complex from S. cerevisiae in transcription and DNA repair. *Cell*, Vol. 75, No. 7, pp. (1379-1387)

Festa, F., Kumar, R., Sanyal, S., Unden, B., Nordfors, L., Lindholm, B., Snellman, E., Schalling, M., Forsti, A. & Hemminki, K. (2005). Basal cell carcinoma and variants in genes coding for immune response, DNA repair, folate and iron metabolism. *Mutat.Res.*, Vol. 574, No. 1-2, pp. (105-111)

Forsti, A., Angelini, S., Festa, F., Sanyal, S., Zhang, Z., Grzybowska, E., Pamula, J., Pekala, W., Zientek, H., Hemminki, K. & Kumar, R. (2004). Single nucleotide polymorphisms in breast cancer. *Oncol.Rep.*, Vol. 11, No. 4, pp. (917-922)

Georgiadis, P., Topinka, J., Stoikidou, M., Kaila, S., Gioka, M., Katsouyanni, K., Sram, R., Autrup, H. & Kyrtopoulos, S.A. (2001). Biomarkers of genotoxicity of air pollution (the AULIS project): bulky DNA adducts in subjects with moderate to low exposures to airborne polycyclic aromatic hydrocarbons and their relationship to environmental tobacco smoke and other parameters. *Carcinogenesis*, Vol. 22, No. 9, pp. (1447-1457)

Goldberg, M.S., Parent, M.E., Siemiatycki, J., Desy, M., Nadon, L., Richardson, L., Lakhani, R., Latreille, B. & Valois, M.F. (2001). A case-control study of the relationship between the risk of colon cancer in men and exposures to occupational agents. *Am J Ind Med*, Vol. 39, No. 6, pp. (531-546)

Gomaa, E.A., Gray, J.I., Rabie, S., Lopez-Bote, C. & Booren, A.M. (1993). Polycyclic aromatic hydrocarbons in smoked food products and commercial liquid smoke flavourings. *Food Addit.Contam*, Vol. 10, No. 5, pp. (503-521)

Goodman, J.E., Mechanic, L.E., Luke, B.T., Ambs, S., Chanock, S. & Harris, C.C. (2006). Exploring SNP-SNP interactions and colon cancer risk using polymorphism interaction analysis. *Int.J.Cancer*, Vol. 118, No. 7, pp. (1790-1797)

Guillen, M.D., Sopelana, P. & Partearroyo, M.A. (1997). Food as a source of polycyclic aromatic carcinogens. *Rev.Environ.Health*, Vol. 12, No. 3, pp. (133-146)

Guzder, S.N., Sung, P., Prakash, L. & Prakash, S. (1998). Affinity of yeast nucleotide excision repair factor 2, consisting of the Rad4 and Rad23 proteins, for ultraviolet damaged DNA. *J.Biol.Chem.*, Vol. 273, No. 47, pp. (31541-31546)

Han, J., Colditz, G.A., Liu, J.S. & Hunter, D.J. (2005). Genetic variation in XPD, sun exposure, and risk of skin cancer. *Cancer Epidemiol.Biomarkers Prev.*, Vol. 14, No. 6, pp. (1539-1544)

Hanawalt, P.C. (2002). Subpathways of nucleotide excision repair and their regulation. *Oncogene*, Vol. 21, No. 58, pp. (8949-8956)

Hansen, R.D., Sorensen, M., Tjonneland, A., Overvad, K., Wallin, H., Raaschou-Nielsen, O. & Vogel, U. (2007). XPA A23G, XPC Lys939Gln, XPD Lys751Gln and XPD Asp312Asn polymorphisms, interactions with smoking, alcohol and dietary factors, and risk of colorectal cancer. *Mutat.Res.*, Vol. 619, No. 1-2, pp. (68-80)

Hansen, R.D., Sorensen, M., Tjonneland, A., Overvad, K., Wallin, H., Raaschou-Nielsen, O. & Vogel, U. (2008). A haplotype of polymorphisms in ASE-1, RAI and ERCC1 and the effects of tobacco smoking and alcohol consumption on risk of colorectal cancer: a Danish prospective case-cohort study. *BMC.Cancer*, Vol. 8, pp. (54)

Hecht, S.S. (2003). Tobacco carcinogens, their biomarkers and tobacco-induced cancer. *Nat.Rev.Cancer*, Vol. 3, No. 10, pp. (733-744)

Hecht, S.S. & Hoffmann, D. (1988). Tobacco-specific nitrosamines, an important group of carcinogens in tobacco and tobacco smoke. *Carcinogenesis*, Vol. 9, No. 6, pp. (875-884)

Hemminki, K., Dickey, C., Karlsson, S., Bell, D., Hsu, Y., Tsai, W.Y., Mooney, L.A., Savela, K. & Perera, F.P. (1997). Aromatic DNA adducts in foundry workers in relation to exposure, life style and CYP1A1 and glutathione transferase M1 genotype. *Carcinogenesis*, Vol. 18, No. 2, pp. (345-350)

Hemminki, K., Grzybowska, E., Chorazy, M., Twardowska-Saucha, K., Sroczynski, J.W., Putman, K.L., Randerath, K., Phillips, D.H., Hewer, A., Santella, R.M. & . (1990a).

DNA adducts in human environmentally exposed to aromatic compounds in an industrial area of Poland. *Carcinogenesis,* Vol. 11, No. 7, pp. (1229-1231)

Hemminki, K., Randerath, K., Reddy, M.V., Putman, K.L., Santella, R.M., Perera, F.P., Young, T.L., Phillips, D.H., Hewer, A. & Savela, K. (1990b). Postlabeling and immunoassay analysis of polycyclic aromatic hydrocarbons--adducts of deoxyribonucleic acid in white blood cells of foundry workers. *Scand J Work Environ Health,* Vol. 16, No. 3, pp. (158-162)

Hemminki, K., Soderling, J., Ericson, P., Norbeck, H.E. & Segerback, D. (1994). DNA adducts among personnel servicing and loading diesel vehicles. *Carcinogenesis,* Vol. 15, No. 4, pp. (767-769)

Hotchkiss, J.H. (1989). Preformed N-nitroso compounds in foods and beverages. *Cancer Surv.,* Vol. 8, No. 2, pp. (295-321)

Hou, S.M., Falt, S., Angelini, S., Yang, K., Nyberg, F., Lambert, B. & Hemminki, K. (2002). The XPD variant alleles are associated with increased aromatic DNA adduct level and lung cancer risk. *Carcinogenesis,* Vol. 23, No. 4, pp. (599-603)

Hu, Z., Wei, Q., Wang, X. & Shen, H. (2004). DNA repair gene XPD polymorphism and lung cancer risk: a meta-analysis. *Lung Cancer,* Vol. 46, No. 1, pp. (1-10)

Hu, Z., Xu, L., Shao, M., Yuan, J., Wang, Y., Wang, F., Yuan, W., Qian, J., Ma, H., Wang, Y., Liu, H., Chen, W., Yang, L., Jing, G., Huo, X., Chen, F., Jin, L., Wei, Q., Wu, T., Lu, D., Huang, W. & Shen, H. (2006). Polymorphisms in the two helicases ERCC2/XPD and ERCC3/XPB of the transcription factor IIH complex and risk of lung cancer: a case-control analysis in a Chinese population. *Cancer Epidemiol.Biomarkers Prev.,* Vol. 15, No. 7, pp. (1336-1340)

Huang, W.Y., Berndt, S.I., Kang, D., Chatterjee, N., Chanock, S.J., Yeager, M., Welch, R., Bresalier, R.S., Weissfeld, J.L. & Hayes, R.B. (2006). Nucleotide excision repair gene polymorphisms and risk of advanced colorectal adenoma: XPC polymorphisms modify smoking-related risk. *Cancer Epidemiol.Biomarkers Prev.,* Vol. 15, No. 2, pp. (306-311)

Huang, W.Y., Chow, W.H., Rothman, N., Lissowska, J., Llaca, V., Yeager, M., Zatonski, W. & Hayes, R.B. (2005). Selected DNA repair polymorphisms and gastric cancer in Poland. *Carcinogenesis,* Vol. 26, No. 8, pp. (1354-1359)

Hunting, D.J., Gowans, B.J. & Dresler, S.L. (1991). DNA polymerase delta mediates excision repair in growing cells damaged with ultraviolet radiation. *Biochem.Cell Biol.,* Vol. 69, No. 4, pp. (303-308)

International Agency for Research on Cancer (IARC). Polynuclear aromatic compounds. Part I: Chemical, environmental and experimental data. IARC monograph on the evaluation of carcinogenic risk of chemical to man. (32). 1983. Lyon, France, IARC.

International Agency for Research on Cancer (IARC) (2003). *Fruit and vegetables* IARC Press, Lyon

Iversen, L.H., Pedersen, L., Riis, A., Friis, S., Laurberg, S. & Sorensen, H.T. (2005). Population-based study of short- and long-term survival from colorectal cancer in Denmark, 1977-1999. *Br.J.Surg.,* Vol. 92, No. 7, pp. (873-880)

Jaspers, N.G., Raams, A., Silengo, M.C., Wijgers, N., Niedernhofer, L.J., Robinson, A.R., Giglia-Mari, G., Hoogstraten, D., Kleijer, W.J., Hoeijmakers, J.H. & Vermeulen, W. (2007). First reported patient with human ERCC1 deficiency has cerebro-oculo-facio-skeletal syndrome with a mild defect in nucleotide excision repair and severe developmental failure. *Am.J.Hum.Genet.*, Vol. 80, No. 3, pp. (457-466)

Jemal, A., Siegel, R., Ward, E., Murray, T., Xu, J., Smigal, C. & Thun, M.J. (2006). Cancer statistics, 2006. *CA Cancer J.Clin.*, Vol. 56, No. 2, pp. (106-130)

Ji, B.T., Weissfeld, J.L., Chow, W.H., Huang, W.Y., Schoen, R.E. & Hayes, R.B. (2006). Tobacco smoking and colorectal hyperplastic and adenomatous polyps. *Cancer Epidemiol.Biomarkers Prev.*, Vol. 15, No. 5, pp. (897-901)

Jo, H., Kang, S., Kim, S.I., Kim, J.W., Park, N.H., Song, Y.S., Park, S.Y., Kang, S.B. & Lee, H.P. (2007). The C19007T Polymorphism of ERCC1 and Its Correlation with the Risk of Epithelial Ovarian and Endometrial Cancer in Korean Women. A Case Control Study. *Gynecol.Obstet.Invest*, Vol. 64, No. 2, pp. (84-88)

Johnson, I.T. & Lund, E.K. (2007). Review article: nutrition, obesity and colorectal cancer. *Aliment.Pharmacol.Ther.*, Vol. 26, No. 2, pp. (161-181)

Jorgensen, T.J., Visvanathan, K., Ruczinski, I., Thuita, L., Hoffman, S. & Helzlsouer, K.J. (2007). Breast cancer risk is not associated with polymorphic forms of xeroderma pigmentosum genes in a cohort of women from Washington County, Maryland. *Breast Cancer Res.Treat.*, Vol. 101, No. 1, pp. (65-71)

Joshi, A.D., Corral, R., Siegmund, K.D., Haile, R.W., Le, M.L., Martinez, M.E., Ahnen, D.J., Sandler, R.S., Lance, P. & Stern, M.C. (2009). Red meat and poultry intake, polymorphisms in the nucleotide excision repair and mismatch repair pathways and colorectal cancer risk. *Carcinogenesis*, Vol. 30, No. 3, pp. (472-479)

Justenhoven, C., Hamann, U., Pesch, B., Harth, V., Rabstein, S., Baisch, C., Vollmert, C., Illig, T., Ko, Y.D., Bruning, T. & Brauch, H. (2004). ERCC2 genotypes and a corresponding haplotype are linked with breast cancer risk in a German population. *Cancer Epidemiol.Biomarkers Prev.*, Vol. 13, No. 12, pp. (2059-2064)

Khan, S.G., Metter, E.J., Tarone, R.E., Bohr, V.A., Grossman, L., Hedayati, M., Bale, S.J., Emmert, S. & Kraemer, K.H. (2000). A new xeroderma pigmentosum group C poly(AT) insertion/deletion polymorphism. *Carcinogenesis*, Vol. 21, No. 10, pp. (1821-1825)

Kiyohara, C. & Yoshimasu, K. (2007). Genetic polymorphisms in the nucleotide excision repair pathway and lung cancer risk: a meta-analysis. *Int.J.Med.Sci.*, Vol. 4, No. 2, pp. (59-71)

Kobayashi, K., Karran, P., Oda, S. & Yanaga, K. (2005). Involvement of mismatch repair in transcription-coupled nucleotide excision repair. *Hum.Cell*, Vol. 18, No. 3, pp. (103-115)

Kyrtopoulos, S.A., Georgiadis, P., Autrup, H., Demopoulos, N.A., Farmer, P., Haugen, A., Katsouyanni, K., Lambert, B., Ovrebo, S., Sram, R., Stephanou, G. & Topinka, J. (2001). Biomarkers of genotoxicity of urban air pollution. Overview and descriptive data from a molecular epidemiology study on populations exposed to moderate-to-low levels of polycyclic aromatic hydrocarbons: the AULIS project. *Mutat Res*, Vol. 496, No. 1-2, pp. (207-228)

Laska, M.J., Nexo, B.A., Vistisen, K., Poulsen, H.E., Loft, S. & Vogel, U. (2005). Polymorphisms in RAI and in genes of nucleotide and base excision repair are not associated with risk of testicular cancer. *Cancer Lett.*, Vol. 225, No. 2, pp. (245-251)

Lee, K.M., Choi, J.Y., Kang, C., Kang, C.P., Park, S.K., Cho, H., Cho, D.Y., Yoo, K.Y., Noh, D.Y., Ahn, S.H., Park, C.G., Wei, Q. & Kang, D. (2005). Genetic polymorphisms of selected DNA repair genes, estrogen and progesterone receptor status, and breast cancer risk. *Clin.Cancer Res.*, Vol. 11, No. 12, pp. (4620-4626)

Lewin, M.H., Bailey, N., Bandaletova, T., Bowman, R., Cross, A.J., Pollock, J., Shuker, D.E. & Bingham, S.A. (2006). Red meat enhances the colonic formation of the DNA adduct O6-carboxymethyl guanine: implications for colorectal cancer risk. *Cancer Res.*, Vol. 66, No. 3, pp. (1859-1865)

Lewtas, J., Walsh, D., Williams, R. & Dobias, L. (1997). Air pollution exposure-DNA adduct dosimetry in humans and rodents: evidence for non-linearity at high doses. *Mutat Res*, Vol. 378, No. 1-2, pp. (51-63)

Li, C., Hu, Z., Liu, Z., Wang, L.E., Strom, S.S., Gershenwald, J.E., Lee, J.E., Ross, M.I., Mansfield, P.F., Cormier, J.N., Prieto, V.G., Duvic, M., Grimm, E.A. & Wei, Q. (2006a). Polymorphisms in the DNA repair genes XPC, XPD, and XPG and risk of cutaneous melanoma: a case-control analysis. *Cancer Epidemiol.Biomarkers Prev.*, Vol. 15, No. 12, pp. (2526-2532)

Li, H., Yao, L., Ouyang, T., Li, J., Wang, T., Fan, Z., Fan, T., Dong, B., Lin, B., Li, J. & Xie, Y. (2006b). Association of p73 G4C14-to-A4T14 (GC/AT) polymorphism with breast cancer survival. *Carcinogenesis,*

Lovatt, T., Alldersea, J., Lear, J.T., Hoban, P.R., Ramachandran, S., Fryer, A.A., Smith, A.G. & Strange, R.C. (2005). Polymorphism in the nuclear excision repair gene ERCC2/XPD: association between an exon 6-exon 10 haplotype and susceptibility to cutaneous basal cell carcinoma. *Hum.Mutat.*, Vol. 25, No. 4, pp. (353-359)

Masutani, C., Sugasawa, K., Yanagisawa, J., Sonoyama, T., Ui, M., Enomoto, T., Takio, K., Tanaka, K., van der Spek, P.J., Bootsma, D. & . (1994). Purification and cloning of a nucleotide excision repair complex involving the xeroderma pigmentosum group C protein and a human homologue of yeast RAD23. *EMBO J.*, Vol. 13, No. 8, pp. (1831-1843)

Matullo, G., Palli, D., Peluso, M., Guarrera, S., Carturan, S., Celentano, E., Krogh, V., Munnia, A., Tumino, R., Polidoro, S., Piazza, A. & Vineis, P. (2001). XRCC1, XRCC3, XPD gene polymorphisms, smoking and (32)P-DNA adducts in a sample of healthy subjects. *Carcinogenesis,* Vol. 22, No. 9, pp. (1437-1445)

Meador, J.P., Stein, J.E., Reichert, W.L. & Varanasi, U. (1995). Bioaccumulation of polycyclic aromatic hydrocarbons by marine organisms. *Rev.Environ.Contam Toxicol.*, Vol. 143, pp. (79-165)

Mechanic, L.E., Millikan, R.C., Player, J., de Cotret, A.R., Winkel, S., Worley, K., Heard, K., Heard, K., Tse, C.K. & Keku, T. (2006). Polymorphisms in nucleotide excision repair genes, smoking and breast cancer in African Americans and whites: a population-based case-control study. *Carcinogenesis,* Vol. 27, No. 7, pp. (1377-1385)

Melikian, A.A., Sun, P., Prokopczyk, B., El-Bayoumy, K., Hoffmann, D., Wang, X. & Waggoner, S. (1999). Identification of benzo[a]pyrene metabolites in cervical mucus and DNA adducts in cervical tissues in humans by gas chromatography-mass spectrometry. *Cancer Lett,* Vol. 146, pp. (127-134)

Metsola, K., Kataja, V., Sillanpaa, P., Siivola, P., Heikinheimo, L., Eskelinen, M., Kosma, V.M., Uusitupa, M. & Hirvonen, A. (2005). XRCC1 and XPD genetic polymorphisms, smoking and breast cancer risk in a Finnish case-control study. *Breast Cancer Res.*, Vol. 7, No. 6, pp. (R987-R997)

Miller, K.L., Karagas, M.R., Kraft, P., Hunter, D.J., Catalano, P.J., Byler, S.H. & Nelson, H.H. (2006). XPA, haplotypes, and risk of basal and squamous cell carcinoma. *Carcinogenesis*, Vol. 27, No. 8, pp. (1670-1675)

Millikan, R.C., Hummer, A., Begg, C., Player, J., de Cotret, A.R., Winkel, S., Mohrenweiser, H., Thomas, N., Armstrong, B., Kricker, A., Marrett, L.D., Gruber, S.B., Culver, H.A., Zanetti, R., Gallagher, R.P., Dwyer, T., Rebbeck, T.R., Busam, K., From, L., Mujumdar, U. & Berwick, M. (2006). Polymorphisms in nucleotide excision repair genes and risk of multiple primary melanoma: the Genes Environment and Melanoma Study. *Carcinogenesis*, Vol. 27, No. 3, pp. (610-618)

Mirvish, S.S. (1995). Role of N-nitroso compounds (NOC) and N-nitrosation in etiology of gastric, esophageal, nasopharyngeal and bladder cancer and contribution to cancer of known exposures to NOC. *Cancer Lett.*, Vol. 93, No. 1, pp. (17-48)

Moller, P., Wallin, H., Vogel, U., Autrup, H., Risom, L., Hald, M.T., Daneshvar, B., Dragsted, L.O., Poulsen, H.E. & Loft, S. (2002). Mutagenicity of 2-amino-3-methylimidazo[4,5-f]quinoline in colon and liver of Big Blue rats: role of DNA adducts, strand breaks, DNA repair and oxidative stress. *Carcinogenesis*, Vol. 23, No. 8, pp. (1379-1385)

Mooney, L.A., Bell, D.A., Santella, R.M., Van Bennekum, A.M., Ottman, R., Paik, M., Blaner, W.S., Lucier, G.W., Covey, L., Young, T.L., Cooper, T.B., Glassman, A.H. & Perera, F.P. (1997). Contribution of genetic and nutritional factors to DNA damage in heavy smokers. *Carcinogenesis*, Vol. 18, No. 3, pp. (503-509)

Mooney, L.A., Madsen, A.M., Tang, D., Orjuela, M.A., Tsai, W.Y., Garduno, E.R. & Perera, F.P. (2005). Antioxidant vitamin supplementation reduces benzo(a)pyrene-DNA adducts and potential cancer risk in female smokers. *Cancer Epidemiol Biomarkers Prev*, Vol. 14, No. 1, pp. (237-242)

Moreno, V., Gemignani, F., Landi, S., Gioia-Patricola, L., Chabrier, A., Blanco, I., Gonzalez, S., Guino, E., Capella, G. & Canzian, F. (2006). Polymorphisms in genes of nucleotide and base excision repair: risk and prognosis of colorectal cancer. *Clin.Cancer Res.*, Vol. 12, No. 7 Pt 1, pp. (2101-2108)

Mort, R., Mo, L., McEwan, C. & Melton, D.W. (2003). Lack of involvement of nucleotide excision repair gene polymorphisms in colorectal cancer. *Br J Cancer*, Vol. 89, No. 2, pp. (333-337)

Mu, D., Hsu, D.S. & Sancar, A. (1996). Reaction mechanism of human DNA repair excision nuclease. *J.Biol.Chem.*, Vol. 271, No. 14, pp. (8285-8294)

Nelson, H.H., Christensen, B. & Karagas, M.R. (2005). The XPC poly-AT polymorphism in non-melanoma skin cancer. *Cancer Lett.*, Vol. 222, No. 2, pp. (205-209)

Niedernhofer, L.J., Essers, J., Weeda, G., Beverloo, B., de, W.J., Muijtjens, M., Odijk, H., Hoeijmakers, J.H. & Kanaar, R. (2001). The structure-specific endonuclease Ercc1-Xpf is required for targeted gene replacement in embryonic stem cells. *EMBO J.*, Vol. 20, No. 22, pp. (6540-6549)

Nielsen, P.S., Andreassen, A., Farmer, P.B., Ovrebo, S. & Autrup, H. (1996a). Biomonitoring of diesel exhaust-exposed workers. DNA and hemoglobin adducts and urinary 1-hydroxypyrene as markers of exposure. *Toxicol Lett*, Vol. 86, No. 1, pp. (27-37)

Nielsen, P.S., de Pater, N., Okkels, H. & Autrup, H. (1996b). Environmental air pollution and DNA adducts in Copenhagen bus drivers--effect of GSTM1 and NAT2 genotypes on adduct levels. *Carcinogenesis*, Vol. 17, No. 5, pp. (1021-1027)

Nielsen, P.S., Okkels, H., Sigsgaard, T., Kyrtopoulos, S. & Autrup, H. (1996c). Exposure to urban and rural air pollution: DNA and protein adducts and effect of glutathione-S-transferase genotype on adduct levels. *Int Arch Occup Environ Health*, Vol. 68, No. 3, pp. (170-176)

Norat, T., Bingham, S., Ferrari, P., Slimani, N., Jenab, M., Mazuir, M., Overvad, K., Olsen, A., Tjonneland, A., Clavel, F., Boutron-Ruault, M.C., Kesse, E., Boeing, H., Bergmann, M.M., Nieters, A., Linseisen, J., Trichopoulou, A., Trichopoulos, D., Tountas, Y., Berrino, F., Palli, D., Panico, S., Tumino, R., Vineis, P., Bueno-de-Mesquita, H.B., Peeters, P.H., Engeset, D., Lund, E., Skeie, G., Ardanaz, E., Gonzalez, C., Navarro, C., Quiros, J.R., Sanchez, M.J., Berglund, G., Mattisson, I., Hallmans, G., Palmqvist, R., Day, N.E., Khaw, K.T., Key, T.J., San Joaquin, M., Hemon, B., Saracci, R., Kaaks, R. & Riboli, E. (2005). Meat, fish, and colorectal cancer risk: the European Prospective Investigation into cancer and nutrition. *J.Natl.Cancer Inst.*, Vol. 97, No. 12, pp. (906-916)

Palli, D., Masala, G., Peluso, M., Gaspari, L., Krogh, V., Munnia, A., Panico, S., Saieva, C., Tumino, R., Vineis, P. & Garte, S. (2004). The effects of diet on DNA bulky adduct levels are strongly modified by GSTM1 genotype: a study on 634 subjects. *Carcinogenesis*, Vol. 25, No. 4, pp. (577-584)

Palli, D., Masala, G., Vineis, P., Garte, S., Saieva, C., Krogh, V., Panico, S., Tumino, R., Munnia, A., Riboli, E. & Peluso, M. (2003). Biomarkers of dietary intake of micronutrients modulate DNA adduct levels in healthy adults. *Carcinogenesis*, Vol. 24, No. 4, pp. (739-746)

Palli, D., Russo, A., Masala, G., Saieva, C., Guarrera, S., Carturan, S., Munnia, A., Matullo, G. & Peluso, M. (2001). DNA adduct levels and DNA repair polymorphisms in traffic-exposed workers and a general population sample. *Int J Cancer*, Vol. 94, No. 1, pp. (121-127)

Palli, D., Vineis, P., Russo, A., Berrino, F., Krogh, V., Masala, G., Munnia, A., Panico, S., Taioli, E., Tumino, R., Garte, S. & Peluso, M. (2000). Diet, metabolic polymorphisms and dna adducts: the EPIC-Italy cross-sectional study. *Int J Cancer*, Vol. 87, No. 3, pp. (444-451)

Park, C.H., Mu, D., Reardon, J.T. & Sancar, A. (1995). The general transcription-repair factor TFIIH is recruited to the excision repair complex by the XPA protein independent of the TFIIE transcription factor. *J.Biol.Chem.*, Vol. 270, No. 9, pp. (4896-4902)

Park, C.H. & Sancar, A. (1994). Formation of a ternary complex by human XPA, ERCC1, and ERCC4(XPF) excision repair proteins. *Proc.Natl.Acad.Sci.U.S.A*, Vol. 91, No. 11, pp. (5017-5021)

Park, J.Y., Park, S.H., Choi, J.E., Lee, S.Y., Jeon, H.S., Cha, S.I., Kim, C.H., Park, J.H., Kam, S., Park, R.W., Kim, I.S. & Jung, T.H. (2002). Polymorphisms of the DNA repair gene xeroderma pigmentosum group A and risk of primary lung cancer. *Cancer Epidemiol.Biomarkers Prev.*, Vol. 11, No. 10 Pt 1, pp. (993-997)

Peluso, M., Merlo, F., Munnia, A., Valerio, F., Perrotta, A., Puntoni, R. & Parodi, S. (1998). 32P-postlabeling detection of aromatic adducts in the white blood cell DNA of nonsmoking police officers. *Cancer Epidemiol Biomarkers Prev*, Vol. 7, No. 1, pp. (3-11)

Perera, F.P., Dickey, C., Santella, R., O'Neill, J.P., Albertini, R.J., Ottman, R., Tsai, W.Y., Mooney, L.A., Savela, K. & Hemminki, K. (1994). Carcinogen-DNA adducts and gene mutation in foundry workers with low-level exposure to polycyclic aromatic hydrocarbons. *Carcinogenesis,* Vol. 15, No. 12, pp. (2905-2910)

Perera, F.P., Hemminki, K., Young, T.L., Brenner, D., Kelly, G. & Santella, R.M. (1988). Detection of polycyclic aromatic hydrocarbon-DNA adducts in white blood cells of foundry workers. *Cancer Res,* Vol. 48, No. 8, pp. (2288-2291)

Peters, U., Sinha, R., Bell, D.A., Rothman, N., Grant, D.J., Watson, M.A., Kulldorff, M., Brooks, L.R., Warren, S.H. & DeMarini, D.M. (2004). Urinary mutagenesis and fried red meat intake: influence of cooking temperature, phenotype, and genotype of metabolizing enzymes in a controlled feeding study. *Environ Mol Mutagen,* Vol. 43, No. 1, pp. (53-74)

Phillips, D.H. (2002). Smoking-related DNA and protein adducts in human tissues. *Carcinogenesis,* Vol. 23, No. 12, pp. (1979-2004)

Pierre, F., Tache, S., Petit, C.R., Van der, M.R. & Corpet, D.E. (2003). Meat and cancer: haemoglobin and haemin in a low-calcium diet promote colorectal carcinogenesis at the aberrant crypt stage in rats. *Carcinogenesis,* Vol. 24, No. 10, pp. (1683-1690)

Poirier, M.C., Weston, A., Schoket, B., Shamkhani, H., Pan, C.F., McDiarmid, M.A., Scott, B.G., Deeter, D.P., Heller, J.M., Jacobson-Kram, D. & Rothman, N. (1998). Biomonitoring of United States Army soldiers serving in Kuwait in 1991. *Cancer Epidemiol Biomarkers Prev,* Vol. 7, No. 6, pp. (545-551)

Popanda, O., Schattenberg, T., Phong, C.T., Butkiewicz, D., Risch, A., Edler, L., Kayser, K., Dienemann, H., Schulz, V., Drings, P., Bartsch, H. & Schmezer, P. (2004). Specific combinations of DNA repair gene variants and increased risk for non-small cell lung cancer. *Carcinogenesis,* Vol. 25, No. 12, pp. (2433-2441)

Preston-Martin, S. & Mack, W. (1991). Gliomas and meningiomas in men in Los Angeles County: investigation of exposures to N-nitroso compounds. *IARC Sci.Publ.,* No. 105, pp. (197-203)

Qiao, Y., Spitz, M.R., Shen, H., Guo, Z., Shete, S., Hedayati, M., Grossman, L., Mohrenweiser, H. & Wei, Q. (2002). Modulation of repair of ultraviolet damage in the host-cell reactivation assay by polymorphic XPC and XPD/ERCC2 genotypes. *Carcinogenesis,* Vol. 23, No. 2, pp. (295-299)

Reardon, J.T., Mu, D. & Sancar, A. (1996). Overproduction, purification, and characterization of the XPC subunit of the human DNA repair excision nuclease. *J.Biol.Chem.,* Vol. 271, No. 32, pp. (19451-19456)

Reardon, J.T. & Sancar, A. (2002). Molecular anatomy of the human excision nuclease assembled at sites of DNA damage. *Mol.Cell Biol.,* Vol. 22, No. 16, pp. (5938-5945)

Ritchey, J.D., Huang, W.Y., Chokkalingam, A.P., Gao, Y.T., Deng, J., Levine, P., Stanczyk, F.Z. & Hsing, A.W. (2005). Genetic variants of DNA repair genes and prostate cancer: a population-based study. *Cancer Epidemiol.Biomarkers Prev.,* Vol. 14, No. 7, pp. (1703-1709)

Rothman, N., Correa-Villasenor, A., Ford, D.P., Poirier, M.C., Haas, R., Hansen, J.A., O'Toole, T. & Strickland, P.T. (1993). Contribution of occupation and diet to white blood cell polycyclic aromatic hydrocarbon-DNA adducts in wildland firefighters. *Cancer Epidemiol Biomarkers Prev,* Vol. 2, No. 4, pp. (341-347)

Rothman, N., Poirier, M.C., Baser, M.E., Hansen, J.A., Gentile, C., Bowman, E.D. & Strickland, P.T. (1990). Formation of polycyclic aromatic hydrocarbon-DNA adducts in peripheral white blood cells during consumption of charcoal-broiled beef. *Carcinogenesis,* Vol. 11, No. 7, pp. (1241-1243)

Sanyal, S., Festa, F., Sakano, S., Zhang, Z., Steineck, G., Norming, U., Wijkstrom, H., Larsson, P., Kumar, R. & Hemminki, K. (2004). Polymorphisms in DNA repair and metabolic genes in bladder cancer. *Carcinogenesis,* Vol. 25, No. 5, pp. (729-734)

Schabath, M.B., Delclos, G.L., Grossman, H.B., Wang, Y., Lerner, S.P., Chamberlain, R.M., Spitz, M.R. & Wu, X. (2005). Polymorphisms in XPD exons 10 and 23 and bladder cancer risk. *Cancer Epidemiol.Biomarkers Prev.,* Vol. 14, No. 4, pp. (878-884)

Schaeffer, L., Roy, R., Humbert, S., Moncollin, V., Vermeulen, W., Hoeijmakers, J.H., Chambon, P. & Egly, J.M. (1993). DNA repair helicase: a component of BTF2 (TFIIH) basic transcription factor. *Science,* Vol. 260, No. 5104, pp. (58-63)

Schottenfeld, D. & Winawer, S.J. (1996). Cancers of the large intestine, In: *Cancer Epidemiology and Prevention,* Schottenfeld, D. & Fraumeni J.F.Jr., pp. (813-840), Oxford University Press, New York

Shivji, K.K., Kenny, M.K. & Wood, R.D. (1992). Proliferating cell nuclear antigen is required for DNA excision repair. *Cell,* Vol. 69, No. 2, pp. (367-374)

Shivji, M.K., Podust, V.N., Hubscher, U. & Wood, R.D. (1995). Nucleotide excision repair DNA synthesis by DNA polymerase epsilon in the presence of PCNA, RFC, and RPA. *Biochemistry,* Vol. 34, No. 15, pp. (5011-5017)

Sijbers, A.M., de Laat, W.L., Ariza, R.R., Biggerstaff, M., Wei, Y.F., Moggs, J.G., Carter, K.C., Shell, B.K., Evans, E., de Jong, M.C., Rademakers, S., de, R.J., Jaspers, N.G., Hoeijmakers, J.H. & Wood, R.D. (1996). Xeroderma pigmentosum group F caused by a defect in a structure-specific DNA repair endonuclease. *Cell,* Vol. 86, No. 5, pp. (811-822)

Sinha, R., Peters, U., Cross, A.J., Kulldorff, M., Weissfeld, J.L., Pinsky, P.F., Rothman, N. & Hayes, R.B. (2005). Meat, meat cooking methods and preservation, and risk for colorectal adenoma. *Cancer Res.,* Vol. 65, No. 17, pp. (8034-8041)

Sinha, R., Rothman, N., Brown, E.D., Salmon, C.P., Knize, M.G., Swanson, C.A., Rossi, S.C., Mark, S.D., Levander, O.A. & Felton, J.S. (1995). High concentrations of the carcinogen 2-amino-1-methyl-6-phenylimidazo- [4,5-b]pyridine (PhIP) occur in chicken but are dependent on the cooking method. *Cancer Res.,* Vol. 55, No. 20, pp. (4516-4519)

Skjelbred, C.F., Sabo, M., Nexo, B.A., Wallin, H., Hansteen, I.L., Vogel, U. & Kure, E.H. (2006a). Effects of polymorphisms in ERCC1, ASE-1 and RAI on the risk of colorectal carcinomas and adenomas: A case control study. *BMC.Cancer,* Vol. 6, No. 1, pp. (175)

Skjelbred, C.F., Saebo, M., Wallin, H., Nexo, B.A., Hagen, P.C., Lothe, I.M., Aase, S., Johnson, E., Hansteen, I.L., Vogel, U. & Kure, E.H. (2006b). Polymorphisms of the XRCC1, XRCC3 and XPD genes and risk of colorectal adenoma and carcinoma, in a Norwegian cohort: a case control study. *BMC.Cancer,* Vol. 6, pp. (67)

Skog, K., Steineck, G., Augustsson, K. & Jagerstad, M. (1995). Effect of cooking temperature on the formation of heterocyclic amines in fried meat products and pan residues. *Carcinogenesis,* Vol. 16, No. 4, pp. (861-867)

Spitz, M.R., Wu, X., Wang, Y., Wang, L.E., Shete, S., Amos, C.I., Guo, Z., Lei, L., Mohrenweiser, H. & Wei, Q. (2001). Modulation of nucleotide excision repair capacity by XPD polymorphisms in lung cancer patients. *Cancer Res.*, Vol. 61, No. 4, pp. (1354-1357)

Starinsky, S., Figer, A., Ben Asher, E., Geva, R., Flex, D., Fidder, H.H., Zidan, J., Lancet, D. & Friedman, E. (2005). Genotype phenotype correlations in Israeli colorectal cancer patients. *International Journal of Cancer*, Vol. 114, No. 1, pp. (58-73), 0020-7136

Stern, M.C., Butler, L.M., Corral, R., Joshi, A.D., Yuan, J.M., Koh, W.P. & Yu, M.C. (2009). Polyunsaturated fatty acids, DNA repair single nucleotide polymorphisms and colorectal cancer in the Singapore Chinese Health Study. *J.Nutrigenet.Nutrigenomics.*, Vol. 2, No. 6, pp. (273-279)

Stern, M.C., Conti, D.V., Siegmund, K.D., Corral, R., Yuan, J.M., Koh, W.P. & Yu, M.C. (2007). DNA repair single-nucleotide polymorphisms in colorectal cancer and their role as modifiers of the effect of cigarette smoking and alcohol in the Singapore Chinese Health Study. *Cancer Epidemiol.Biomarkers Prev.*, Vol. 16, No. 11, pp. (2363-2372)

Stern, M.C., Johnson, L.R., Bell, D.A. & Taylor, J.A. (2002). XPD codon 751 polymorphism, metabolism genes, smoking, and bladder cancer risk. *Cancer Epidemiol.Biomarkers Prev.*, Vol. 11, No. 10 Pt 1, pp. (1004-1011)

Stern, M.C., Siegmund, K.D., Conti, D.V., Corral, R. & Haile, R.W. (2006). XRCC1, XRCC3, and XPD polymorphisms as modifiers of the effect of smoking and alcohol on colorectal adenoma risk. *Cancer Epidemiol.Biomarkers Prev.*, Vol. 15, No. 12, pp. (2384-2390)

Stigger, E., Drissi, R. & Lee, S.H. (1998). Functional analysis of human replication protein A in nucleotide excision repair. *J.Biol.Chem.*, Vol. 273, No. 15, pp. (9337-9343)

Sugasawa, K., Ng, J.M., Masutani, C., Iwai, S., van der Spek, P.J., Eker, A.P., Hanaoka, F., Bootsma, D. & Hoeijmakers, J.H. (1998). Xeroderma pigmentosum group C protein complex is the initiator of global genome nucleotide excision repair. *Mol.Cell*, Vol. 2, No. 2, pp. (223-232)

Terry, M.B., Gammon, M.D., Zhang, F.F., Eng, S.M., Sagiv, S.K., Paykin, A.B., Wang, Q., Hayes, S., Teitelbaum, S.L., Neugut, A.I. & Santella, R.M. (2004). Polymorphism in the DNA repair gene XPD, polycyclic aromatic hydrocarbon-DNA adducts, cigarette smoking, and breast cancer risk. *Cancer Epidemiol.Biomarkers Prev.*, Vol. 13, No. 12, pp. (2053-2058)

Tomkinson, A.E. & Levin, D.S. (1997). Mammalian DNA ligases. *Bioessays*, Vol. 19, No. 10, pp. (893-901)

van Maanen, J.M., Moonen, E.J., Maas, L.M., Kleinjans, J.C. & Van Schooten, F.J. (1994). Formation of aromatic DNA adducts in white blood cells in relation to urinary excretion of 1-hydroxypyrene during consumption of grilled meat. *Carcinogenesis*, Vol. 15, No. 10, pp. (2263-2268)

Viguier, J., Boige, V., Miquel, C., Pocard, M., Giraudeau, B., Sabourin, J.C., Ducreux, M., Sarasin & A., Praz, F. (2005). ERCC1 codon 118 polymorphism is a predictive factor for the tumor response to oxaliplatin/5-fluorouracil combination chemotherapy in patients with advanced colorectal cancer. *Clin.Cancer.Res.*, Vol. 11, pp. (6212-6217)

Vogel, U., Olsen, A., Wallin, H., Overvad, K., Tjonneland, A. & Nexo, B.A. (2005a). Effect of polymorphisms in XPD, RAI, ASE-1 and ERCC1 on the risk of basal cell carcinoma among Caucasians after age 50. *Cancer Detect.Prev.*, Vol. 29, No. 3, pp. (209-214)

Vogel, U., Overvad, K., Wallin, H., Tjonneland, A., Nexo, B.A. & Raaschou-Nielsen, O. (2005b). Combinations of polymorphisms in XPD, XPC and XPA in relation to risk of lung cancer. *Cancer Lett.*, Vol. 222, No. 1, pp. (67-74)

Volker, M., Mone, M.J., Karmakar, P., van, H.A., Schul, W., Vermeulen, W., Hoeijmakers, J.H., van, D.R., Van Zeeland, A.A. & Mullenders, L.H. (2001). Sequential assembly of the nucleotide excision repair factors in vivo. *Mol.Cell*, Vol. 8, No. 1, pp. (213-224)

Wakasugi, M. & Sancar, A. (1998). Assembly, subunit composition, and footprint of human DNA repair excision nuclease. *Proc.Natl.Acad.Sci.U.S.A*, Vol. 95, No. 12, pp. (6669-6674)

Wakasugi, M. & Sancar, A. (1999). Order of assembly of human DNA repair excision nuclease. *J.Biol.Chem.*, Vol. 274, No. 26, pp. (18759-18768)

Wang, J., Zhao, Y., Jiang, J., Gajalakshmi, V., Kuriki, K., Nakamura, S., Akasaka, S., Ishikawa, H., Suzuki, S., Nagaya, T. & Tokudome, S. (2010). Polymorphisms in DNA repair genes XRCC1, XRCC3 and XPD, and colorectal cancer risk: a case-control study in an Indian population. *J.Cancer Res.Clin.Oncol.*, Vol. 136, No. 10, pp. (1517-1525)

Weiss, J.M., Weiss, N.S., Ulrich, C.M., Doherty, J.A. & Chen, C. (2006). Nucleotide excision repair genotype and the incidence of endometrial cancer: effect of other risk factors on the association. *Gynecol.Oncol.*, Vol. 103, No. 3, pp. (891-896)

Weiss, J.M., Weiss, N.S., Ulrich, C.M., Doherty, J.A., Voigt, L.F. & Chen, C. (2005). Interindividual variation in nucleotide excision repair genes and risk of endometrial cancer. *Cancer Epidemiol.Biomarkers Prev.*, Vol. 14, No. 11 Pt 1, pp. (2524-2530)

Wrensch, M., Kelsey, K.T., Liu, M., Miike, R., Moghadassi, M., Sison, J.D., Aldape, K., McMillan, A., Wiemels, J. & Wiencke, J.K. (2005). ERCC1 and ERCC2 polymorphisms and adult glioma. *Neuro.Oncol.*, Vol. 7, No. 4, pp. (495-507)

Wu, X., Gu, J., Grossman, H.B., Amos, C.I., Etzel, C., Huang, M., Zhang, Q., Millikan, R.E., Lerner, S., Dinney, C.P. & Spitz, M.R. (2006). Bladder cancer predisposition: a multigenic approach to DNA-repair and cell-cycle-control genes. *Am.J.Hum.Genet.*, Vol. 78, No. 3, pp. (464-479)

Wu, X., Zhao, H., Wei, Q., Amos, C.I., Zhang, K., Guo, Z., Qiao, Y., Hong, W.K. & Spitz, M.R. (2003). XPA polymorphism associated with reduced lung cancer risk and a modulating effect on nucleotide excision repair capacity. *Carcinogenesis*, Vol. 24, No. 3, pp. (505-509)

Xing, D., Tan, W., Wei, Q. & Lin, D. (2002). Polymorphisms of the DNA repair gene XPD and risk of lung cancer in a Chinese population. *Lung Cancer*, Vol. 38, No. 2, pp. (123-129)

Yang, K., Fang, J.L. & Hemminki, K. (1998). Abundant lipophilic DNA adducts in human tissues. *Mutat Res*, Vol. 422, No. 2, pp. (285-295)

Yeh, C.C., Sung, F.C., Tang, R., Chang-Chieh, C.R. & Hsieh, L.L. (2005). Polymorphisms of the XRCC1, XRCC3, & XPD genes, and colorectal cancer risk: a case-control study in Taiwan. *BMC.Cancer*, Vol. 5, pp. (12)

Yeh, C.C., Sung, F.C., Tang, R., Chang-Chieh, C.R. & Hsieh, L.L. (2007). Association between polymorphisms of biotransformation and DNA-repair genes and risk of colorectal cancer in Taiwan. *J.Biomed.Sci.*, Vol. 14, No. 2, pp. (183-193)

Yin, J., Vogel, U., Ma, Y., Guo, L., Wang, H. & Qi, R. (2006). Polymorphism of the DNA repair gene ERCC2 Lys751Gln and risk of lung cancer in a northeastern Chinese population. *Cancer Genet.Cytogenet.*, Vol. 169, No. 1, pp. (27-32)

Yokoi, M., Masutani, C., Maekawa, T., Sugasawa, K., Ohkuma, Y. & Hanaoka, F. (2000). The xeroderma pigmentosum group C protein complex XPC-HR23B plays an important role in the recruitment of transcription factor IIH to damaged DNA. *J.Biol.Chem.*, Vol. 275, No. 13, pp. (9870-9875)

You, J.S., Wang, M. & Lee, S.H. (2003). Biochemical analysis of the damage recognition process in nucleotide excision repair. *J.Biol.Chem.*, Vol. 278, No. 9, pp. (7476-7485)

Zienolddiny, S., Campa, D., Lind, H., Ryberg, D., Skaug, V., Stangeland, L., Phillips, D.H., Canzian, F. & Haugen, A. (2006). Polymorphisms of DNA repair genes and risk of non-small cell lung cancer. *Carcinogenesis*, Vol. 27, No. 3, pp. (560-567)

# 7

# Low Penetrance Genetic Variations in DNA Repair Genes and Cancer Susceptibility

Ravindran Ankathil
*Human Genome Center, School of Medical Sciences*
*University Sains Malaysia, Health Campus*
*Kubang Kerian, Kelantan*
*Malaysia*

## 1. Introduction

The genetic material, DNA, which encodes genes needed for the production of essential proteins, is vulnerable to damage in a number of ways. Human DNA is assaulted on a daily basis by a variety of exogenous factors including UV light, cigarette smoke, dietary factors, and other carcinogens all of which can cause varying degrees of DNA damage and can lead to mutations. Similarly, endogenous factors such as undue DNA replication which can cause mismatches, hydrolysis leading to spontaneous DNA depurination, replication form collapse which can result in strand breaks, loss of bases because of spontaneous disintegration of chemical bonds, and DNA damage secondary to endogenous reactants such as alkyl groups, metal cations, and reactive oxygen species (ROS) which can induce base oxidation and DNA breaks also contribute to DNA damage (Branzei & Foiani, 2008; Capella et al., 2008)

When DNA is damaged, an intertwined network of surveillance mechanisms will act including

- Sensing and recognizing DNA damage by activation of cell cycle checkpoints, pause that permit assessment and complete of DNA processing , either DNA damage repair or processing of DNA intermediates
- Up regulation a large number of genes
- Programmed cell death or apoptosis when the cell is unable to repair the damage sustained and
- Elicitation of multiple distinct DNA repair responses

DNA damages are repaired by enzymes coded by one or more DNA repair pathways according to their structure, or their location in the cellular genome. DNA repair enzymes can be characterized as cellular proteins acting directly on damaged DNA in an attempt to restore the correct DNA sequence and structure. These relatively specialized enzymes appear to undertake the initial stages of recognition and repair of specific forms of DNA damage. Since there are various kinds of DNA damage, a variety of repair mechanisms are essential. Cells integrate DNA repair process with transcription and apoptosis through a network known as the DNA damage response (DDR) which is orchestrated by the checkpoint proteins.

## 2. DNA repair pathways

Damages in DNA are repaired by various DNA repair genes belonging to distinct pathways. Each pathway is recognized for efficient repair of specific types of DNA damage. To date, more than 150 human DNA repair genes have been identified, which can be categorized into at least 5 distinct pathways: Base Excision Repair (BER), Nucleotide Excision Repair (NER), Mismatch Repair (MMR), Double Strand Break Repair (DSBR), and Transcription Coupled Repair (TCR) (Wood et al., 2005). The Base Excision Repair Pathway operates on small lesions such as oxidized or reduced bases, fragmented or nonbulky adduct, and adducts produced by methylating agents. The Nucleotide Excision repair (NER) pathway repairs bulky lesions such as pyramidine dimmers, other products of phytochemical reactions, large chemical adducts and DNA crosslinks. For Double Strand Break Repair (DSBR), at least two pathways exist: homologous recombination and non homologous end joining. Replication errors such as base-base or insert-deletion mismatches caused by the DNA polymerase are repaired by Mismatch Repair (MMR) pathway genes. Finally, the suicide enzyme methyl-guanine-DNA, methyl transferase, is an additional category of DNA repair pathway that directly removes the alkylated bases.

The repair gene products operate in a co-ordinated fashion to form repair pathways that control restitution of specific types of DNA damage. Repair pathways are further co-ordinated with other metabolic processes, such as cell cycle control, to optimize the prospects of successful repair. During the cell cycle, checkpoint mechanisms ensure that a cell's DNA is intact before permitting DNA replication and cell division to occur. Failures in these checkpoints can lead to an accumulation of damage, which in turn leads to mutations.

Repair of damaged DNA is of paramount importance and is essential to prevent loss of or in correct transmission of genetic information, to prevent genetic damage from propagating and accumulating, to maintain genome integrity and stability of cells, and also to prevent mutations The failure of the cell to adequately repair the acquired damage and to undergo apoptosis may lead to further errors which can cause developmental abnormalities and neoplastic transformation of the cell and finally to carcinogenesis.

## 3. Genetic susceptibility to cancer - High and low penetrance DNA repair genes

Genetic susceptibility to cancer result from variations in the genetic code that alter either protein expression, function or localization. Susceptibility to cancer is determined by two types of genes – high penetrance genes and low penetrance genes. High penetrance genes are genes with allelic variants that confer a high degree of risk to the individual. Relatively few individuals in the population carry risk –increasing genotypes at these loci. The proportion of cancer in the population that may be explained by these genotypes will be low. Therefore the population attributable risk also will be low. But high penetrance genes have a large magnitude of effects on cancer risk and usually follow a mendelian autosomal domiant pattern of inheritance and involve multiple cancer sites that form a cancer syndrome. High penetrance genes with an attendant high likelihood of causing cancer, account for only a small proportion of cancer cases. In humans, high penetrance DNA repair genes that cause family or hereditary cancer syndromes can have substantial impact in affected families (eg: BRCA1 and BRCA2 genes in hereditary breast cancer, DNA Mismatch Repair (MMR) genes in Hereditary Non Polyposis Colorectal Cancer (HNPCC), p53 in

LiFraumeni Syndrome). But these genes affect only a small portion of cancer cases and a small percent age of the population.

Loss of function mutations in a significant number of DNA damage response genes predispose to a variety of familial cancers (Spry et al., 2007). There are several examples such as mutations in BRCA1 and BRCA2 belonging to homologous recombination pathways predispose to breast and ovarian cancer (Bertwistle & Ashworth, 2000). So also, mutations in other double strand break repair genes such as ATM predispose to the familial tumorigenic condition ataxia telangiectasia (Lavin & Shiloh 1996) and breast cancer (Renwick et al., 2006). Mutations in NBSI have been reported to predispose to Nijwegen breakage syndrome (Matsuura et al., 2004). Somatic mutations in another DSB repair gebe, ATR, correlate with sporadic microsatellite (MSI) positive stomach cancer (Menoyo et al., 2001). PALB2 gene, which encodes a BRCA2 indicating protein, has also been identified as a breast cancer susceptibility gene (Rahman et al., 2007). Mutations in a group of DNA mismatch repair (MMR) gene predispose to hereditary non-poplyposis colorectal cancer and other cancers in the extra colonic sites in Lynch syndrome (Jacob & Praz, 2002). Biallelic germline mutations of the base excision repair gene MUTYH have been identified in patients with autosomal recessive form of hereditary multiple colorectal adenoma and carcinoma (Jones et al., 2002). Defect in the nucleotide excision repair pathway genes predispose to xeroderma pimentosum (XP), Cockayne syndrome (CS) and Trichothiodystrophy (TTD), which are all autosomal recessive syndromes (Leibeling et al., 2006).

In contrast, in the remaining major portion of sporadic cases, genetic variations in the form of low to moderate penetrance alleles may predispose individuals to cancer in combination with environmental factors and thus affect a large segment of the population. Low penetrance genes, also referred to as modifier genes, are genes in which subtle sequence variants may be associated with a small to moderate increased relative risk for sporadic cancers.

## 4. Single Nucleotide Polymorohisms (SNPs) in low penetrance genes

Genetic variations seen in human genome includes insertion/deletion of one or more nucleotides (indels) the copy number variations (CNNs) that can involve DNA sequences of a few kilobases up to millions of bases and single nucleotide polymorphisms (SNPs) which are the substitution of a single nucleotide along the DNA. With an estimated number of more than 10 million to be present in the human genome, SNPs are the most common form of genetic variation (Miller et al., 2005).

Variations in several classes of low penetrance genes known as Single nucleotide polymorphism (SNPs) are very common in the population. SNPs are DNA sequence variations that occur when a single nucleotide (A, T, C, or G) in the genome sequence is altered. For a variation to be considered as SNP, it must be present in at least 1% of the population. SNPs are relatively common in the population and as such may be associated with a much higher attributable risk in the population as a whole than the rare high penetrance genes. Therefore, variants in low penetrance genes could explain a greater proportion of sporadic cancers than the high penetrance genes. SNPs acting together with environmental factors are well documented candidates for cancer susceptibility. Even though, SNPs in these low penetrance genes have only small effect when considered singly, they may produce a high risk profile when acting together with other shared genetic variants and environmental factors (Gary et al., 1999). On the basis of biological plausibility,

SNPs in low penetrance genes whose protein products would affect a pathway involved in carcinogenesis have been documented as cancer predisposition or susceptibility risk factors. Low penetrance candidates are found in a wide variety of pathways ranging from metabolism and detoxification or environmental carcinogens to DNA damage repair.

The recognition that carcinogens can also be mutagens that change the DNA sequence gave impetus to the relevance of DNA damage and repair to carcinogenesis. All the effects of exogenous factors and endogenous factors on tumor production could be accounted for by the DNA damage that they cause and by the errors introduced into DNA during the cell's efforts to repair this damage. According to the mutator phenotype hypothesis, cancer phenotypes result from mutations in genes that maintain genetic stability in normal cells. Mutations in genetic stability genes can cause mutations in other genes that govern genetic stability, initiating a cascade of mutations throughout the genome. So, the prompt response of the cells to repair genetic injury and its ability to maintain genomic stability by means of a variety of DNA repair mechanisms are therefore essential in preventing tumor initiation and progression.

Genetic variants or mutations in high penetrance genes are disease causing whereas genetic variations in low penetrance genes are insufficient to cause cancer, but may influence cancer risk. So genetic variants in low penetrance genes are disease risk associated. Individual low penetrance risk alleles are insufficient to cause cancer, but influence cancer risk. Low penetrance genes, with an attendant increased risk of causing cancer, albeit, less likely than high penetrance genes (Ponder, 2001; Shields & Harris, 2000) predispose individuals to cancer upon interacting with environmental factors.

## 5. SNPs in DNA damage repair genes

DNA repair mechanisms are controlled by specific set of genes encoding the enzymes that catalyze cellular response to DNA damage. It is well documented that loss of repair function, or alteration of the control of repair process, can have very serious consequences for cells and individuals and can lead to development of cancer. Several genes involved in DNA repair pathways are considered to be low penetrance genes. A link between failure of DNA repair and carcinogens was suggested when individuals with chromosome breakage syndrome such as Xeroderma Pigmentosum, Fanconi Anemia, Bloom Syndrome, Ataxia telangiectasia who have inherited genetic defects in certain DNA repair systems were recognized to be at an increased susceptibility to development of certain cancers.

Because DNA damage is associated with cancer development, it was hypothesized that genes involved in DNA damage repair may influence cancer susceptibility. Polymorphisms in DNA repair genes may be associated with differences in the DRC of DNA damage and may influence an individual's risk for cancer, because the variant genotype in those polymorphisms might destroy or alter repair function. A large number of SNPs have been determined among individuals in DNA repair genes. It has been documented that genetic variations in DNA damage repair genes could result in variations in efficacy and accuracy of DNA repair enzymes and could have effect on the sensitivity of the organism to environmental genotoxins.

Genetic variation in DNA repair genes in each of the five DNA repair pathways has been implicated in cancer susceptibility (Berwick & Vineis 2000; Goode et al., 2002). Genetic variations such as SNPs in DNA repair genes are associated with reduced function of their encoded proteins, rather than absence of function and may alter an individual's capability to repair damaged DNA. This may result in gene product (protein) not being formed, or that the protein is less active, or that it is formed in an uncontrolled fashion, may be at the wrong time, or in the wrong amount. Some minor genetic alterations may not affect protein activity, or interactions, whereas others may significantly disrupt cellular function. It is also possible that since certain proteins work in a number of different processes or complexes, the loss or impairment of one type of protein can affect several different functions of the cell and organism. Deficiency or impairment in DNA repair genes which results in alteration of the key gene expression may have an influence on DNA repairs functions and could lead to altered cancer risk. The importance of these mechanisms in cancer prevention is evident from the increased cancer risk associated with disruption of these pathways (Digweed, 2003). So studies on DNA repair as a susceptibility factor for cancer are increasing exponentially. Majority of cancer susceptibility studies have focused on the identification of low-penetrance disease susceptibility alleles applying candidate gene pathway studies and genome wide association studies. Genetic association studies and genome wide association scans have identified a number of polymorphisms in several low penetrance genes and their role in etiology of several cancers, through risk modification (Tomlinson et al., 2008).

DNA double strand breaks (DSBs) which can result from a variety of factors including ionizing radiation, free radicals, replication errors, telomere dysfunction are one of the most severe types of DNA damage (Khanna & Jackson, 2001). Unpaired or misrepaired DSBs can lead to cell death, genomic instability and oncogenic transformation (Jeggo & Jackson, 2001). Homologous recombination (HR) and nonhomologous enjoining (NHEJ) are the two major DSB repair pathways in mammalian cells. Reports are available suggesting that several Single Nucleotide Polymorphisms in the NHEJ genes may be relevant to modify the risk of multiple myloma (Roddam et al., 2002), glioma (Liu et al., 2008) and, breast cancer (Garcia-Closas et al., 2006). Another study by (Tseng et al., 2009), showed significant association between the XRCC4 and LIGH genotypes with non-small cell lung cancer (NSCLC) risk in an analysis of individual polymorphism associations, and the risk of NSCLC increased further in a combined analysis of multiple polymorphisms.

# 6. XRCC3

The X-ray repair cross-complementing group 3 (XRCC3), the DNA repair gene which codes for a protein participating in homologous recombination repair (HRR) of double strand breaks (DSB), has been of considerable interest as a candidate gene for cancer susceptibility. The variant allele of the Thr241Met had been reported to have relatively high DNA adduct levels in lymphocyte DNA and hence with relatively low DNA repair capacity (Matullo et al., 2001). Several molecular epidemiologic studies have been performed to evaluate the role of XRCC3 polymorphisms such as XRCC3 4541 A>G, XRCC3 17893 A>G, XRCC3 Thr 241 Met on various neoplasms, such as cancer of breast, lung, bladder, colorectal, head and neck, skin etc (Han et al., 2004; Shen et al., 2004; Ritchey et al., 2005; Jin et al., 2005; Matullo et al.,2005; Garcia-Closas et al., 2006; Zienoldding et al., 2006; Yi et al., 2006). But rather than conclusive, the results from these studies remain fairly conflicting. Ahmd Aizat (2011) reported lack of association of XRCC3 Thr 241 Met with sporadic colorectal cancer susceptibility in Malaysian population. (Han et al., 2006) performed a meta –analysis on XRCC3 polymorphism and cancer risk involving 48 case-control studies including 24,975 cancer patients and 34,209 controls. From the analysis results, (Han et al., 2006) reported that individuals carrying the XRCC3 Met/Met genotype showed a. small cancer risk under a recessive genetic model. Specifically, the XRCC3 Met/Met genotype showed significantly increased risk of breast cancer, but not significant risk of cancer for head and neck, bladder, and non-melanoma skin cancer. This meta analysis results support that the XRCC3 might represent a low penetrance susceptible gene especially for cancer of breast, bladder, head and neck, and non-melanoma skin cancer.

## 7. XRCC1

The X-Ray Cross Complementing group I XRCC1 gene belongs to The Base Excision Repair IBER) pathway. The XRCC1 gene product plays an important role in the BER pathway by acting as a scolfold for the other DNA repair proteins, such as DNA polymerase B (Kubota et al., 1996), and DNA ligase III (Caldecott, 2003). Few common single nucleotide polymoprhisms of the XRCC1 gene have been identified at codon 194 (G>T substitution at position 26304, exon 6, Arg to Trp), codon 280 (G>A substitution at position 27466, exon 9, Arg to His) and 399 (G>A substitution at position 28152, exon 10, Arg to Gln). The individuals carrying XRCC1 399 variants have been shown to have higher levels of DNA adduct (Lunn et al., 2000) and to be at greater risk for tobacco related DNA damage (Lei et al., 2002). Few studies reported XRCC1 399AA genotype to be significantly associated with lung cancer risk in Caucasian population (Divine et al., 2001; Zhou et al., 2003), Korean population (Park et al., 2002), and Indian population (Sreeja et al., 2008).

## 8. XPD (ERCC2)

Xeroderma Pigmentosum group D (XPD) also known as ERCC2 (Excision Repair Cross Completing group 2) gene encodes a helicase, a major DNA repair protein, which is involved in transcription-coupled NER and in the removal of a variety of structurally unrelatedas DNA lesions (Lehmann, 2001) including those induced by tobacco carcinogens (Leadon & Cooper, 1993), (Tang et al., 2002). The normal functioning XPD protein plays an essential role in NER and participates in the unwinding of DNA at the site of deleterious DNA lesions (Hoeijmakers et al., 1996). Several studies have reported association between A751C variant of XPD and increased risk of lung cancer (Hou et al., 2002; Spitz et al 2003; Ramachandran et al., 2006). Hou et al., 2003 reported a marginally increased risk for those carrying heterozygous A>C transversions, compared to those with wildtype homozygous, indicating that heterozygosity also carry the risk. In a Northeastern Chinese population, XPD 751 AC heterozygous genotype carriers were at 2.7 fold higher risk of lung cancer than carrier of AA genotype (Yin et al., 2006). A significant association of XPD variants in modulating NSCLC risk was reported by Zienolddiny et al (2006) in Norwegian lung cancer population. So also, in an Indian population, Sreeja et al., (2008) also reported significant association of XPD heterozygous variants in modulating Non small cell lung cancer risk. SNPs in genes involved in nucleotide excision repair (ERCC1, XPD, XPC, XPA, XPF and XPG) and mismatch repair genes (MLH1 and MSH2) in 577 colorectal cancer cases and 307 case-affected sibling controls were examined by Joshi et al., (2009). Their results showed that consumption of red meat, heavily brown on the outside or inside, increased colorectal cancer risk only among subjects with XPD codon 751 Lys/Lys or XPD codon 312 Asp/Asp genotypes.

## 9. P53

The P53 gene plays a critical role in cell cycle control, the initiation of apoptosis, and maintenance of genomic stability and in DNA repair (Levine, 1997). TP53 is highly polymorphic in coding and non coding regions and some of these polymorphisms have been shown to increase cancer susceptibility and modify cancer phenotypes in TP53 mutation carriers (Whibley et al., 2009). Over 80 TP53 polymorphisms have been identified and validated in human populations. (IARC TP53 Database, R13). Nearly 90% are located in introns, outside splice sites, or in non coding exons. Among the P53 polymorphism, the codon 72 polymorphism (Arg72 Pro) in exon 4 of TP53 is the most extensively studied, both in experimental and population studies. Codon 72 is located within a proline rich region and Arg72 has been reported to be more effective in inducing apoptosis than Pro72. The Arg/Pro polymorphism at codon 72 of the P53 gene alters the ability of the P53 protein to induce apoptosis, influences the behaviour of mutant P53, decreases the DNA repair capacity and has been linked to with an increased risk of cancer, especially lung cancer. Several studies have examined the associations between P53 codon 72 (Arg72Pro) polymorphism and risk of different cancers, but with in consistent results. Few studies reported higher risk for lung cancer in individuals with the Arg/Pro or Pro/Pro genotype and especially Pro/Pro genotype with smoking induced lung cancer (Weston et al., 1992; Jin et al., 1995 ; Fan et al., 2000; Zhou et al., 2001). In a Chilean population, Fan et al., (2000) investigated the influence of polymorphic genotype TP53 on lung cancer susceptibility and the Pro/Pro genotype of TP53 was found to contribute significantly to lung cancer susceptibility risk [ OR 3.88 (95% CI 1.16 – 13.39) ]. The study by Alexandrov et al (2002) were consistent with the hypothesis that Benzo(a)pyrene (Polycyclic aromatic hydrocarbon, PAH) induce G : C to T : A transverse mutations in the hotspot codons of TP53 and are hence involved in the malignant transformation of the lung tissue of smokers. In an Indian case – control study involving 211 lung cancer cases and 211 controls, Sreeja et al (2007) reported an OR of 2.5 (95% CI 1.470 4.302 , p= 0.001) for the TP53 Pro/Pro variant genotype for lung cancer susceptibility and the risk tended to be higher for women [ OR =2.4 , p=0.003 ] . Recently, Ahmd Aizat (2011) reported a significant association of Pro/Pro homozygous variant of p53 with sporadic colorectal cancer susceptibility (OR = 1.886, CI: 1,046 – 3.399 , p= 0,035) and suggested that p53 Pro72Pro genotype carriers might be having a higher risk for Colorectal cancer susceptibility in Malaysian population (personal communication, unpublished data) However, meta-analysis on the risk association of TP53 Arg72Pro polymorphism with lung cancer (Matakidou et al., 2003) and breast cancer (Schmidt et al., 2007) do not support a significant role for this polymorphism in susceptibility.

## 10. MMR genes

In the maintenance of genomic stability, the DNA mismatch repair (MMR) systems comprising of various MMR genes play a key role. MMR genes mediate DNA repair through removal of mismatched nucleotide pairs and insertion/ deletion heterologies generated during DNA replication. Germ line mutations as well as hypermethylation in MMR genes have been reported in familial/hereditary forms of colorectal cancer. So, it was hypothesized that common variants in relevant genes encoding DNA MMR enzymes might impact the risk of sporadic form of CRC and studies have been carried out to explore this possibility. Even though the functional relevance of majority of polymorphisms in the genes involved in MMR is not known, recent studies suggest an influence of SNPs or biochemical interaction between components of the MMR pathways or on epigenetic mediated functional regulation (Chen et al., 2007).

Several common polymorphisms in DNA repair genes representing different repair pathways have been reported. Many studies have been carried out to elucidate the association between DNA repair gene polymorphisms and cancer susceptibility. But studies have shown inconsistent associations. The impact of many these polymorphisms on repair phenotype and cancer susceptibility remain uncertain (Berwick & Vineis, 2000; Au et al., 2004). In a study on 5 DNA repair genes (XRCC1 Arg194Trp and Arg399gln, PARP Val762Ala and Lys940Arg, XPD Asp312Asn and Lys751Gln, OGG1 Ser326Cys MGMT Leu84Phe) in Singaporean Chinese population, Stern et al., (2007) provided support to the hypothesis that selected variants in DNA repair genes may contribute to colorectal cancer risk and may modify the effects of relevant life style risk factors that have been inconsistently associated with the disease. This study which reported the overall effects of PARP on colorectal cancer risk and XRCC1 SNPs as modifiers of the effects of smoking and alcohol on colorectal cancer risk, also highlighted the role of the base excision repair pathway in colorectal carcinogenesis. Vinies et al (2009) conducted meta-analyses of 241 associations between variants in DNA repair genes and cancer and had found sparse association signals with strong epidemiological credibility. Using 1087 datasets and publicly available data from genome wide association platforms, meta-analysis using dominant and recessive models were performed on 241 associations between individual variants and specific cancer types that had been tested in two or more independent studies. Thirty one nominally statistically significant (P<0.05 without adjustment for multiple comparisons) associations were recorded for 16 genes in dominant and/or recessive model analyses (BRCA2, CCND1, ERCC1, ERCC2, ERCC4, ERCC5, MGMT, NBN, PARP1, POL1, TP53, XPA, XRCC1, XRCC2, XRCC3 and XRCC4). XRCC1, XRCC2, TP53, and ERCC2 variants were each nominally associated with several types of cancer. Three associations were graded as having "strong" credibility, another four had "modest" credibility and 24 had "weak" credibility based on Vinies criteria. Requiring more stringent P values to account for multiplicity of comparisons, only the associations of ERCC2 codon 751 (recessive model) and of XRCC1-77 T>C (dominant-model) with lung cancer had P≤ 0.0001 and retained P≤ 0.001 even when the first published studies on the respective associations were excluded. The analyses suggested that the vast majority of postulated associations between DNA repair alleles and cancer risk have not been replicated sufficiently to give them strong credibility. This meta-analysis implies that larger scale studies would be necessary to establish specific associations of genetic variants in DNA repair and cancer and that the added risk conferred by single variants in DNA repair genes may be small. In another recent meta analysis, (Kiyohara et al.) found XPA G23A, OGG1 Ser326Cys and XPD Lys751Gln polymorphisms were associated with lung cancer risk .

## 11. Limitations and future prospectives

In SNP association studies, the most important critical point is associated with often too small size of cohort of cases and controls, resulting in a low statistical power and false, by chance, positive or negative outcomes. Another important aspect concerns inclusion of different ethnic groups. Different results may be expected due to intrinsic difference in genetic background among Caucasians, Asians, Afro Americans and other ethnic groups.

There is wide population variability in repair capability phenotype on account of the variation in the polymorphic allele frequencies of DNA repair genes between different ethnic groups. Even through majority of SNPs are common to at least three historic human populations (Caucasians, Africans, Asians), some SNPs are specific to different ethnicities. These differences among human populations has highlighted the need to consider ethnic genetic differences while conducting genetic association studies evaluating disease risk, treatment response and outcome studies. This could also be accounting for a several fold variation in cancer risk and significant heterogeneity across all included studies. So susceptibility factor in one population may not be a factor in another population. Thus, different study designs, differences in the prevalence of genetic polymorphisms and linkage disequilibrium in different ethnic populations are possible explanations for the varying results obtained in different studies across the world. Effect modification by environmental or other genetic risk factors that differ between study populations are also alternative causes. This warrants the need to undertake large studies on homogeneous populations to avoid such influences. .

It is hoped that in future, advances in genotyping utilizing high throughput genotyping methods could facilitate the analysis of multiple polymorphisms within DNA repair genes and also the analysis of multiple genes within DNA repair pathways. Data generated from multiple polymorphisms within a gene can be combined to create haplotypes , the set of multiple alleles on a single chromosome. Because of higher heterozygosity and tighter linkage disequilibrium within disease causing mutations, haplotype analysis can increase the power to detect disease associations. Haplotype analysis also allows for the possibility of an ungenotyped functional variant to be in linkage disequilibrium with the genotyped polymorphisms. Investigations on gene-gene interactions or pathway analysis also would provide more comprehensive insight into the role of low penetrance genetic variants of DNA repair genes in cancer susceptibility. The identification of common, moderate or low penetrance genes for cancer will potentially be of great benefit , because it allows screening to be targeted to those at greatest risk which in turn will help in implementing cancer prevention strategies.

## 12. Acknowledgements

The author would like to express his sincere thanks and gratitude to Mr. Ahmad Aizat and Mr. Au Zian Liang, postgraduate students in Human Genome Center, University Sains Malaysia, Malaysia for their valuable help in word processing and reference collection related to this work.

## 13. References

Ahmad Aizat AA, Siti Nurfatimah MS, Aminudin MB, Biswa BM, Venkatesh RN, Zaidi Z, Ahmad Shanwani MS, Ankathil R. Polymorphism Thr241Met of the XRCC3 Gene and Lack of Association with Colorectal Cancer Susceptibility Risk among Malaysian Population –A Preliminary Report (Accepted to publish in International Medical Journal in next issue)

Ahmad Aizat AA , Siti Nurfatimah MS, Aminudin MM, Zaidi Z , Ahmad Shanwani MS , Muhammad Radzi AH , Ankathil R . Increased risk of sporadic Colorectal Cancer in Pro/Pro 72 genotype carriers of TP53 Arg72Pro polymorphism- A Malaysian case control study (In press)

Alexandrov, K. Cascorbi, I. Rojas, M. Bouvier, G. Kriek, E. & Bartsch, H. (2002). CYP1A1 and GSTM1 genotypes affect benzo[a]pyrene DNA adducts in smokers' lung: comparison with aromatic/hydrophobic adduct formation. Carcinogenesis. 23(12): 1969-1977.

Au, W. W., Navasumrit, P. & Ruchirawat, M. (2004) Use of biomarkers to characterize functions of polymorphic DNA repair genotypes. *Int J Hyg Environ Health,* 207, 301-13.

Bertwistle, D. & Ashworth, A. (2000) BRCA1 and BRCA2. *Curr Biol,* 10, R582.

Berwick, M. & Vineis, P. (2000) Markers of DNA repair and susceptibility to cancer in humans: an epidemiologic review. *J Natl Cancer Inst,* 92, 874-97.

Branzei, D. & Foiani, M. (2008) Regulation of DNA repair throughout the cell cycle. *Nat Rev Mol Cell Biol,* 9, 297-308.

Caldecott, K. W. (2003) XRCC1 and DNA strand break repair. *DNA Repair (Amst),* 2, 955-69.

Chen, H., Taylor, N. P., Sotamaa, K. M., Mutch, D. G., Powell, M. A., Schmidt, A. P., Feng, S., Hampel, H. L., De La Chapelle, A. & Goodfellow, P. J. (2007) Evidence for heritable predisposition to epigenetic silencing of MLH1. *Int J Cancer,* 120, 1684-8.

Digweed, M. (2003) Response to environmental carcinogens in DNA-repair-deficient disorders. *Toxicology,* 193, 111-24.

Divine, K. K., Gilliland, F. D., Crowell, R. E., Stidley, C. A., Bocklage, T. J., Cook, D. L. & Belinsky, S. A. (2001) The XRCC1 399 glutamine allele is a risk factor for adenocarcinoma of the lung. *Mutat Res,* 461, 273-8.

Fan, R., Wu, M. T., Miller, D., Wain, J. C., Kelsey, K. T., Wiencke, J. K. & Christiani, D. C. (2000) The p53 codon 72 polymorphism and lung cancer risk. *Cancer Epidemiol Biomarkers Prev,* 9, 1037-42.

Garcia-Closas, M., Egan, K. M., Newcomb, P. A., Brinton, L. A., Titus-Ernstoff, L., Chanock, S., Welch, R., Lissowska, J., Peplonska, B., Szeszenia-Dabrowska, N., Zatonski, W., Bardin-Mikolajczak, A. & Struewing, J. P. (2006) Polymorphisms in DNA double-strand break repair genes and risk of breast cancer: two population-based studies in USA and Poland, and meta-analyses. *Hum Genet,* 119, 376-88.

Gary, R., Park, M. S., Nolan, J. P., Cornelius, H. L., Kozyreva, O. G., Tran, H. T., Lobachev, K. S., Resnick, M. A. & Gordenin, D. A. (1999) A novel role in DNA metabolism for the binding of Fen1/Rad27 to PCNA and implications for genetic risk. *Mol Cell Biol,* 19, 5373-82.

Gilad, S., Bar-Shira, A., Harnik, R., Shkedy, D., Ziv, Y., Khosravi, R., Brown, K., Vanagaite, L., Xu, G., Frydman, M., Lavin, M. F., Hill, D., Tagle, D. A. & Shiloh, Y. (1996) Ataxia-telangiectasia: founder effect among north African Jews. *Hum Mol Genet,* 5, 2033-7.

Goode, E. L., Ulrich, C. M. & Potter, J. D. (2002) Polymorphisms in DNA repair genes and associations with cancer risk. *Cancer Epidemiol Biomarkers Prev,* 11, 1513-30.

Han, J., Colditz, G. A., Samson, L. D. & Hunter, D. J. (2004) Polymorphisms in DNA double-strand break repair genes and skin cancer risk. *Cancer Res,* 64, 3009-13.

Han, S., Zhang, H. T., Wang, Z., Xie, Y., Tang, R., Mao, Y. & Li, Y. (2006) DNA repair gene XRCC3 polymorphisms and cancer risk: a meta-analysis of 48 case-control studies. *Eur J Hum Genet,* 14, 1136-44.

Hoeijmakers, J. H., Egly, J. M. & Vermeulen, W. (1996) TFIIH: a key component in multiple DNA transactions. *Curr Opin Genet Dev, 6*, 26-33.

Hou, S. M., Falt, S., Angelini, S., Yang, K., Nyberg, F., Lambert, B. & Hemminki, K. (2002) The XPD variant alleles are associated with increased aromatic DNA adduct level and lung cancer risk. *Carcinogenesis, 23*, 599-603.

Hou, S. M., Ryk, C., Kannio, A., Angelini, S., Falt, S., Nyberg, F. & Husgafvel-Pursiainen, K. (2003) Influence of common XPD and XRCC1 variant alleles on p53 mutations in lung tumors. *Environ Mol Mutagen, 41*, 37-42.

Jacob, S. & Praz, F. (2002) DNA mismatch repair defects: role in colorectal carcinogenesis. *Biochimie, 84*, 27-47.

Jeggo, P. A. & Lobrich, M. (2006) Contribution of DNA repair and cell cycle checkpoint arrest to the maintenance of genomic stability. *DNA Repair (Amst), 5*, 1192-8.

Jin, X., Wu, X., Roth, J. A., Amos, C. I., King, T. M., Branch, C., Honn, S. E. & Spitz, M. R. (1995) Higher lung cancer risk for younger African-Americans with the Pro/Pro p53 genotype. *Carcinogenesis, 16*, 2205-8.

Jones, S., Emmerson, P., Maynard, J., Best, J. M., Jordan, S., Williams, G. T., Sampson, J. R. & Cheadle, J. P. (2002) Biallelic germline mutations in MYH predispose to multiple colorectal adenoma and somatic G:C-->T:A mutations. *Hum Mol Genet, 11*, 2961-7.

Joshi, A. D., Corral, R., Siegmund, K. D., Haile, R. W., Le Marchand, L., Martinez, M. E., Ahnen, D. J., Sandler, R. S., Lance, P. & Stern, M. C. (2009) Red meat and poultry intake, polymorphisms in the nucleotide excision repair and mismatch repair pathways and colorectal cancer risk. *Carcinogenesis, 30*, 472-9.

Khanna, K. K. & Jackson, S. P. (2001) DNA double-strand breaks: signaling, repair and the cancer connection. *Nat Genet, 27*, 247-54.

Kiyohara, C., Takayama, K. & Nakanishi, Y. (2010) Lung cancer risk and genetic polymorphisms in DNA repair pathways: a meta-analysis. *J Nucleic Acids, ,* 701-760.

Kubota, Y., Nash, R. A., Klungland, A., Schar, P., Barnes, D. E. & Lindahl, T. (1996) Reconstitution of DNA base excision-repair with purified human proteins: interaction between DNA polymerase beta and the XRCC1 protein. *EMBO J, 15*, 6662-70.

Lavin, M. F. & Shiloh, Y. (1996). Ataxia telangiectasia: a multifaceted genetic disorder associated with defective signal transduction. Curr Opin immunol. 8: 459-464.

Leadon, S. A. & Cooper, P. K. (1993) Preferential repair of ionizing radiation-induced damage in the transcribed strand of an active human gene is defective in Cockayne syndrome. *Proc Natl Acad Sci U S A, 90*, 10499-503.

Lehmann, A. R. (2001) The xeroderma pigmentosum group D (XPD) gene: one gene, two functions, three diseases. *Genes Dev, 15*, 15-23.

Lei, Y. C., Hwang, S. J., Chang, C. C., Kuo, H. W., Luo, J. C., Chang, M. J. & Cheng, T. J. (2002) Effects on sister chromatid exchange frequency of polymorphisms in DNA repair gene XRCC1 in smokers. *Mutat Res, 519*, 93-101.

Leibeling, D., Laspe, P. & Emmert, S. (2006) Nucleotide excision repair and cancer. *J Mol Histol, 37*, 225-38.

Levine, A. J. (1997) p53, the cellular gatekeeper for growth and division. *Cell, 88*, 323-31.

Liu, Y., Zhou, K., Zhang, H., Shugart, Y. Y., Chen, L., Xu, Z., Zhong, Y., Liu, H., Jin, L., Wei, Q., Huang, F., Lu, D. & Zhou, L. (2008) Polymorphisms of LIG4 and XRCC4 involved in the NHEJ pathway interact to modify risk of glioma. *Hum Mutat,* 29, 381-9.

Lunn, R. M., Helzlsouer, K. J., Parshad, R., Umbach, D. M., Harris, E. L., Sanford, K. K. & Bell, D. A. (2000) XPD polymorphisms: effects on DNA repair proficiency. *Carcinogenesis,* 21, 551-5.

Matakidou, A., Eisen, T. & Houlston, R. S. (2003) TP53 polymorphisms and lung cancer risk: a systematic review and meta-analysis. *Mutagenesis,* 18, 377-85.

Matsuura, S., Kobayashi, J., Tauchi, H. & Komatsu, K. (2004) Nijmegen breakage syndrome and DNA double strand break repair by NBS1 complex. *Adv Biophys,* 38, 65-80.

Matullo, G., Palli, D., Peluso, M., Guarrera, S., Carturan, S., Celentano, E., Krogh, V., Munnia, A., Tumino, R., Polidoro, S., Piazza, A. & Vineis, P. (2001) XRCC1, XRCC3, XPD gene polymorphisms, smoking and (32)P-DNA adducts in a sample of healthy subjects. *Carcinogenesis,* 22, 1437-45.

Menoyo, A., Alazzouzi, H., Espin, E., Armengol, M., Yamamoto, H. & Schwartz, S., Jr. (2001) Somatic mutations in the DNA damage-response genes ATR and CHK1 in sporadic stomach tumors with microsatellite instability. *Cancer Res,* 61, 7727-30.

Miller, R. D., Phillips, M. S., Jo, I., Donaldson, M. A., Studebaker, J. F., Addleman, N., Alfisi, S. V., Ankener, W. M., Bhatti, H. A., Callahan, C. E., Carey, B. J., Conley, C. L., Cyr, J. M., Derohannessian, V., Donaldson, R. A., Elosua, C., Ford, S. E., Forman, A. M., Gelfand, C. A., Grecco, N. M., Gutendorf, S. M., Hock, C. R., Hozza, M. J., Hur, S., In, S. M., Jackson, D. L., Jo, S. A., Jung, S. C., Kim, S., Kimm, K., Kloss, E. F., Koboldt, D. C., Kuebler, J. M., Kuo, F. S., Lathrop, J. A., Lee, J. K., Leis, K. L., Livingston, S. A., Lovins, E. G., Lundy, M. L., Maggan, S., Minton, M., Mockler, M. A., Morris, D. W., Nachtman, E. P., Oh, B., Park, C., Park, C. W., Pavelka, N., Perkins, A. B., Restine, S. L., Sachidanandam, R., Reinhart, A. J., Scott, K. E., Shah, G. J., Tate, J. M., Varde, S. A., Walters, A., White, J. R., Yoo, Y. K., Lee, J. E., Boyce-Jacino, M. T. & Kwok, P. Y. (2005) High-density single-nucleotide polymorphism maps of the human genome. *Genomics,* 86, 117-26.

Park, J. Y., Lee, S. Y., Jeon, H. S., Bae, N. C., Chae, S. C., Joo, S., Kim, C. H., Park, J. H., Kam, S., Kim, I. S. & Jung, T. H. (2002) Polymorphism of the DNA repair gene XRCC1 and risk of primary lung cancer. *Cancer Epidemiol Biomarkers Prev,* 11, 23-7.

Ponder, B. A. (2001) Cancer genetics. *Nature,* 411, 336-41.

Rahman, N., Seal, S., Thompson, D., Kelly, P., Renwick, A., Elliott, A., Reid, S., Spanova, K., Barfoot, R., Chagtai, T., Jayatilake, H., Mcguffog, L., Hanks, S., Evans, D. G., Eccles, D., Easton, D. F. & Stratton, M. R. (2007) PALB2, which encodes a BRCA2-interacting protein, is a breast cancer susceptibility gene. *Nat Genet,* 39, 165-7.

Ramachandran, S., Ramadas, K., Hariharan, R., Rejnish Kumar, R. & Radhakrishna Pillai, M. (2006) Single nucleotide polymorphisms of DNA repair genes XRCC1 and XPD and its molecular mapping in Indian oral cancer. *Oral Oncol,* 42, 350-62.

Renwick, A., Thompson, D., Seal, S., Kelly, P., Chagtai, T., Ahmed, M., North, B., Jayatilake, H., Barfoot, R., Spanova, K., Mcguffog, L., Evans, D. G., Eccles, D., Easton, D. F., Stratton, M. R. & Rahman, N. (2006) ATM mutations that cause ataxia-telangiectasia are breast cancer susceptibility alleles. *Nat Genet,* 38, 873-5.

Ritchey, J. D., Huang, W. Y., Chokkalingam, A. P., Gao, Y. T., Deng, J., Levine, P., Stanczyk, F. Z. & Hsing, A. W. (2005) Genetic variants of DNA repair genes and prostate cancer: a population-based study. *Cancer Epidemiol Biomarkers Prev,* 14, 1703-9.

Roddam, P. L., Rollinson, S., O'driscoll, M., Jeggo, P. A., Jack, A. & Morgan, G. J. (2002) Genetic variants of NHEJ DNA ligase IV can affect the risk of developing multiple myeloma, a tumour characterised by aberrant class switch recombination. *J Med Genet,* 39, 900-5.

Schmidt, M. K., Reincke, S., Broeks, A., Braaf, L. M., Hogervorst, F. B., Tollenaar, R. A., Johnson, N., Fletcher, O., Peto, J., Tommiska, J., Blomqvist, C., Nevanlinna, H. A., Healey, C. S., Dunning, A. M., Pharoah, P. D., Easton, D. F., Dork, T. & Van't Veer, L. J. (2007) Do MDM2 SNP309 and TP53 R72P interact in breast cancer susceptibility? A large pooled series from the breast cancer association consortium. *Cancer Res,* 67, 9584-90.

Shen, H., Wang, X., Hu, Z., Zhang, Z., Xu, Y., Hu, X., Guo, J. & Wei, Q. (2004) Polymorphisms of DNA repair gene XRCC3 Thr241Met and risk of gastric cancer in a Chinese population. *Cancer Lett,* 206, 51-8.

Shields, P. G. & Harris, C. C. (2000) Cancer risk and low-penetrance susceptibility genes in gene-environment interactions. *J Clin Oncol,* 18, 2309-15.

Spry, M., Scott, T., Pierce, H. & D'orazio, J. A. (2007) DNA repair pathways and hereditary cancer susceptibility syndromes. *Front Biosci,* 12, 4191-207.

Sreeja L, Syamala V, Raveendran PB, Santhi S, Madhavan J, Ankathil R., (2008). P53 Arg72Pro polymorphism Predicts Survival Outcome in Lung Cancer Patients in Indian Population. *Cancer Investigation,* 26:41-46.

Sreeja, L., Syamala, V. S., Syamala, V., Hariharan, S., Raveendran, P. B., Vijayalekshmi, R. V., Madhavan, J. & Ankathil, R. (2007) Prognostic importance of DNA repair gene polymorphisms of XRCC1 Arg399Gln and XPD Lys751Gln in lung cancer patients from India. *J Cancer Res Clin Oncol,* 134, 645-52.

Stern, M. C., Conti, D. V., Siegmund, K. D., Corral, R., Yuan, J. M., Koh, W. P. & Yu, M. C. (2007) DNA repair single-nucleotide polymorphisms in colorectal cancer and their role as modifiers of the effect of cigarette smoking and alcohol in the Singapore Chinese Health Study. *Cancer Epidemiol Biomarkers Prev,* 16, 2363-72.

Tang, D., Cho, S., Rundle, A., Chen, S., Phillips, D., Zhou, J., Hsu, Y., SCHNABEL, F., ESTABROOK, A. & PERERA, F. P. (2002) Polymorphisms in the DNA repair enzyme XPD are associated with increased levels of PAH-DNA adducts in a case-control study of breast cancer. *Breast Cancer Res Treat,* 75, 159-66.

Tomlinson, I. P., Webb, E., Carvajal-Carmona, L., Broderick, P., Howarth, K., Pittman, A. M., Spain, S., Lubbe, S., Walther, A., Sullivan, K., Jaeger, E., Fielding, S., Rowan, A., Vijayakrishnan, J., Domingo, E., Chandler, I., Kemp, Z., Qureshi, M., Farrington, S. M., Tenesa, A., Prendergast, J. G., Barnetson, R. A., Penegar, S., Barclay, E., Wood, W., Martin, L., Gorman, M., Thomas, H., Peto, J., Bishop, D. T., Gray, R., Maher, E. R., Lucassen, A., Kerr, D., Evans, D. G., Schafmayer, C., Buch, S., Volzke, H., Hampe, J., Schreiber, S., John, U., Koessler, T., Pharoah, P., Van Wezel, T., Morreau, H., Wijnen, J. T., Hopper, J. L., Southey, M. C., Giles, G. G., Severi, G., Castellvi-Bel, S., Ruiz-Ponte, C., Carracedo, A., Castells, A., Forsti, A., Hemminki, K., Vodicka, P., Naccarati, A., Lipton, L., Ho, J. W., Cheng, K. K., Sham, P. C., Luk, J., Agundez, J. A., Ladero, J. M., De La Hoya, M., Caldes, T., Niittymaki, I., Tuupanen, S., Karhu, A., Aaltonen, L., Cazier, J. B., Campbell, H., Dunlop, M. G. & Houlston, R. S. (2008) A genome-wide association study identifies colorectal cancer susceptibility loci on chromosomes 10p14 and 8q23.3. *Nat Genet,* 40, 623-30.

Tseng, R. C., Hsieh, F. J., Shih, C. M., Hsu, H. S., Chen, C. Y. & Wang, Y. C. (2009) Lung cancer susceptibility and prognosis associated with polymorphisms in the nonhomologous end-joining pathway genes: a multiple genotype-phenotype study. *Cancer*, 115, 2939-48.

Weston, A., Perrin, L. S., Forrester, K., Hoover, R. N., Trump, B. F., Harris, C. C. & Caporaso, N. E. (1992) Allelic frequency of a p53 polymorphism in human lung cancer. *Cancer Epidemiol Biomarkers Prev*, 1, 481-3.

Whibley, C., Pharoah, P. D. & Hollstein, M. (2009) p53 polymorphisms: cancer implications. *Nat Rev Cancer*, 9, 95-107.

Wood, R. D., Mitchell, M. & Lindahl, T. (2005) Human DNA repair genes. *Mutat Res*, 577, 275-83.

Yin, J., Vogel, U., Ma, Y., Guo, L., Wang, H. & Qi, R. (2006) Polymorphism of the DNA repair gene ERCC2 Lys751Gln and risk of lung cancer in a northeastern Chinese population. *Cancer Genet Cytogenet*, 169, 27-32.

Zhou, G., Zhang, X., Lui, S., Lin, D., Liang, C. & Yang, X. (2001) [Pilot study on mutations of p53 gene in laryngeal carcinoma]. *Hua Xi Yi Ke Da Xue Xue Bao*, 32, 359-60, 434.

Zhou, W., Liu, G., Miller, D. P., Thurston, S. W., Xu, L. L., Wain, J. C., Lynch, T. J., Su, L. & Christiani, D. C. (2003) Polymorphisms in the DNA repair genes XRCC1 and ERCC2, smoking, and lung cancer risk. *Cancer Epidemiol Biomarkers Prev*, 12, 359-65.

Zienolddiny, S., Campa, D., Lind, H., Ryberg, D., Skaug, V., Stangeland, L., Phillips, D. H., Canzian, F. & Haugen, A. (2006) Polymorphisms of DNA repair genes and risk of non-small cell lung cancer. *Carcinogenesis*, 27, 560-7.

# 8

# Variants and Polymorphisms of DNA Repair Genes and Neurodegenerative Diseases

Fabio Coppedè

*Contract Professor at Faculty of Medicine, University of Pisa*
*Italy*

## 1. Introduction

Oxidative DNA damage is one of the earliest detectable events in several neurodegenerative diseases, often preceding the onset of the clinical symptoms. Moreover, neurons in the adult human brain can re-enter the cell division cycle, likely allowing DNA repair. Impairments of DNA repair pathways are reported in neurons of patients suffering from one of several neurodegenative diseases and might result in the accumulation of mutations critical for neurodegeneration. Current investigation aims at understanding the causes of such impairment (Coppedè & Migliore, 2010). One of the most robust set of data that demonstrates association between DNA repair and neurodegenerative diseases comes from studies on early onset ataxia with ocular motor apraxia and hypoalbuminemia/ataxia with oculomotor apraxia type 1 (EAOH/AOA1), an autosomal recessive form of cerebellar ataxia caused by mutations in the aprataxin (*APTX*) gene. It was shown that aprataxin participates in DNA repair suggesting that genes involved in DNA repair pathways might have a role in neurodegeneration (Hirano et al., 2007; Takahashi et al., 2007). Also parkin, encoded by one of the causative genes of Parkinson's disease (PD), seems to contribute to DNA repair (Kao, 2009). Variants and polymorphisms of DNA repair genes, particularly DNA base excision repair (BER) genes, have been investigated as possible risk factors for Alzheimer's disease (AD), Parkinson's disease, amyotrophic lateral sclerosis (ALS), and other neurodegenerative diseases (Coppedè & Migliore, 2010). There is also evidence that BER could contribute to CAG repeat expansion in Huntington's disease (HD) (Kovtun et al., 2007). Most of the genetic association studies have been performed in the last few years and gave often conflicting or inconclusive results, their power was limited by the sample size of case-control groups, gene-gene interactions were missing, and only common polymorphisms have been included with little or no attention paid to rare gene variants (Coppedè, 2011). In this chapter I discuss the current knowledge on DNA repair gene variants and polymorphisms and major neurodegenerative disorders.

## 2. DNA repair pathways

A brief overview of the major DNA repair pathways in mammals is shown in Table 1. It is estimated that our cells are subjected to a daily average of about one million lesions that, if not properly repaired, can drive mutagenesis, disrupt normal gene expression or create aberrant protein products. Cells have therefore developed several repair systems that can be generally divided into single stand break (SSB) and double strand break (DSB) repair pathways (Table 1).

| Pathway | Type of repair | Type of damage |
|---|---|---|
| Base excision repair (BER) | SSB | Modifications of DNA bases due to oxidation, alkylation, and deamination |
| Nucleotide excision repair (NER) | SSB | Repair of UV photoproducts, DNA crosslinks, and bulky lesions. |
| Mismatch repair (MMR) | SSB | Repair of mismatches and small insertions or deletions during replication. |
| Homologous recombination (HR) | DSB | Repair of DNA DSBs, such as those caused by ionizing radiations, through recombination with regions of homology (usually a sister chromatid) during late S or G2 phases of the cell cycle |
| Non homologous end joining (NHEJ) | DSB | Repair of DNA DSBs, such as those induced by radiations, without recombination with regions of homology; it occurs during G0, G1, and early S phases of the cell cycle |

Table 1. Major DNA repair pathways in mammalian cells

## 2.1 Base excision repair (BER)

The DNA base excision repair pathway deserves a detailed description since it is believed to be the major pathway for repairing DNA base modifications caused by oxidation, deamination and alkylation. DNA glycosylases catalyze the first step in the BER process by cleaving the N-glycosylic bond between a damaged base and the sugar moiety; after the cleavage the damaged base is released resulting in the formation of an abasic site which is then cleaved by an AP lyase activity or by the major mammalian apurinic/apyrimidinic endonuclease (APEX1). Repair can then proceed through short or long-patch BER. In short-patch BER, which is the most common sub-pathway, a single nucleotide is incorporated into the gap by DNA polymerase β (Pol β) and ligated by the DNA ligase III/ X-ray repair cross-complementing group 1 (XRCC1) complex. In long-patch BER several nucleotides (two to seven-eight) are incorporated, followed by cleavage of the resulting 5′ flap structure and ligation. It has been suggested that after Pol β adds the first nucleotide into the gap, it is substituted by Pol δ/ε which continues long-patch BER. DNA ligase I completes the long-patch pathway. Several other proteins, including the proliferating cell nuclear antigen (PCNA), the RPA protein, and the 5′-flap endonuclease (FEN-1) participate in long-patch BER. Recent evidence suggests that XRCC1 acts as a scaffold protein in short-patch BER, regulating and coordinating the whole process. XRCC1 recruits DNA Pol β and DNA ligase III required for filling and sealing the damaged strand. Moreover, it also interacts with DNA glycosylases and APEX1, mediating their exchange at the damaged site. XRCC1 also interacts with PARP-1, which is one of the cellular sensors of DNA SSBs and DSBs. BER takes places either in nuclei and mitochondria, and mitochondria have independent BER machinery encoded by nuclear genes. Indeed, several BER enzymes have been identified which have both nuclear and mitochondrial forms. The gaps generated by the action of AP endonucleases/lyases are filled in by Pol γ in the mitochondria, and ligation is mediate by ligase III. To date, there is no evidence of long-patch BER in mitochondria (Weissman et al., 2007).

## 2.2 Nucleotide excision repair (NER)

The nucleotide excision repair pathway (NER) is required for the removal of a wide variety of forms of DNA damage, including UV induced photoproducts, DNA crosslinks, and other bulky lesions. NER involves at least 20-30 proteins or complexes of proteins, and is divided into global genome repair (GGR) and transcription coupled repair (TCR). The two pathways mainly differ in the initial steps that recognize the DNA lesion, and different initial recognition factors are involved. NER senses the presence of a lesion through the distortion it causes to the DNA structure. In GGR DNA damage recognition requires the xeroderma pigmentosum (XP) complementing protein XPC-HR23B-centrin complex. The DNA damage is verified by opening of the DNA strands surrounding the lesion by the transcription factor TFIIH. This is followed by recruitment of XPA and other components of the transcription factor TFIIH to the lesion site. In TCR the recognition step is initiated when a RNA polymerase stalls at a lesion site and requires the Cockayne's syndrome proteins CSA and CSB. After a correct assembly of the NER complex, a fragment of 24-32 nucleotides is incised and removed from the damaged strand by the simultaneous action of the DNA excision repair cross complementing (ERCC) proteins ERCC5 (XPG; 3' endonuclease) and ERCC4 (XPF; 5' endonuclease) complexed with ERCC1. Repair is completed by new DNA synthesis mediated by DNA Pol δ/ε, DNA Pol κ, and ligation (DNA ligase I, DNA ligase III) of the nascent DNA to the parental strands using the undamaged strand as a template. The GGR pathway removes damages overall in the genome irrespective of genome location and point in the cell cycle, whereas TCR is required for the specific repair of bulky lesions in the transcribed strand of active genes. Mitochondria have been shown to lack NER, which operates in the nucleus removing the majority of DNA lesions (Fleck & Nielsen, 2004; Subba Rao, 2007).

## 2.3 Mismatch repair (MMR)

Mismatch repair (MMR) corrects mismatches and small insertions or deletions during DNA replication, thus eliminating potentially pre-mutagenic bases. Repair involves recognition of the mismatch by MutSα (MSH2 and MSH6 proteins), or by MutSβ (MSH2 and MSH3 proteins) in the case of small insertions/deletions (1-10 nucleotides). MutLα (a heterodimer of MLH1 and PMS2 proteins) is then recruited and serves to coordinate the process that involves, among others, the PCNA protein for strand discrimination and exonuclease 1, DNA Pol δ and a DNA ligase, for DNA repair (Kunkel & Erie, 2005).

## 2.4 Homologous recombination (HR) and non homologous end joining (NHEJ)

Non homologous end joining (NHEJ) is the major pathway for the repair of DSBs because it can function throughout the cell cycle and does not require a homologous chromosome. Rather, NHEJ involves rejoining of what remains of the two DNA ends, tolerating nucleotide loss or addition at the rejoining site. When a DSB occurs during G0, G1, and early

S phase, the Ku heterodimer (Ku70/Ku80) recognizes DSB ends, aligns them, protects them from excessive degradation, and ultimately prepares them for ligation. The Ku heterodimer is capable of interacting with the nuclease (Artemis-DNA-PKcs) complex, the polymerases (μ and λ), and the ligase (XLF-XRCC4-DNA ligase IV) complex. If complementary ends are not present at the break, the Artemis-DNA-PKcs complex resects some of the overhangs to create single-strand overhangs with short stretches of micro-homology. When necessary, polymerization of missing nucleotides is performed by DNA polymerases. Then, the XLF-XRCC4-LigaseIV complex seals the DSB. When homologous recombination (HR) is used for repair, it is promoted by the recombinase RAD51, the human homolog of the *E. coli* RecA protein, which binds to 3'-tailed single strands at the end of DSBs in a helical fashion and promotes pairing with homologous DNA sequences (usually the sister chromatid) as a prelude to strand invasion and repair of the DSBs. Strand invasion is the invasion of the 3' end of the single-stranded DNA overhang into the region of complementarity in the intact sister chromatid. The process is directed by RAD51 which forms a nucleoprotein filament that directs homology search, strand pairing, and invasion of the homologous chromosome. Rad51 is assisted in this process by several RAD family members (RAD51B, RAD51C, RAD51D, XRCC2, XRCC3, RAD54, and RAD52). During strand invasion, RAD51 creates a four-stranded Holliday junction intermediate. Then, the invading strand is extended by DNA polymerase η and the Holliday junction is resolved by a RAD51C and XRCC3 directed mechanism. Several nucleases and helicases, such as the RecQ family members, also participate in resolving Holliday junctions. Since eukaryotic genomes contain dispersed repeated DNA, repair of DSBs by HR can occur not only through an interaction with the sister chromatid or the homolog chromosome, but also with repeats on non-homolog chromosomes. Numerous factors affect the decision to repair a DSB via these pathways, and accumulating evidence suggests these major repair pathways both cooperate and compete with each other at double-strand break sites to facilitate efficient repair and promote genomic integrity (Kass & Jasin, 2010).

## 3. Polymorphisms of DNA repair genes and Alzheimer's disease

Alzheimer's disease is a complex multi-factorial neurodegenerative disorder and represents the most common form of dementia in the elderly. In 2006, the worldwide prevalence of AD was 26.6 million. It has been estimated that following the global aging of the world's population this number will quadruple by 2050, suggesting that 1 in 85 persons worldwide will be living with the disease, which is clinically characterized by a progressive neurodegeneration in selected brain regions, including the temporal and parietal lobes and restricted regions within the frontal cortex and the cingulate gyrus, resulting in gross atrophy of the affected regions and leading to memory loss accompanied by changes of behaviour and personality severe enough to affect work, lifelong hobbies or social life (Brookmeyer et al., 2007). Increasing evidence reports oxidative DNA damage in affected brain regions of AD patients, paralleled by a decrease in DNA repair activities, particularly concerning the BER pathway (Lovell et al., 2000; Weissman et al., 2007). This has driven current research to focus on common polymorphisms of BER genes as candidate AD risk factors. Studied genes are those encoding for 8-oxoguanine DNA glycosylase (OGG1), APEX1 and XRCC1. Particularly, we screened 178 Italian late onset AD patients and 146 matched controls for the presence of the *OGG1* Ser326Cys gene polymorphism (rs1052133), observing no difference in allele and genotype frequencies between patients and controls (Coppedè et al., 2007a). Subsequently, 91 sporadic Turkish AD patients and 93 matched controls have been genotyped for the presence of *OGG1* Ser326Cys, *APEX1* Asp148Glu (rs1130409), *XRCC1* Arg280His (rs25489) and *XRCC1* Arg399Gln (rs25487) polymorphisms, but none of them was associated with increased AD risk (Parildar-Karpuzoğlu et al. 2008). Also a small case-control study performed in Poland with 41 AD patients and 51 controls failed to find significant differences in *OGG1* Ser326Cys allele frequencies between groups (Dorszewska et al., 2009). A borderline association with AD risk ($P$ =0.06) was observed for the *XRCC1* Arg194Trp (rs1799782) polymorphism in a group of 98 Turkish AD patients and 95 healthy subjects (Doğru-Abbasoğlu et al., 2007), but a recent study failed to replicate this association in a larger case-control group of over 200 Chinese AD patients (Quian et al., 2010). Overall, five common functional polymorphisms of BER genes have been investigated as possible AD risk factors, but none of them resulted significantly associated with increased AD risk (Table 2). Also polymorphisms of NER genes have been evaluated as candidate AD risk factors. Particularly, two common polymorphisms of the *XPD (ERCC2)* gene (namely, rs238406 and rs13181), and a silent mutation in exon 11 (T>C at codon 824) of the *XPF (XRCC4)* gene have been investigated in 97 Turkish AD patients and in 101 matched controls, but none of them resulted to be associated with AD risk (Doğru-Abbasoğlu et al., 2006). Poly-ADP-ribose polymerase-1 (PARP-1) is a zinc-finger DNA binding protein that is activated by DNA SSBs or DSBs. The primary function of PARP-1 is in DNA repair processes through the detection of DNA damage and the prevention of chromatide exchanges. PARP-1 poly-ADP-ribosylates several proteins involved in DNA repair including histones, thus inducing local relaxation of the chromatin structure and facilitating the access of repair proteins to damaged DNA. There is evidence for widespread DNA SSBs and DSBs in AD brains, as well as increased PARP-1 activity (Love et al., 1999). Two independent groups evaluated *PARP-1* gene polymorphisms as putative AD risk factors. Infante and coworkers screened 263 Spanish AD patients and 293 matched controls for the presence of two *PARP-1* promoter polymorphisms (–410 and –1672). If evaluated independently, nor *PARP-1* –410 neither *PARP-1* –1672 resulted associated with increased AD risk (Table 2). However, *PARP-1* –410 and *PARP-1* –1672 polymorphisms resulted in linkage disequilibrium and some haplotypes were associated with increased AD risk. Particularly, haplotypes 2-1 and 1-2 were significantly overrepresented in AD individuals and associated with an increased risk for the disease with an adjusted OR of 1.42 and 5.38, respectively (Infante et al., 2007). More recently two *PARP-1* exonic polymorphisms, 414C>T (rs1805404) and 2456T>C (rs1136410), have been evaluated in 120 Chinese AD patients and 111 matched controls (Liu et al., 2010). Again, none of the polymorphisms resulted independently associated with increased AD risk (Table 2). However, authors found that the distributions of haplotype 3-TT and haplotype 4-CC were significantly associated with an increased risk of AD, whereas the haplotype 1-TC showed a protective effect, with OR of 12.2 and 0.52, respectively (Liu et al., 2010). Overall, both studies support the hypothesis that *PARP1* haplotypes might affect AD risk.

### 3.1 Searching for BER gene variants in DNA extracted from post-mortem AD brain

Mao and colleagues extracted nuclear DNA from post-mortem brain specimens of 14 late stage AD patients and 10 neurologically healthy controls. They identified and characterized novel *OGG1* mutations (a single base deletion C796del, and two base substitutions leading

| Reference | Polymorphism | Number of subjects AD/Controls | Variant allele frequency AD/Controls | Odds Ratio (95% CI) |
|---|---|---|---|---|
| Coppedè et al., 2007a | OGG1 Ser326Cys | 178/146 | 0.19/0.18 | 1.04 (0.70-1.55) |
| Parildar-Karpuzoğlu et al. 2008 | OGG1 Ser326Cys | 91/93 | 0.29/0.23 | 1.32 (0.83-2.11) |
| Dorszewska et al., 2009 | OGG1 Ser326Cys | 41/51 | 0.29/0.21 | 1.60 (0.81-3.14) |
| Parildar-Karpuzoğlu et al. 2008 | APEX1 Asp148Glu | 91/93 | 0.33/0.31 | 1.08 (0.70-1.68) |
| Parildar-Karpuzoğlu et al. 2008 | XRCC1 Arg280His | 91/93 | 0.06/0.10 | 0.53 (0.24-1.14) |
| Parildar-Karpuzoğlu et al. 2008 | XRCC1 Arg399Gln | 91/93 | 034/0.33 | 1.05 (0.68-1.63) |
| Doğru-Abbasoğlu et al., 2007 | XRCC1 Arg194Trp | 98/95 | 0.11/0.06 | 2.06 (0.97-4.37) |
| Quian et al., 2010 | XRCC1 Arg194Trp | 212/203 | 0.31/0.31 | 1.04 (0.70-1.52) |
| Doğru-Abbasoğlu et al., 2006 | ERCC2 rs238406 (XPD exon 6) | 97/101 | 0.40/0.42 | 0.94 (0.63-1.41) |
| Doğru-Abbasoğlu et al., 2006 | ERCC2 rs13181 (XPD exon 23) | 97/101 | 0.41/0.36 | 1.24 (0.83-1.86) |
| Doğru-Abbasoğlu et al., 2006 | ERCC4 (XPF exon 11) | 97/101 | 0.37/0.35 | 1.09 (0.72-1.64) |
| Infante et al., 2007a | PARP-1 (-410) | 263/293 | 0.35/0.33 | 1.08 (0.84-1.38) |
| Infante et al., 2007a | PARP-1 (-1672) | 263/293 | 0.17/0.18 | 0.94 (0.69-1.27) |
| Liu et al., 2010 | PARP-1 (rs1805404) | 120/111 | 0.53/0.59 | 0.76 (0.53-1.11) |
| Liu et al., 2010 | PARP-1 (rs1136410) | 120/111 | 0.52/0.59 | 0.75 (0.52-1.09) |

Table 2. DNA repair gene polymorphisms and risk of Alzheimer's disease

to Ala53Thr or Ala288Val amino acidic changes, respectively) in 4 of 14 AD subjects. Particularly, two AD patients carried the C796 deletion, one patient had the Ala53Thr substitution, and another patient carried the Ala288Val substitution. No mutations were found in any of 10 studied age-matched controls (Mao et al., 2007). This study is not an association study for risk assessment but a genetic screening performed on brain DNA specimens searching for novel *OGG1* variants. The authors created the mutant proteins by site-directed mutagenesis observing that the C796del mutant OGG1 lacks glycosylase activity, whereas both Ala53Thr and Ala288Val substitutions result in 40–50% reduced activity (Mao et al., 2007). Therefore, we cannot exclude that the activity of the OGG1 protein might be partially impaired by rare gene variants in some AD subjects. However, given the limited sample-size of the studied case-control group, further studies are required to confirm this hypothesis.

## 4. Polymorphisms of DNA repair genes and amyotrophic lateral sclerosis

Amyotrophic lateral sclerosis (ALS), also known as motor neuron disease (MND), is one of the major neurodegenerative diseases alongside AD and PD. It is a progressive disorder characterized by the degeneration of motor neurons of the motor cortex, brainstem and spinal cord. The incidence of the disease is similar worldwide and ranges from 1 to 3 cases per 100,000 individuals per year, with the exception of some high-risk areas around the Pacific Rim. Several studies report increased oxidative DNA damage and a compromised DNA repair activity, particularly BER activity, in spinal cords and other tissues of ALS patients (Bogdanov et al., 2000; Ferrante et al., 1997; Kikuchi et al., 2002; Kisby et al., 1997). Missense mutations in the gene encoding *APEX1* were found in DNA obtained from 8 of 11 ALS patients, including the common *APEX1* Asp148Glu polymorphism (Hayward et al., 1999), that was subsequently associated with increased ALS risk in a Scottish cohort of 117 ALS patients and 58 controls, and in an Irish group of 105 ALS individuals and 82 controls (Greenway et al., 2004). The analysis of 88 English ALS patients and 88 matched controls still revealed an increased frequency of the variant allele in the ALS cohort, even if not statistically significant (Tomkins et al. 2000). We have recently performed the largest case-control study aimed at clarifying the role of *APEX1* Asp148Glu in sporadic ALS pathogenesis. No difference in *APEX1* Asp148Glu allele and genotype frequencies was found between 134 ALS patients and 129 controls of Italian origin, nor was the polymorphism associated with disease age or site of onset, or duration of the disease, suggesting that it might not play a major role in ALS pathogenesis in the Italian population (Coppedè et al., 2010a). The ALSGene database (www.alsgene.org) is a public database containing all the ALS genetic association studies, genome-wide association studies and updated meta-analyses of the literature. A meta-analysis of the four studies described above revealed a significant increased frequency of the variant 148Glu allele in ALS cases with respect to controls, suggesting a protective role for the wild type 148Asp variant with an OR = 0.78 (95%CI=0.62-0.97) (www.alsgene.org). Our analysis of the *OGG1* Ser326Cys polymorphism in 136 ALS patients and 129 matched controls of Italian origin revealed a significant association of the variant allele with increased ALS risk (Coppedè et al., 2007b) (Table 3). At best of our knowledge this study is the first in the literature addressing this issue, still pending replication in other populations. More recently, we screened over 400 individuals, including 206 ALS patients and 203 matched controls of Italian origin for the presence of *XRCC1* Arg194Trp, Arg280His and Arg399Gln polymorphisms, observing a significant increased frequency of the 399Gln variant allele and a borderline significant decreased frequency of the 194Trp allele in ALS patients with respect to controls (Coppedè et al., 2010b). Interestingly, others have evaluated the same *XRCC1* polymorphisms and two additional ones (rs939461 and rs915927) in 108 ALS patients and 39 controls from New-England, observing that rs939461 was associated with reduced ALS risk, and Arg399Gln with a borderline significant reduced risk (Fang et al., 2010) (Table 3). Overall, even if still inconclusive, the results of both studies suggest that additional investigation is required to clarify the role of *XRCC1* polymorphisms and haplotypes in ALS pathogenesis.

## 4.1 Less frequent BER gene variants and polymorphisms

Alongside with common BER gene polymorphisms, less frequent gene variants or polymorphisms have been observed in the DNA of both ALS subjects and matched controls, but with very low allele frequencies and no significant difference between groups. Some examples are *APEX1* 1835C/A (Intron3), *APEX1* 2712A/T (3'UTR), *APEX1* 459C/T (Exon1), and *APEX1* rs1048945 (Q51H) (Hayward et al., 1999; Tomkins et al., 2000).

| Reference | Polymorphism | Number of subjects ALS/Controls | Variant allele frequency ALS/Controls | Odds Ratio (95% CI) |
|---|---|---|---|---|
| Coppedè et al., 2007b | OGG1 Ser326Cys | 136/129 | 0.26/0.18 | 1.62 (1.07-2.45) |
| Hayward et al. 1999 | APEX1 Asp148Glu | 117/58 | 0.62/0.49 | 1.66 (1.06-2.60) |
| Tomkins et al. 2000 | APEX1 Asp148Glu | 88/88 | 0.51/0.45 | 1.28 (0.85-1.95) |
| Greenway et al. 2004 | APEX1 Asp148Glu | 105/82 | 0.60/0.51 | 1.46 (0.97-2.21) |
| Coppedè et al. 2010a | APEX1 Asp148Glu | 134/129 | 0.44/0.45 | 0.99 (0.70-1.40) |
| Coppedè et al. 2010b | XRCC1 Arg194Trp | 206/195 | 0.05/0.08 | 0.58 (0.32-1.05) |
| Coppedè et al. 2010b | XRCC1 Arg280His | 205/203 | 0.09/0.08 | 1.25 (0.76-2.04) |
| Coppedè et al. 2010b | XRCC1 Arg399Gln | 197/194 | 0.39/0.28 | 1.39 (1.05-1.85) |
| Fang et al. 2010 | XRCC1 Arg194Trp | 108/39 | 0.06/0.03 | 2.4 (0.5-2.2)[a] |
| Fang et al. 2010 | XRCC1 Arg280His | 108/39 | 0.05/0.03 | 2.0 (0.4-2.0)[a] |
| Fang et al. 2010 | XRCC1 Arg399Gln | 108/39 | 0.35/0.47 | 0.4 (0.2-1.0)[a] |
| Fang et al. 2010 | XRCC1 rs915927 | 108/39 | 0.45/0.33 | 2.4 (0.5-2.2)[a] |
| Fang et al. 2010 | XRCC1 rs939461 | 108/39 | 0.06/0.15 | 0.4 (0.1-0.9)[a] |

Table 3. DNA repair gene polymorphisms and risk of Amyotrophic Lateral sclerosis[a] OR are derived from the original paper and referred to (heterozygous+minor homozygous) vs major homozygous.

## 5. Polymorphisms of DNA repair genes and Parkinson's disease

Parkinson's disease is the second most common neurodegenerative disorder after AD, affecting 1–2% of the population over the age of 50 years, and is characterized by progressive and profound loss of neuromelanin containing dopaminergic neurons in the *substantia nigra* (SN) resulting in resting tremor, rigidity, bradykinesia, and postural instability. The majority of PD cases are sporadic idiopathic forms, resulting from three interactive events: an individual's inherited genetic susceptibility, subsequent exposure to environmental risk factors, and aging (Bekris et al., 2010). However, in a minority of the cases PD is inherited as a Mendelian trait. Parkin is an E3 ubiquitin ligase that acts on a variety of substrates, resulting in polyubiquitination and degradation by the proteasome or monoubiquitination and regulation of biological activity. Mutation of *parkin* is one of the most prevalent causes of autosomal recessive familial PD and a recent study has shown that parkin is essential for optimal repair of DNA damage. Particularly, DNA damage induces nuclear translocation of parkin leading to interactions with PCNA and possibly other nuclear proteins involved in DNA repair (Kao, 2009). Moreover, parkin protects mitochondrial genome integrity and supports mitochondrial DNA (mtDNA) repair (Rothfuss et al., 2009). DNA polymerase gamma (POLG1) participates in mtDNA replication and repair, thus playing a fundamental role in mtDNA maintenance. Missense mutations in *POLG1* co-segregate with a phenotype that includes progressive external ophthalmoplegia and parkinsonism (Hudson et al., 2007). Moreover, missense mutations in *POLG1* have been reported in case studies, in which parkinsonism was part of the clinical symptoms (Davidzon et al., 2006; Remes et al., 2008). *POLG1* mutations and polymorphisms have been also investigated in sporadic idiopathic PD, among them a polyglutamine (poly-Q) located in the N-terminal of POLG1, encoded by a CAG repeat in exon 2. The poly-Q tract normally consists of 10Q (frequency >80%), followed by 11Q (frequency > 6-12%), whereas non-10Q/11Q alleles are considered as less frequent alleles. Several authors investigated whether or not non-10Q alleles are more frequent in PD cases than in matched controls (Hudson et al., 2009; Luoma et al., 2007; Taanman & Shapira, 2005; Tiangyou et al., 2006). Eerola and coworkers recently screened 641 PD patients and 292 controls from USA and performed a pooled analysis of their data with those available in the literature (Hudson et al., 2009; Luoma et al., 2007; Taanman & Shapira, 2005; Tiangyou et al., 2006) for a total of 1163 sporadic PD patients and 1214 controls observing that variant alleles defined as non-10Q were significantly increased in PD patients than in controls (16.3%vs.13.4%, $p$ = 0.005) (Eerola et al., 2010). A few months later Anvret and coworkers screened 243 PD patients and 279 matched controls from Sweden, observing that non10Q/11Q alleles were more frequent in PD cases than in controls with an OR of 2.0 (1.3-3.1, 95%CI) strengthening the evidence that non frequent *POLG1* alleles might be more frequent in sporadic PD patients than in controls, thus representing a PD risk factor (Anvret et al., 2010) (Table 4). We screened 139 sporadic PD patients and 211 healthy matched controls for the presence of the *OGG1* Ser326Cys polymorphism. The Cys326 allele frequency was similar between the groups (0.20 in PD patients and 0.19 in controls), and no difference in genotype frequencies was observed. Moreover, the *OGG1* Ser326Cys polymorphism was not associated with PD age at onset (Coppedè et al., 2010c). In human cells the oxidized purine nucleoside triphosphatase MTH1 efficiently hydrolyzes oxidized purines such as 8-oxo-guanine in the nucleotide pools, thus avoiding their incorporation into DNA or RNA. A Val83Met polymorphism of the *MTH1* gene was studied in 73

Japanese patients with sporadic PD and 151 age-matched controls but was not associated with sporadic PD risk (Satoh & Kuroda, 2000). Another *MTH1* polymorphism (Ile45Thr) was investigated in 106 PD patients and 135 unrelated controls from China. The variant allele frequency resulted borderline increased in PD males (Jiang et al., 2008). This finding is pending replication in other populations. *PARP1* promoter polymorphisms (-410C/T, -1672G/A, and a (CA)n microsatellite) have been investigated in 146 Spanish PD cases and 161 matched controls. A protective effect against PD was found for heterozygosity at -410 (OR = 0.44) and (CA)n microsatellite (OR = 0.53) polymorphisms, and heterozygosity at -1672 polymorphism delayed by 4 years on the onset age of PD (Infante et al., 2007). Also these findings are original and waiting for replication in additional case-control groups (Table 4).

| Reference | Polymorphism | Number of subjects PD/Controls | Variant allele frequency PD/Controls | Odds Ratio (95% CI) |
|---|---|---|---|---|
| Eerola et al., 2010 | POLG1 Poly-Q tract | 641/292 | 0.17/0.12 | OR = n.a. P = 0.004 |
| Eerola et al., 2010 | POLG1 Poly-Q tract | 1163/1214[a] | 0.16/0.13 | OR = n.a. P = 0.005 |
| Anvret et al., 2010 | POLG1 Poly-Q tract | 243/279 | 0.11/0.06 | 2.0 (1.3-3.1) |
| Coppedè et al., 2010c | OGG1 Ser326Cys | 139/211 | 0.20/0.19 | 1.05 (0.72-1.53) |
| Satoh & Kuroda, 2000 | MTH1 Val83Met | 73/151 | 0.07/0.11 | OR = n.a. P = 0.219 |
| Jiang et al., 2008 | MTH1 Ile45Thr | 106/135 | 0.05/0.02 | OR = n.a. P = 0.08[b] |
| Infante et al., 2007b | PARP-1 (-410) | 146/161 | 0.35/0.53[c] | 0.44 (0.26-0.75) |
| Infante et al., 2007b | PARP-1 (-1672)[d] | 146/161 | 0.29/0.27[c] | 0.87 (0.50-1.52) |
| Infante et al., 2007b | PARP-1 (CA)n | 146/161 | 0.36/0.50[c] | 0.53 (0.31-0.90) |

Table 4. DNA repair gene polymorphisms and risk of Parkinson's disease. [a] = pooled-analysis of (Eerola et al., 2010, Hudson et al., 2009; Luoma et al., 2007; Taanman & Shapira, 2005; Tiangyou et al., 2006). [b] = Allele frequency difference (PD/Controls) approached significance in the male subgroup (0.07/0.02, P = 0.05). [c] = Heterozygous genotype frequency. [d] = Associated with PD age at onset

## 5.1 Other mutations and polymorphisms
As previously observed, several *POLG1* mutations have been observed to co-segregate in families with parkinsonism. For a detailed description I suggest a recent review by Orsucci and coworkers (Orsucci et al., 2010).

# 6. Other neurodegenerative diseases

## 6.1 Spinocerebellar ataxias

Hereditary ataxias are a heterogeneous group of diseases with different patterns of inheritance. Some of them are caused by recessive mutations in genes involved in DNA repair pathways that likely predispose the affected individuals to neurodegeneration. Spinocerebellar ataxia with axonal neuropathy 1 (SCAN1) is caused by autosomal recessive mutations in the gene encoding tyrosyl-DNA phosphodiesterase 1 (TDP1), a protein required for the repair of DNA SSBs that arise independent of DNA replication from abortive topoisomerase 1 activity or oxidative stress. Ataxia-telangiectasia (AT), ataxia-telangiectasia-like disorder (ATLD), ataxia oculomotor apraxia type 1 (AOA1) and ataxia oculomotor apraxia type 2A (AOA2) are a subgroup of the autosomal recessive spinocerebellar ataxias characterized by cerebellar atrophy and oculomotor apraxia. The progressive neurodegeneration described in AT and ATLD is due to mutations in genes encoding for ATM and Mre11, respectively. ATM recognizes and signals DNA DSBs to the cell cycle checkpoints and the DNA repair machinery. The Mre11 DNA repair complex, composed of Rad50, Mre11 and Nbs1 proteins, is involved in DNA damage recognition, DNA repair, and initiating cell cycle checkpoints. ATM and the Mre11 complex combine to recognize and signal DNA DSBs. AOA1 is caused by mutations in the gene encoding aprataxin (APTX), a nuclear protein that interacts with several DNA repair proteins, including XRCC1, Polβ, DNA ligase III, PARP-1, and p53. It functions in the endprocessing of DNA SSBs removing 3'-phosphate, 5'-phosphate, and 3'-phosphoglycolate ends. AOA2 is caused by autosomal recessive mutations in the gene encoding senataxin (SETX). SETX is a member of the superfamily I DNA/RNA helicases, likely involved in oxidative DNA damage response. *SETX* mutations have been also linked to juvenile ALS. Overall, spinocerebellar ataxias deficient in DNA damage responses represent the most robust set of data linking mutations in DNA repair genes to neurodegeneration (Gueven et al., 2007; Martin, 2008).

## 6.2 Huntington's disease

Huntington's disease (HD) is a progressive neurodegenerative disorder resulting in cognitive impairment, choreiform movements and death which usually occurs 15–20 years after the onset of the symptoms. The disease is also characterized by psychiatric and behavioural disturbances. HD is an autosomal dominant disorder caused by a CAG repeat expansion within exon 1 of the gene encoding for huntingtin (*IT15*) on chromosome 4. In the normal population the number of CAG repeats is maintained below 35, while in individuals affected by HD it ranges from 35 to more than 100, resulting in an expanded polyglutamine segment in the protein. Age at onset of the disease is inversely correlated with the CAG repeat length; moreover the length of the expanded polyglutamine segment seems to be related to the rate of clinical progression of neurological symptoms and to the progression of motor impairment, but not to psychiatric symptoms. Somatic CAG repeat expansion in the gene encoding for huntingtin has been observed in several HD tissues, including the striatum which is the region most affected by the disease and the OGG1 protein has been involved in somatic CAG repeat expansion in HD, suggesting that it might contribute to disease age at onset (Kovtun et al. 2007). We recently observed a weak borderline association between the *OGG1* Ser326Cys polymorphism and HD age at onset in a small group of 91 HD subjects (Coppedè et al. 2010d). However, replication of the study in a cohort of more than 400 HD individuals failed to confirm the association between *OGG1* Ser326Cys and HD age at onset (Taherzadeh-Fard et al., 2010).

### 6.3 Multiple sclerosis

Multiple sclerosis (MS) has been classically regarded as an inflammatory demyelinating disease of the central nervous system. In recent years, it is also becoming increasingly apparent that there is a significant neurodegenerative component in the disease (Moore, 2010). MS is a complex autoimmune disease with a prominent genetic component. The primary genetic risk factor is the human leukocyte antigen *(HLA)-DRB1*1501* allele; however, much of the remaining genetic contribution to MS remains to be elucidated. Briggs and collaborators screened 1,343 MS cases and 1,379 healthy controls of European ancestry for a total of 485 single nucleotide polymorphisms within 72 genes related to DNA repair pathways. Only a single nucleotide polymorphism (rs1264307) within the general transcription factor IIH polypeptide 4 gene (*GTF2H4*), a nucleotide excision repair gene, was significantly associated with MS risk (OR = 0.7) after correcting for multiple testing. However, using a nonparametric approach comprising the Random Forests and CART algorithms, authors observed evidence for a predictive relation for MS based on 9 variants in nucleotide excision repair (rs4134860, rs2974754, rs7783714, rs4134813, rs2957873 and rs4150454), homologous recombination (rs9562605), and nonhomologous end-joining genes (rs9293329 and rs1231201). Specifically, variants within nucleotide excision repair genes were most prominent among predictors of MS (Briggs et al., 2010). Variants of DNA repair genes, particularly *BRCA2* (rs1801406) and *XRCC5* (rs207906), might also increase the risk for the development of secondary acute promyelocytic leukemia in MS patients (Hasan et al., 2011).

### 6.4 Diseases caused by mutations of NER genes

Xeroderma pigmentosum (XP), Cockayne's syndrome (CS) and trichothiodystropy (TTD) represent a clinically heterogeneous group of progeroid syndromes characterized by defects in NER proteins. A subset of these patients exhibits neurological dysfunction and neurodegeneration, and many XP patients have high cancer predisposition, thus linking DNA repair defects to premature aging, cancer and neurodegeneration. Several studies performed in mice, as well as in cell cultures, suggest that neurodegeneration in XP and CS patients might arise as a consequence of impaired repair of oxidative DNA lesions caused by mutations of NER genes. Details are provided in our recent updated review (Coppedè & Migliore, 2010)

## 7. Conclusions

The present chapter describes the current knowledge concerning DNA repair genes and neurodegeneration. Studies in ataxias (section 6.1) have undoubtedly linked genes involved in DNA repair to neurodegeneration. These observations, alongside with evidence of increased DNA damage in affected brain regions, have driven researchers to search for variant and polymorphisms of DNA repair genes in major neurodegenerative diseases such as AD, ALS and PD. Studies in sporadic late onset AD patients (Section 3) suggest that common polymorphisms of BER genes, namely *OGG1* Ser326Cys, *APEX1* Asp148Glu, and *XRCC1* (Arg194Trp, Arg280His and Arg399Gln) are unlikely to represent major AD risk factors. However, further studies are required to replicate and clarify the associations observed between *PARP-1* haplotypes and disease risk. Moreover, the power of these studies was limited by the sample size of case-control groups (Table 2), gene-gene interactions were missing, and only common polymorphisms have been included with little or no attention paid to rare gene variants. Concerning ALS, although results are still inconclusive, some studies performed in northern Europe suggest a possible association between the *APEX1* Asp148Glu polymorphism and disease risk, the *OGG1* Ser326Cys polymorphism was associated with increased ALS risk in Italy, and *XRCC1* variants gave conflicting results in different populations (Table 3). Overall, these studies (Section 4) suggest the need of further investigation aimed at addressing the contribution of haplotypes, gene-gene and gene-environment interactions. There is evidence for a contribution of *POLG1* mutations in PD, and parkin seems to be involved in mtDNA repair, thus strengthening the contribution of mtDNA mutations to disease pathogenesis (Section 5). Increasing evidence suggests that BER proteins might be involved in CAG repeat expansion in somatic cells of HD individuals (Section 6.2), however studies aimed at addressing the possible contribution of variant of BER genes to disease age at onset are still in their beginnings. Recent evidence also suggests a possible contribution of NER genes in MS (Section 6.3), and the impaired ability to repair oxidative DNA damage might cause neurodegeneration observed in progeroid syndromes caused by mutations of NER genes (Section 6.4). In summary, increasing evidence supports a role for DNA repair genes in neurodegeneration, making this field a promising area for further investigation.

# 8. References

Anvret, A., Westerlund, M., Sydow, O., Willows, T., Lind, C., Galter, D. & Belin, A.C. (2010) Variations of the CAG trinucleotide repeat in DNA polymerase gamma (POLG1) is associated with Parkinson's disease in Sweden. *Neurosci Lett.* 485(2): 117-20.

Bekris, L.M., Mata, I.F. & Zabetian, C.P. (2010) The genetics of Parkinson disease. *J Geriatr Psychiatry Neurol.* 23(4): 228-42.

Bogdanov, M., Brown, R.H., Matson, W., Smart, R., Hayden, D., O'Donnell, H., Beal, F.M. & Cudkowicz, M. (2000) Increased oxidative damage to DNA in ALS patients. *Free Radic Biol Med.* 29(7): 652-8.

Briggs, F.B., Goldstein, B.A., McCauley, J.L., Zuvich, R.L., De Jager, P.L., Rioux, J.D., Ivinson, A.J., Compston, A., Hafler, D.A., Hauser, S.L., Oksenberg, J.R., Sawcer, S.J., Pericak-Vance, M.A., Haines, J.L., Barcellos, L.F. & International Multiple Sclerosis Genetics Consortium (2010) Variation within DNA repair pathway genes and risk of multiple sclerosis. *Am J Epidemiol.* 172(2): 217-24.

Brookmeyer, R., Johnson, E., Ziegler-Graham, K. & Arrighi, H.M. (2007) Forecasting the global burden of Alzheimer's disease. *Alzheimers Dement.* 3(3): 186-191.

Coppedè, F., Mancuso, M., Lo Gerfo, A., Manca, M.L., Petrozzi, L., Migliore, L., Siciliano, G., & Murri, L. (2007a) A Ser326Cys polymorphism in the DNA repair gene hOGG1 is not associated with sporadic Alzheimer's disease. *Neurosci Lett.* 414(3): 282-5.

Coppedè, F., Mancuso, M., Lo Gerfo, A., Carlesi, C., Piazza, S., Rocchi, A., Petrozzi, L., Nesti, C., Micheli, D., Bacci, A., Migliore, L., Murri, L. & Siciliano, G. (2007b) Association of the hOGG1 Ser326Cys polymorphism with sporadic amyotrophic lateral sclerosis. *Neurosci Lett.* 420(2): 163-8.

Coppedè, F. & Migliore, L. (2010) DNA repair in premature aging disorders and neurodegeneration. *Curr Aging Sci* 3 (1): 3-19.

Coppedè, F., Lo Gerfo, A., Carlesi, C., Piazza, S., Mancuso, M., Pasquali, L., Murri, L., Migliore, L. & Siciliano, G. (2010a) Lack of association between the APEX1 Asp148Glu polymorphism and sporadic amyotrophic lateral sclerosis. *Neurobiol Aging.* 31(2): 353-5.

Coppedè, F., Migheli, F., Lo Gerfo, A., Fabbrizi, M.R., Carlesi, C., Mancuso, M., Corti, S., Mezzina, N., del Bo, R., Comi, G.P., Siciliano, G. & Migliore, L. (2010b) Association study between XRCC1 gene polymorphisms and sporadic amyotrophic lateral sclerosis. *Amyotroph Lateral Scler.* 11(1-2): 122-4.

Coppedè, F., Ceravolo, R., Migheli, F., Fanucchi, F., Frosini, D., Siciliano, G., Bonuccelli, U. & Migliore, L. (2010c) The hOGG1 Ser326Cys polymorphism is not associated with sporadic Parkinson's disease. *Neurosci Lett.* 473(3): 248-51.

Coppedè, F., Migheli, F., Ceravolo, R., Bregant, E., Rocchi, A., Petrozzi, L., Unti, E., Lonigro, R., Siciliano, G. & Migliore, L. (2010d) The hOGG1 Ser326Cys polymorphism and Huntington's disease. Toxicology 278(2):199-203.

Coppedè, F. (2011) Variants and polymorphisms of DNA base excision repair genes and Alzheimer's disease. *J Neurol Sci.* 300(1-2): 200-1.

Davidzon, G., Greene, P., Mancuso, M., Klos, K.J., Ahlskog, J.E., Hirano, M. & DiMauro, S. (2006) Early-onset familial parkinsonism due to POLG mutations. *Ann Neurol.* 59(5): 859-62

Dogru-Abbasoglu, S., Inceoglu, M., Parildar-Karpuzoglu, H., Hanagasi, H.A., Karadag, B., Gurvit, H., Emre, M., Aykac-Toker, G. & Uysal, M. (2006) Polymorphisms in the DNA repair genes XPD (ERCC2) and XPF (ERCC4) are not associated with sporadic late-onset Alzheimer's disease. *Neurosci Lett.* 404(3):258-61.

Doğru-Abbasoğlu, S., Aykaç-Toker, G., Hanagasi, H.A., Gürvit, H., Emre, M. & Uysal, M. (2007) The Arg194Trp polymorphism in DNA repair gene XRCC1 and the risk for sporadic late-onset Alzheimer's disease. *Neurol Sci.* 28(1):31-4.

Dorszewska, J., Kempisty, B., Jaroszewska-Kolecka, J., Rózycka, A., Florczak, J., Lianeri, M., Jagodziński, P.P. & Kozubski, W. (2009) Expression and polymorphisms of gene 8-oxoguanine glycosylase 1 and the level of oxidative DNA damage in peripheral blood lymphocytes of patients with Alzheimer's disease. *DNA Cell Biol.* 28(11): 579-88.

Eerola, J., Luoma, P.T., Peuralinna, T., Scholz, S., Paisan-Ruiz, C., Suomalainen, A., Singleton ,A.B. & Tienari, P.J. (2010) POLG1 polyglutamine tract variants associated with Parkinson's disease. *Neurosci Lett.* 477(1): 1-5.

Fang, F., Umbach, D.M., Xu, Z., Ye, W., Sandler, D.P., Taylor, J.A. & Kamel, F. (2010) No association between DNA repair gene XRCC1 and amyotrophic lateral sclerosis. *Neurobiol Aging.* Epub Aug 16.

Ferrante, R.J., Browne, S.E., Shinobu, L.A., Bowling, A.C., Baik, M.J., MacGarvey, U., Kowall, N.W., Brown, R.H. Jr. & Beal, M.F. Evidence of increased oxidative damage in both sporadic and familial amyotrophic lateral sclerosis. *J Neurochem.* 69(5): 2064-74.

Fleck, O & Nielsen, O. (2004) DNA repair. *J Cell Sci* 117(Pt4): 515-7.

Greenway, M.J., Alexander, M.D., Ennis, S., Traynor, B.J., Corr, B., Frost, E., Green, A. & Hardiman, O. (2004) A novel candidate region for ALS on chromosome 14q11.2. *Neurology.* 63(10): 1936-8.

Gueven, N., Chen, P., Nakamura, J., Becherel, O.J., Kijas, A.W., Grattan-Smith, P. & Lavin, M.F. (2007) A subgroup of spinocerebellar ataxias defective in DNA damage responses. *Neuroscience.* 145(4): 1418-25.

Hasan, S.K., Buttari, F., Ottone, T., Voso, M.T., Hohaus, S., Marasco, E., Mantovani, V., Garagnani, P., Sanz, M.A., Cicconi, L., Bernardi ,G., Centonze, D. & Lo-Coco, F. (2011) Risk of acute promyelocytic leukemia in multiple sclerosis: Coding variants of DNA repair genes. *Neurology.* 76(12): 1059-65.

Hayward, C., Colville, S., Swingler, R.J. & Brock, D.J. (1999) Molecular genetic analysis of the APEX nuclease gene in amyotrophic lateral sclerosis. *Neurology.* 52(9): 1899-901.

Hirano, M., Yamamoto, A., Mori, T., Lan, L., Iwamoto, T.A., Aoki, M., Shimada, K., Furiya, Y., Kariya, S., Asai, H., Yasui, A., Nishiwaki, T., Imoto, K., Kobayashi, N., Kiriyama, T., Nagata, T., Konishi, N., Itoyama, Y. & Ueno S. (2007) DNA single-strand break repair is impaired in apraxian-related ataxia. *Ann Neurol* 61(2): 162-74.

Hudson, G., Schaefer, A.M., Taylor, R.W., Tiangyou, W., Gibson, A., Venables, G., Griffiths, P., Burn, D.J., Turnbull, D.M. & Chinnery, P.F. (2007) Mutation of the linker region of the polymerase gamma-1 (POLG1) gene associated with progressive external ophthalmoplegia and Parkinsonism. *Arch Neurol.* 64(4): 553-7.

Hudson, G., Tiangyou, W., Stutt, A., Eccles, M., Robinson, L., Burn, D.J. & Chinnery, P.F. (2009) No association between common POLG1 variants and sporadic idiopathic Parkinson's disease. *Mov Disord.* 24(7): 1092-4.

Infante, J., Llorca, J., Mateo, I., Rodríguez-Rodríguez, E., Sánchez-Quintana, C., Sánchez-Juan, P., Fernández-Viadero, C., Peña, N., Berciano, J. & Combarros, O. (2007) Interaction between poly(ADP-ribose) polymerase 1 and interleukin 1A genes is associated with Alzheimer's disease risk. *Dement Geriatr Cogn Disord.* 23(4): 215-8.

Infante, J., Sánchez-Juan, P., Mateo, I., Rodríguez-Rodríguez, E., Sánchez-Quintana, C., Llorca, J., Fontalba, A., Terrazas, J., Oterino, A., Berciano, J. & Combarros O. (2007) Poly (ADP-ribose) polymerase-1 (PARP-1) genetic variants are protective against Parkinson's disease. *J Neurol Sci.* 256(1-2): 68-70.

Jiang, G., Xu, L., Wang, L., Song, S. & Zhu, C. (2007) Association study of human MTH1 Ile45Thr polymorphism with sporadic Parkinson's disease. *Eur Neurol.* 59(1-2): 15-7.

Kao, S.Y. (2009) Regulation of DNA repair by parkin. *Biochem Biophys Res Commun* 382(2): 321-5.

Kass, E.M. & Jasin, M. (2010) Collaboration and competition between DNA double-strand break repair pathways. *FEBS Lett.* 584(17): 3703-8.

Kikuchi, H., Furuta, A., Nishioka, K., Suzuki, S.O., Nakabeppu, Y. & Iwaki, T. (2002) Impairment of mitochondrial DNA repair enzymes against accumulation of 8-oxoguanine in the spinal motor neurons of amyotrophic lateral sclerosis. *Acta Neuropathol.* 103(4): 408-14.

Kisby, G.E., Milne, J. & Sweatt, C. (1997) Evidence of reduced DNA repair in amyotrophic lateral sclerosis brain tissue. *Neuroreport.* 8(6): 1337-40.

Kovtun, I.V., Liu, Y., Bjoras, M., Klungland, A., Wilson, S.H. & McMurray, C.T. (2007) OGG1 initiates age-dependent CAG trinucleotide expansion in somatic cells. *Nature* 447 (7143): 447-52.

Kunkel, T.A. & Erie, D.A. (2005) DNA mismatch repair. *Annu Rev Biochem* 74: 681-710.

Liu, H.P., Lin, W.Y., Wu, B.T., Liu, S.H., Wang, W.F., Tsai, C.H., Lee, C.C. & Tsai, F.J. (2010) Evaluation of the poly(ADP-ribose) polymerase-1 gene variants in Alzheimer's disease. *J Clin Lab Anal.* 24(3): 182-6.

Love, S., Barber, R. & Wilcock, G.K. (1999) Increased poly(ADP-ribosyl)ation of nuclear proteins in Alzheimer's disease. *Brain.* 122(Pt 2): 247-53

Lovell, M.A., Xie, C. & Markesbery, W.R. (2000) Decreased base excision repair and increased helicase activity in Alzheimer's disease brain. *Brain Res* 855(1): 116-23 .

Luoma, P.T., Eerola, J., Ahola, S., Hakonen, A.H., Hellström, O., Kivistö, K.T., Tienari, P.J. & Suomalainen, A. (2007) Mitochondrial DNA polymerase gamma variants in idiopathic sporadic Parkinson disease. *Neurology.* 69(11): 1152-9.

Mao, G., Pan, X., Zhu, B.B., Zhang, Y., Yuan, F., Huang, J., Lovell, M.A., Lee, M.P., Markesbery, W.R., Li, G.M. & Gu, L. (2007) Identification and characterization of OGG1 mutations in patients with Alzheimer's disease. *Nucleic Acids Res.* 35(8): 2759-66.

Martin, L.J. (2008) DNA damage and repair: relevance to mechanisms of neurodegeneration. *J Neuropathol Exp Neurol.* 67(5): 377-87.

Moore, G.R. (2010) Current concepts in the neuropathology and pathogenesis of multiple sclerosis. *Can J Neurol Sci.* 37 (Suppl 2) :S5-15.

Orsucci, D., Caldarazzo Ienco, E., Mancuso, M. & Siciliano, G. (2011) POLG1-Related and other "Mitochondrial Parkinsonisms": an Overview. *J Mol Neurosci.* Epub Jan 8.

Parildar-Karpuzoğlu, H., Doğru-Abbasoğlu, S., Hanagasi, H.A., Karadağ, B., Gürvit, H., Emre, M. & Uysal, M. (2008) Single nucleotide polymorphisms in base-excision repair genes hOGG1, APE1 and XRCC1 do not alter risk of Alzheimer's disease. *Neurosci Lett.* 442(3): 287-91.

Qian, Y., Chen, W., Wu, J., Tao, T., Bi, L., Xu, W., Qi, H., Wang, Y. & Guo L. (2010) Association of polymorphism of DNA repair gene XRCC1 with sporadic late-onset Alzheimer's disease and age of onset in elderly Han Chinese. *J Neurol Sci.* 295(1-2): 62-5.

Remes, A.M., Hinttala, R., Kärppä, M., Soini, H., Takalo, R., Uusimaa, J. & Majamaa, K. (2008) Parkinsonism associated with the homozygous W748S mutation in the POLG1 gene. *Parkinsonism Relat Disord.* 14(8): 652-4.

Rothfuss, O., Fischer, H., Hasegawa, T., Maisel, M., Leitner, P., Miesel, F., Sharma, M., Bornemann, A., Berg, D., Gasser, T. & Patenge N. (2009) Parkin protects mitochondrial genome integrity and supports mitochondrial DNA repair. *Hum Mol Genet.* 18(20): 3832-50.

Satoh, J. & Kuroda, Y. (2000) A valine to methionine polymorphism at codon 83 in the 8-oxo-dGTPase gene MTH1 is not associated with sporadic Parkinson's disease. *Eur J Neurol.* 7(6): 673-7.

Subba Rao, K. (2007) Mechanisms of disease: DNA repair defects and neurological disease. *Nat Clin Pract Neurol* 3(3): 162-72.

Taanman, J.W. & Schapira, A.H. (2005) Analysis of the trinucleotide CAG repeat from the DNA polymerase gamma gene (POLG) in patients with Parkinson's disease. *Neurosci Lett.* 376(1): 56-9.

Taherzadeh-Fard, E., Saft, C., Wieczorek, S., Epplen, J.T. & Arning, L. (2010) Age at onset in Huntington's disease: replication study on the associations of ADORA2A, HAP1 and OGG1. *Neurogenetics.* 11(4): 435-9.

Takahashi, T., Tada, M., Igarashi, S., Koyama, A., Date, H., Yokoseki, A., Shiga, A., Yoshida, Y., Tsuji, S., Nishizawa, M. & Onodera O. (2007) Aprataxin, causative gene product for EAOH/AOA1, repairs DNA single-strand breaks with damaged 3'-phosphate and 3'-phosphoglycolate ends. *Nucleic Acids Res* 35(11): 3797-809.

Tiangyou, W., Hudson, G., Ghezzi, D., Ferrari, G., Zeviani, M., Burn, D.J. & Chinnery, P.F. (2006) POLG1 in idiopathic Parkinson disease. Neurology. 67(9):1698-700.

Tomkins, J., Dempster, S., Banner, S.J., Cookson, M.R. & Shaw, P.J. (2000) Screening of AP endonuclease as a candidate gene for amyotrophic lateral sclerosis (ALS). *Neuroreport.* 11(8): 1695-7.

Weissman, L., de Souza-Pinto, N.C., Stevnsner, T. & Bohr, V.A. (2007) DNA repair, mitochondria, and neurodegeneration. *Neuroscience* 145 (12): 1318-29.

Weissman, L., Jo, D.G., Sørensen, M.M., de Souza-Pinto, N.C., Markesbery, W.R., Mattson, M.P. & Bohr V.A. (2007b) Defective DNA base excision repair in brain from individuals with Alzheimer's disease and amnestic mild cognitive impairment. *Nucleic Acids Res* 35(16): 5545-55.

# Part 3

# Telomeres and DNA Repair

# 9

# Roles of DNA Repair Proteins in Telomere Maintenance

Masaru Ueno
*Hiroshima University*
*Japan*

## 1. Introduction

Most eukaryots have specialized protein-DNA complexs, called telomeres at the ends of natural linear chromosomes. Telomeric DNA consists of a tandemly repeated G-rich sequence. The lengths of telomeric DNAs in *S. pombe*, *S. cerevisiae*, and human are ~300 nucleotids, ~350 nucleotides, and ~10 kb, respectively. The ends of the telomeric DNA have 3′ single-stranded overhangs. The protein components of telomeres consists of double-stranded telomere-binding proteins, such as human TRF1 and TRF2, S. pombe Taz1, and single-stranded telomere-binding proteins, such as *S. cerevisiae* Cdc13, *S. pombe* Pot1, and human POT1. DNA double-strand breaks (DSBs) must be repaired to maintain genomic integrity. In contrast, natural chromosome ends should not be recognized as DSBs. The telomere is capped to protect from DNA repair activity. If this capping function is lost, this uncapped telomere is recognized as DNA damage and becomes substrate for DNA repair proteins. The first step in homologous recombination (HR) repair is processing of DNA ends by 5′ to 3′ degradation to create 3′ single-stranded overhangs. The proteins involved in this steps include *S. cerevisiae* Mre11-Rad50-Xrs2 complex (MRX), Sae2, Sgs1, and Dna2. Recent works revealed that proteins involved in the processing of DNA DSB ends are also involved in the processing of capped or uncapped telomere. These facts raised new question of how these proteins are regulated at telomere ends. This chapter will focus on the roles of proteins involved in the processing of DBS ends at capped (functional) and uncapped (dysfunctional) telomere in *S. pombe*, *S. cerevisiae* and human. This chapter will also focus on the functional interactions between telomere-binding proteins and proteins involved in the processing of DBS ends. Resent works revealed that double-stranded and single-stranded telomere-binding proteins play critical roles to control proteins involved in DNA repair at chromosome ends.

## 2. Roles of proteins involved in DNA end-processing in telomere maintenance

DNA DSBs are repaired by HR or non-homologous end-joining (NHEJ) [1]. *S. cerevisiae* MRX is involved in both HR and NHEJ [2]. MRX is suggested to be involved in the processng of DSB ends in HR repair. Recently, several other  proteins involved the processing have been discovered. Some of the these proteins are also involved in the processing of telomere ends. In this section, the roles of these proteins in telomere maintenance will be discussed.

## 2.1 Roles of proteins involved in DNA end-processing at DSB ends

Role of *S. cerevisiae* MRX in HR is well studied both in vivo and in vitro (Mimitou and Symington 2009) (Mimitou and Symington 2008) (Zhu et al. 2008) (Gravel et al. 2008) (Cejka et al. 2010) (Niu et al. 2010). MRX cooperates with Sae2 to initiate 5′ resection at DNA DSB end. Although both MRX and Sae2 have nuclease activities, it remains unclear the contribution of these nucleases to DSB resection. The resultant 3′ single-stranded ovehangs are further resected by two redundant pathways. One is dependent on Sgs1 helicase, a conserved RecQ family member, and the Dna2. Dna2 has both helicase and nuclease domains, but nuclease activity is enough for DSB resection (Zhu et al. 2008). The other is dependent on Exo1 5′-3′ exonuclease. *S. cerevisiae* Yku70-Yku80 heterodimer (Ku) binds to DSB ends and recruits downstream NHEJ factors (Critchlow and Jackson 1998). Ku inhibits 5′ resection by MRX (Mimitou and Symington 2010) (Shim et al. 2010). Similar model is proposed in *S. pombe*(Tomita et al. 2003). *S. pombe* Mre11-Rad50-Nbs1 (S. cerevisiae Xrs2 homologue) complex (MRN) is also suggested to be involved in 5′ resection at DNA DSB end. *S. pombe* Ku also inhibits 5′ resection by MRN. In the absence of MRN, Exo1 can resect DSB ends. Contribution of *S. pombe* RecQ helicase Rqh1 and Dna2 in the resection of DSB ends remains unclear. It has been shown that human BLM, a RecQ helicase family, and DNA2 interact to resect DNA end and helicase activity of BLM and nuclease activity of DNA2 are required for this reaction (Nimonkar et al. 2011). The functional conservation of these proteins from yeast to human suggests that the functions of these proteins in *S. pombe* are also conserved.

## 2.2 Roles of proteins involved in DNA end-processing in telomere maintenance in *S. pombe* and in *S. cerevisiae*

Telomere ends should not be recognized as DSB ends, because telomere ends should no be repaired by HR or NHEJ. However, proteins involved in HR or NHEJ are also involved in telomere maintenance (Longhese et al. 2010). The chromosome end replicated by lagging-strand synthesis has 3′ single-stranded overhangs. In contrast, the chromosome end replicated by leading-strand synthesis is blunt-end. However, most eukaryotes have 3′ single-stranded overhangs at both ends, suggesting that the chromosome end replicated by leading-strand synthesis is resected (Wellinger et al. 1996; Makarov, Hirose, and Langmore 1997). *S. cerevisiae* MRX is suggested to be involved in this resection (Diede and Gottschling 2001). However, MRX independent resection has been suggested, which may be produced at lagging-strand telomere after DNA replication without any nuclease activity (Larrivee, LeBel, and Wellinger 2004). MRX mainly binds to the leading-strand telomere, further suggesting that MRX is involved in this resection at leading-strand telomere (Faure et al. 2010). An inducible short telomere assay revealed that artificial telomere ends is resected by the same DNA repair factors (Bonetti et al. 2009) (Longhese et al. 2010) (Iglesias and Lingner 2009) (Fig. 1). MRX and Sae2 act in the same resection pathway. Concomitant inactivation of Sae2 and Sgs1 abolishes end resection, suggesting that they have redundant function for the resection. Dna2 acts redundantly with Exo1, but not with Sgs1, suggesting that Dna2 supports Sgs1 activity. The lack of Sgs1, Dna2 or Exo1 by itself does not affect the resection, suggesting that Exo1 and Sgs1-Dna2 may less important for the resection than MRX and Sae2. These results were obtained by using artificial telomere, which initially produces blunt-end telomere by nuclease. However, leading-strand synthesis in wild-type cells also produces blunt-end telomere. Consistently, Sae2 and Sgs1 also play redundant functions in natural telomere end-processing (Bonetti et al. 2009), suggesting that an inducible short telomere assay mimic wild-type telomere end. In wild-type *S. cerevisiae* cells, 3' single-stranded overhangs increase in S phase at telomeres (Wellinger, Wolf, and Zakian 1993) (Dionne and Wellinger 1996). In contrast, 3' single-stranded overhangs can be detected at telomeres throughout the cell cycle in the absence of *S. cerevisiae* Ku, suggesting that Ku inhibits resection at telomere (Gravel et al. 1998) (Polotnianka, Li, and Lustig 1998). This function of Ku is conserved in *S. pombe* Ku (Kibe et al. 2003). However, proteins involved in the resection of telomere ends are not well studied in *S. pombe*. In *S. pombe*, Dna2 is involved in the resection of telomere ends (Tomita et al. 2004).

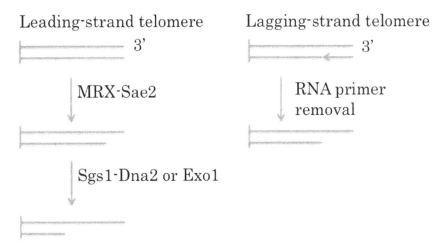

Fig. 1. Model for DNA end-processing at *S. cerevisiae* telomere. DNA replication will create blunt-end at leading-strand telomere and 3' single-stranded overhangs at lagging-strand telomere after removal of the last RNA primer. Similar to the case at DSB ends, MRX and Sae2 play a major role to produce 3' single-stranded overhangs at telomeres. Sgs1-Dna2 and Exo1 can provide compensatory activities to produce 3' single-stranded overhangs.

### 2.3 Proteins involved in DNA end-processing in *S. pombe taz1Δ* cells

*S. pombe* Taz1 binds telomeric double-stranded DNA (Cooper, Watanabe, and Nurse 1998). Deletion of *taz1* causes massive telomere elongation. Asynchronous wild-type *S. pombe* cells have small amount of 3' single-stranded overhangs (Kibe et al. 2003). In contrast, *taz1* disruptant has very long 3' single-stranded overhangs (Tomita et al. 2003). In this mutant background, roles of MRN, Ku, Dna2, and Exo1 are studied (Fig. 2). MRN and Dna2 are responsible for the production of 3' single-stranded overhangs (Tomita et al. 2004). But, 3' single-stranded overhangs are produced by concomitant deletion of Ku and MRN, suggesting that unknown nuclease can produce the overhangs in the absence of both MRN and Ku in *taz1* disruptant. Exo1 is not involved in this activity. Telomere ends in *taz1* disruptant is partially unprotected. Indeed, RPA foci and Rad22[Rad52] foci are produced at telomere in *taz1* disruptant (Carneiro et al. 2010). Therefore, proteins involved in the resection in *taz1* disruptant may not be same as that in wild-type cells. However, Dna2 is involved in the resection in both wild-type and *taz1Δ* background, suggesting that some of the proteins involved in the resection in *taz1* disruptant are also involved in the resection in wild-type cells.

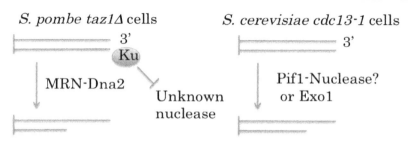

Fig. 2. Model for DNA end-processing at dysfunctional telomere. 3' single-stranded overhangs are produced by MRN and Dna2 in *S. pombe taz1Δ* cells (Left). Ku inhibits unknown nuclease, but not nuclease activity depending on MRN-Dna2. 3' single-stranded overhangs are produced by Pif1 or Exo1 in *S. cerevisiae cdc13-1* cells (Right). Unknown nuclease is suggested to function together with Pif1 helicase.

### 2.4 Proteins involved in DNA end-processing in *S. cerevisiae cdc13-1* cells
*S. cerevisiae* Cdc13 binds telomeric single-stranded DNA (Garvik, Carson, and Hartwell 1995). *cdc13-1* temperature sensitive mutant is used to study proteins that are involved in the resection at uncapped telomeres (Lydall 2009). These studies revealed that the single-stranded DNA at telomeres in *cdc13-1* mutants resembles a DSB end. However, there are some differences between these ends (Fig. 2). In *cdc13-1* mutants at high temperature, Pif1 helicase and Exo1 are redundantly involved in the resection of uncapped telomere (Dewar and Lydall 2010). It remains unclear how Pif1 contribute to the resection. As Pif1 has no nuclease activity, involvement of the unknown nuclease is suggested to cleave single-stranded DNA unwound by Pif1 helicase. Sgs1 also contributes to resection of telomeres in *cdc13-1* mutants (Ngo and Lydall 2010). However, unlike *pif1 exo1* double mutant, resection of telomeres in *cdc13-1* mutant background occurs in *sgs1 exo1* double mutant, demonstrating that Pif1 and Exo1 play major roles in the resection of uncapped telomere at high temperature.

## 3. Roles of RecQ helicase in telomere maintenance

RecQ helicase is conserved from *E. coli*. to human and play a critical role in genome stability (Bernstein, Gangloff, and Rothstein 2010). Werner Syndrome (WS) is a premature aging syndrome resulting from loss of function of one of the human RecQ helicase WRN. The roles of *S. cerevisiae* RecQ helicase Sgs1 in homologous recombination are well studied. RecQ helicase is also involved in telomere maintenance especially at dysfunctional telomere. In this section, roles of RecQ helicase in telomere maintenance will be discussed. Functional interaction between RecQ helicase and POT1 in *S. pombe* and in human will be also discussed.

### 3.1 Roles of RecQ helicase in DNA repair
*S. cerevisiae* RecQ helicase Sgs1 is involved in several steps in HR (Ashton and Hickson 2010). As discussed above, Sgs1 is involved in the resection of DSB ends. Genetic and in vitro studies also suggest that Sgs1 inhibits unscheduled recombinogenic events, but promotes the resolution of recombination intermediates. Strains deleted for *SGS1* display hyperrecombination phenotype, but are defective in DNA damage-induced heteroallilic

recombination (Watt et al. 1996) (Onoda et al. 2001). *S. cerevisiae* Sgs1 and Top3 migrate and disentangle a double Holliday junction (dHJ) to produce non-crossover recombination products in vitro (Cejka et al. 2010). This activity is also detected in human RecQ helicase BLM and human topoisomerase IIIa (Wu and Hickson 2003). Mutant of *S. pombe* RecQ helicase *rqh1* is sensitive to DNA damage and has high frequency of recombination under normal growth conditions and following DNA damage, suggesting that Rqh1 is also involved in HR repair both positively and negatively (Murray et al. 1997) (Stewart et al. 1997) (Doe et al. 2000) (Caspari, Murray, and Carr 2002).

### 3.2 Roles of RecQ helicase in telomere maintenance in *S. cerevisiae*
As mutation of *S. cerevisiae SGS1* does not affect telomere length, Sgs1 has no apparent role in telomere maintenance in the presence of telomerase activity (Watt et al. 1996). However, the double mutant between telomerase RNA component *TLC1* and *SGS1* shorten telomeres at an increased rate per population doubling and Sgs1 affects telomere-telomere recombination in the absence of telomerase, demonstrating that Sgs1 plays roles at telomere in the absence of telomerase activity (Johnson et al. 2001) (Cohen and Sinclair 2001) (Huang et al. 2001). X-shaped structures are accumulated at telomeres in senescing *tlc1 sgs1* double mutants and these structures are suggested to be the recombination intermediates related to hemicatenanes. This result suggests that Sgs1 is required for the efficient resolution of telomere recombination intermediates in the absence of telomerase (Lee et al. 2007; Chavez, Tsou, and Johnson 2009).

### 3.3 Roles of RecQ helicase in telomere maintenance in mammals
Human RecQ helicase WRN binds to telomere in S phase in primary human IMR90 fibroblasts and is required for efficient replication of the G-rich telomeric DNA strand, suggesting that WRN is required for replication of telomeric DNA in telomerase-negative primary human fibroblasts (Crabbe et al. 2004). In Werner syndrome (WS) cells, replication-associated telomere loss results in the chromosome fusions, causing genomic instability (Crabbe et al. 2007). The life span of normal human skin fibroblasts derived from WS patients can be extended by expression of the catalytic subunit human telomerase reverse transcriptase (hTERT) (Wyllie et al. 2000; Ouellette et al. 2000). These facts demonstrate that dysfunctional telomere is a major determinant of the premature aging syndrome and WRN plays important role at dysfunctional telomere and telomerase activity can suppress the defect in WRN deficient cells. Consistently, Wrn-deficient mouse, which has telomerase activity, has no disease phenotype, but telomerase-Wrn double null mouse elicits a Werner-like premature aging syndrome (Chang et al. 2004). Telomere sister chromatid exchange (T-SEC) increases in cells from telomerase-Wrn double null mouse, suggesting that WRN are required to repress inappropriate telomere recombination (Laud et al. 2005) (Multani and Chang 2007). Human WRN and other RecQ helicase BLM co-localizes with telomere in human cancer cells that lack telomerase, ALT cells (Johnson et al. 2001; Opresko et al. 2004; Lillard-Wetherell et al. 2004). As telomeres in ALT cells are maintained by HR, human WRN and BLM are suggested to be involved in the recombination at telomere in ALT cells. Possible roles of WRN in telomere maintenance will be discussed in the next section.

### 3.4 Functional interaction between RecQ helicase and POT1 in *S. pombe* and in human
Pot1 is conserved from *S. pombe* to human and binds to single-stranded telomeric DNA sequence specifically (Baumann and Cech 2001). Deletion of *S. pombe pot1* causes rapid

telomere loss and chromosome circularization and this circularization is mediated by single strand annealing (SSA) (Wang and Baumann 2008). In *S. cerevisiae*, Rad52, Rad1/Rad10 nuclease, RPA, Srs2 helicase, and Sgs1 are involved in SSA (Fishman-Lobell and Haber 1992) (Ivanov and Haber 1995) (Ivanov et al. 1996) (Paques and Haber 1997), (Sugawara, Ira, and Haber 2000; Umezu et al. 1998) (Zhu et al. 2008). Consistently, the double mutants between *S. pombe* homologue of these proteins and *pot1* are synthetically lethal (Wang and Baumann 2008). *S. pombe* telomerase disruptant can survive either by maintaining telomere by HR or chromosome circularization(Nakamura, Cooper, and Cech 1998). In contrast, *pot1* disruptant survives only by chromosome circularization (Baumann and Cech 2001). One possible explanation is that Pot1 is required for prevention of rapid telomere loss, which would lead chromosome circularization dominantly. Recently our group has reported that the double mutant between *rqh1-hd* (helicase dead point mutant) and *pot1* is not synthetically lethal (Takahashi et al. 2011). The chromosome ends of the *pot1 rqh1-hd* double mutant are maintained by HR. There are several possible explanations for this. First, helicase dead Rqh1 may bind to the chromosome ends in *pot1* disruptant to inhibit rapid telomere loss, allowing cells to maintain chromosome ends by HR. Second, helicase activity of the Rqh1 may be involved in the rapid telomere loss in the *pot1* disruptant, because *S. cerevisiae* RecQ helicase is involved in the processing of telomere ends. This will also allow cells to maintain chromosome ends by HR. Third, helicase activity of the Rqh1 may be required for the suppression of recombination at telomere. This will also allow cells to maintain chromosome ends by HR. The exact role of the helicase dead Rqh1 in pot1 disruptant remains unclear. Interestingly, *pot1 rqh1-hd* double mutant is sensitive to anti-microtubule drug thiabendazole (TBZ) (Takahashi et al. 2011). The *pot1 rqh1-hd* double mutant has recombination intermediates even in the M phase at the chromosome ends. This physical link between the sister chromatids in M phase will inhibit chromosome segregation, especially in the presence of TBZ, which would lender cells sensitive to TBZ. Interestingly, concomitant

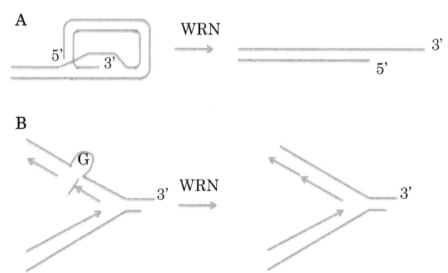

Fig. 3. WRN activities on a telomeric D-loop structure (A) and on a lagging strand telomere (B) during S phase. **A.** The model shows that WRN helicase releases the invading strand during S phase. **B.** WRN resolves G-quartet (G) formed on the lagging telomeric DNA.

inhibition of WRN and POT1 also lender human cells sensitive to anti-microtubule drug vinblastine, implying the functional conservation between human POT1 and WRN and *S. pombe* Pot1 and Rqh1(Takahashi et al. 2011). The other double knockdown experiments of WRN and POT1 in human cells show that human POT1 is required for efficient telomere C-rich strand replication in the absence of WRN (Arnoult et al. 2009). The functional interaction between human POT1 and RecQ helicase WRN is also suggested by in vitro experiment. Purified human POT1 binds to WRN and POT1 binding on telomeric DNA regulates the unwinding activity of WRN (Opresko et al. 2005; Sowd, Lei, and Opresko 2008; Opresko, Sowd, and Wang 2009). Based on these and other data, several possible roles of WRN at telomere are suggested (Rossi, Ghosh, and Bohr 2010) (Fig. 3). Telomere is capped by telomere binding proteins called shelterin and the chromosome end is protected through strand invasion of the duplex telomeric repeat by the 3' single-stranded overhangs, which is called t-loop (Palm and de Lange 2008). As WRN acts to release the 3' invading tail from a telomeric D loop in vitro, WRN may be involved in the regulation of the t-loop (Opresko et al. 2004). Single-stranded overhangs can fold into G-quadruplex DNA, which may inhibit DNA polymerase and telomerase at telomere (Zaug, Podell, and Cech 2005). Therefore, WRN may disrupt telomeric G-quadruplex with POT1 to facilitate DNA replication and/or telomere elongation at telomeres.

## 4. Roles of RPA in telomere maintenance

Replication protein A (RPA) is a heterotrimeric single-stranded non-specific DNA-binding protein consisting of a large (70 kDa), middle (32 kDa) and small (14 kDa) subunit. RPA is conserved from yeast to human and is essential for DNA replication, repair, and recombination (Binz, Sheehan, and Wold 2004). The large subunits of RPA in human, *S, cerevisiae* and *S. pombe* are named as RPA70, Rfa1 and Rad11, respectively. RPA is involved in HR repair by binding the single-stranded DNA generated by DNA end-processing at DSB ends. Single-stranded DNA is also produced at telomere. But RPA is suggested to be excluded from single-stranded telomere overhangs because it will lead to DNA damage checkpoint activation and cell cycle arrest. However, genetic evidences suggest the role of RPA in telomere maintenance. In this section, possible roles of RPA in telomere maintenance will be discussed. The functional relationship between RPA, RecQ helicase, and Taz1 will be also discussed.

### 4.1 Roles of RPA in DNA repair

Mutations in *S, cerevisiare rfa1* lender cells to sensitive to DNA damage and affect recombination efficiency, suggesting the involvement of RPA in recombination and repair processes (Smith and Rothstein 1995; Firmenich, Elias-Arnanz, and Berg 1995; Umezu et al. 1998). *S. pombe rad11* mutants are also sensitive to DNA damage and *rad11-D223Y* mutant is epitatic to *rad50* mutant, suggesting that RPA is involved in the HR repair (Parker et al. 1997; Ono et al. 2003). The roles of RPA in HR repair is well studied by in vitro system using *S. pombe* proteins (Kurokawa et al. 2008; Murayama et al. 2008). These in vitro and other genetic studies suggest that RPA binds to the single-stranded DNA generated by processing at DSB end. Then Rad22 (the *S. pombe* Rad52 homolog) helps Rad51 to displace RPA from single-stranded DNA. RPA bound to the single-stranded DNA recruits DNA damage checkpoint proteins to the DSB site to activate DNA damage checkpoint (Zou and Elledge 2003).

## 4.2 Roles of RPA in telomere maintenance

Telomere ends have single-strand overhangs, which may serve substrates for RPA. However, it is believed that RPA is excluded from telomere to suppress DNA damage checkpoint activation at telomere. Indeed, binding of human and mouse POT1 to telomeric ssDNA inhibits the localization of RPA to telomeres (Barrientos et al. 2008) (Gong and de Lange 2010). However, there are several genetic evidences suggesting that RPA is involved in telomere maintenance. Mutation of *S. cerevisiae RFA1* gene, *rfa1-D228Y* in *Yku70* mutant background causes telomere shortening, demonstration that RPA is required for telomere length regulation at dysfunctional telomere (Smith, Zou, and Rothstein 2000). Moreover, certain mutant alleles of *RFA2* gene, encoding the middle subunit of RPA, in wild-type background causes telomere shortening, demonstration that RPA is required for telomere length regulation (Mallory et al. 2003). In addition, *S. cerevisiae* RPA binds to telomere especially in S phase and cells expressing truncated Rfa2 show impaired binding of the Est1, a component of telomerase (Schramke et al. 2004). Based on these data, they proposed that RPA activates telomerase by loading Est1 onto telomeres during S phase. *S. pombe rad11-D223Y* mutant, which corresponds to the *S.cerevisiae rfa1-D228Y* mutant, has short telomere in wild-type background. Moreover, *S. pombe* RPA binds to telomere especially in S phase (Ono et al. 2003; Moser et al. 2009). A genome-wide screen for *S. pombe* deletion mutants shows that deletion of *ssb3*, the small subunit of RPA, affects telomere length(Liu et al. 2010). These facts suggest that RPA plays important role in telomere maintenance in both *S. cerevisiae* and *S. pombe*. Human RPA is also enriched at telomere during S phase, possibly due to exposure of single-stranded DNA during telomere replication (Verdun and Karlseder 2006). The aspartic acid at position 223 in *S. pombe* Rad11 is important for telomere length regulation, which corresponds to the position 227 in human RPA70 (Ono et al. 2003). Similarly, expression of RPA70-D227Y mutant protein in human fibrosarcoma HT1080 cells causes telomere shortening, suggesting that human RPA also plays role in telomere length regulation (Kobayashi et al. 2010). Possible role of RPA at telomere is the regulation of the processing of telomere ends by controlling accessibility of DNA repair proteins and/or Pot1 to single-stranded overhang (Fig. 4).

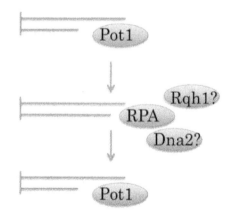

Fig. 4. The model shows that *S. pombe* RPA regulates the localizations and/or activities of proteins involved in the telomere maintenance. RPA may regulate Dna2 and/or Rqh1 during S phase.

### 4.3 Functional interaction between *S. pombe* Taz1, RPA and RecQ helicase

*S. pombe taz1 rad11-D223Y* double mutant lose telomere very rapidly, demonstrating that Taz1 and RPA collaborate to maintain telomere (Kibe et al. 2007). This rapid telomere loss can be suppressed by overexpression of Pot1. One possible explanation for this data is that Taz1 and RPA are required for the function of Pot1 at telomere and overexpression of Pot1 can rescue this defect. The rapid telomere loss of *taz1 rad11-D223Y* double mutant can be also suppressed by deletion of *rqh1*. Sgs1 is involved in the processing of telomere ends in *S. cerevisiae*. Similarly, *S. pombe* Rqh1 may be involved in the rapid telomere loss, possible by degradation of C-rich strand in *taz1 rad11-D223Y* double mutant (Fig. 5). The other functional relationship between Taz1 and Rqh1 is reported by Cooper group. *taz1* disruptant is sensitive to low temperature (Miller and Cooper 2003). Telomere entanglement is suggested to be a reason for this cold sensitivity. They found that unsumoylated Rqh1 mutant can suppress this cold sensitivity (Rog et al. 2009). Trt1 is a catalitic subunit of telomerase in *S. pombe*. *trt1* single mutant loses telomeric DNA gradually (Nakamura, Cooper, and Cech 1998). In contrast, *taz1 trt1* double mutant lose telomere very rapidly (Miller, Rog, and Cooper 2006). The replication fork stalling at the telomeres and resultant DSB is suggested to be a season for the rapid telomere loss in *taz1 trt1* double mutant. Unsumoylated Rqh1 mutant can also suppress this rapid telomere loss. Based on these data, they propose that sumoylated Rqh1 promotes telomere breakage and entanglement in *taz1* disruptant. This data demonstrate that the activity of Rqh1 at telomere is regulated to protect telomere. However, it remains unclear how Rqh1 and other DNA repair proteins are regulated at telomere. The functional interactions between human TRF1/TRF2 (*S. pombe*

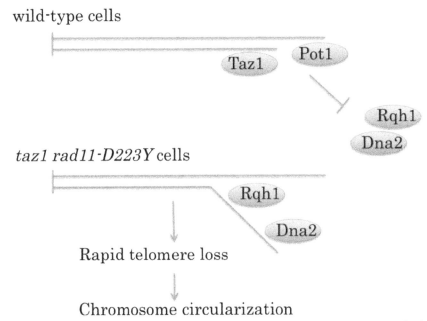

Fig. 5. The model shows that *S. pombe* Taz1 and RPA are required for prevent rapid telomere loss. In *taz1 rad11-D223Y* double mutant, Pot1 can not function properly and Rqh1 and possibly Dna2 resects telomere ends, which causes rapid telomere loss.

Taz1 ortholog) and human RecQ homolog WRN and BLM in telomere maintenance are also suggested (Opresko 2008). TRF2 interacts with WRN and stimulates helicase activity of WRN in vitro (Opresko et al. 2002; Machwe, Xiao, and Orren 2004). Expression of a TRF2 lacking the amino terminal basic domain induces the telomeric circle formations and rapid telomere deletions (Wang, Smogorzewska, and de Lange 2004). These events are dependent on WRN (Li et al. 2008). TRF2 also protects the displacement of Holliday junctions with telomeric arm by WRN in vitro (Nora, Buncher, and Opresko 2010). These facts suggest that the regulation of WRN activity by TRF2 is required to protect telomere.

## 5. Conclusion

This chapter focused on the roles of proteins involved in the processing of DBS ends at functional and dysfunctional telomere in *S. pombe, S. cerevisiae* and human. We found that MRN, Dna2, and possibly RecQ helicase Rqh1 are involved in the processing at telomere ends in *S. pombe*. Lydall group and other group found that Exo1, RecQ helicase Sgs1, Dna2, and Pif1 are involved in the processing at telomere ends in *S. cerevisiae*. Interestingly, most of these proteins were also involved in the processing of DNA double-strand break ends. These facts raise a new question of how these proteins are regulated at telomere ends. This chapter also focused on the functional interactions between telomere capping proteins and proteins involved in the processing of DBS ends mainly in *S. pombe*. We found that Taz1 and RPA collaborate to inhibit DNA end-processing, possibly by RecQ helicase, to prevent telomere loss. We also found that single-stranded telomere-binding protein Pot1 and RecQ helicase Rqh1 collaborate to inhibit homologous recombination at telomere. Cooper group found that RecQ helicase Rqh1 makes *taz1* disruptant sensitive to cold temperature by creating telomere entanglement. From these analyses, we learned that both double-stranded and single-stranded telomere binding proteins play critical roles to control proteins involved in DNA repair at chromosome ends.

## 6. Acknowledgment

I wish to thank all of my collaborators for support on my research. Part of this work was supported by Grants-in-Aid for Scientific Research on Priority Areas from the Ministry of Education, Science, Sports and Culture of Japan to Masaru Ueno.

## 7. References

Arnoult, N., C. Saintome, I. Ourliac-Garnier, J. F. Riou, and A. Londono-Vallejo. 2009. Human POT1 is required for efficient telomere C-rich strand replication in the absence of WRN. *Genes Dev.* 23 (24):2915-24.

Ashton, T. M., and I. D. Hickson. 2010. Yeast as a model system to study RecQ helicase function. *DNA Repair (Amst)* 9 (3):303-14.

Barrientos, K. S., M. F. Kendellen, B. D. Freibaum, B. N. Armbruster, K. T. Etheridge, and C. M. Counter. 2008. Distinct functions of POT1 at telomeres. *Mol. Cell. Biol.* 28 (17):5251-64.

Baumann, P., and T. R. Cech. 2001. Pot1, the putative telomere end-binding protein in fission yeast and humans. *Science* 292 (5519):1171-5.

Bernstein, K. A., S. Gangloff, and R. Rothstein. 2010. The RecQ DNA helicases in DNA repair. *Annu Rev Genet* 44:393-417.

Binz, S. K., A. M. Sheehan, and M. S. Wold. 2004. Replication protein A phosphorylation and the cellular response to DNA damage. *DNA Repair (Amst)* 3 (8-9):1015-24.

Bonetti, D., M. Martina, M. Clerici, G. Lucchini, and M. P. Longhese. 2009. Multiple pathways regulate 3' overhang generation at *S. cerevisiae* telomeres. *Mol. Cell* 35 (1):70-81.

Carneiro, T., L. Khair, C. C. Reis, V. Borges, B. A. Moser, T. M. Nakamura, and M. G. Ferreira. 2010. Telomeres avoid end detection by severing the checkpoint signal transduction pathway. *Nature* 467 (7312):228-32.

Caspari, T., J. M. Murray, and A. M. Carr. 2002. Cdc2-cyclin B kinase activity links Crb2 and Rqh1-topoisomerase III. *Genes Dev.* 16 (10):1195-208.

Cejka, P., E. Cannavo, P. Polaczek, T. Masuda-Sasa, S. Pokharel, J. L. Campbell, and S. C. Kowalczykowski. 2010. DNA end resection by Dna2-Sgs1-RPA and its stimulation by Top3-Rmi1 and Mre11-Rad50-Xrs2. *Nature* 467 (7311):112-6.

Chang, S., A. S. Multani, N. G. Cabrera, M. L. Naylor, P. Laud, D. Lombard, S. Pathak, L. Guarente, and R. A. DePinho. 2004. Essential role of limiting telomeres in the pathogenesis of Werner syndrome. *Nat Genet* 36 (8):877-82.

Chavez, A., A. M. Tsou, and F. B. Johnson. 2009. Telomeres do the (un)twist: helicase actions at chromosome termini. *Biochim Biophys Acta* 1792 (4):329-40.

Cohen, H., and D. A. Sinclair. 2001. Recombination-mediated lengthening of terminal telomeric repeats requires the Sgs1 DNA helicase. *Proc Natl Acad Sci U S A* 98 (6):3174-9.

Cooper, J. P., Y. Watanabe, and P. Nurse. 1998. Fission yeast Taz1 protein is required for meiotic telomere clustering and recombination. *Nature* 392 (6678):828-31.

Crabbe, L., A. Jauch, C. M. Naeger, H. Holtgreve-Grez, and J. Karlseder. 2007. Telomere dysfunction as a cause of genomic instability in Werner syndrome. *Proc. Natl. Acad. Sci. U. S. A.* 104 (7):2205-10.

Crabbe, L., R. E. Verdun, C. I. Haggblom, and J. Karlseder. 2004. Defective telomere lagging strand synthesis in cells lacking WRN helicase activity. *Science* 306 (5703):1951-3.

Critchlow, S. E., and S. P. Jackson. 1998. DNA end-joining: from yeast to man. *Trends Biochem Sci* 23 (10):394-8.

Dewar, J. M., and D. Lydall. 2010. Pif1- and Exo1-dependent nucleases coordinate checkpoint activation following telomere uncapping. *EMBO J* 29 (23):4020-34.

Diede, S. J., and D. E. Gottschling. 2001. Exonuclease activity is required for sequence addition and Cdc13p loading at a de novo telomere. *Curr Biol* 11 (17):1336-40.

Dionne, I., and R. J. Wellinger. 1996. Cell cycle-regulated generation of single-stranded G-rich DNA in the absence of telomerase. *Proc. Natl. Acad. Sci. U. S. A.* 93 (24):13902-7.

Doe, C. L., J. Dixon, F. Osman, and M. C. Whitby. 2000. Partial suppression of the fission yeast *rqh1-* phenotype by expression of a bacterial Holliday junction resolvase. *EMBO J.* 19 (11):2751-62.

Faure, V., S. Coulon, J. Hardy, and V. Geli. 2010. Cdc13 and telomerase bind through different mechanisms at the lagging- and leading-strand telomeres. *Mol Cell* 38 (6):842-52.

Firmenich, A. A., M. Elias-Arnanz, and P. Berg. 1995. A novel allele of Saccharomyces cerevisiae RFA1 that is deficient in recombination and repair and suppressible by RAD52. *Mol Cell Biol* 15 (3):1620-31.

Fishman-Lobell, J., and J. E. Haber. 1992. Removal of nonhomologous DNA ends in double-strand break recombination: the role of the yeast ultraviolet repair gene RAD1. *Science* 258 (5081):480-4.

Garvik, B., M. Carson, and L. Hartwell. 1995. Single-stranded DNA arising at telomeres in cdc13 mutants may constitute a specific signal for the RAD9 checkpoint. *Mol. Cell. Biol.* 15 (11):6128-38.

Gong, Y., and T. de Lange. 2010. A Shld1-controlled POT1a provides support for repression of ATR signaling at telomeres through RPA exclusion. *Mol Cell* 40 (3):377-87.

Gravel, S., J. R. Chapman, C. Magill, and S. P. Jackson. 2008. DNA helicases Sgs1 and BLM promote DNA double-strand break resection. *Genes Dev.* 22 (20):2767-72.

Gravel, S., M. Larrivee, P. Labrecque, and R. J. Wellinger. 1998. Yeast Ku as a regulator of chromosomal DNA end structure. *Science* 280 (5364):741-4.

Huang, P., F. E. Pryde, D. Lester, R. L. Maddison, R. H. Borts, I. D. Hickson, and E. J. Louis. 2001. SGS1 is required for telomere elongation in the absence of telomerase. *Curr Biol* 11 (2):125-9.

Iglesias, N., and J. Lingner. 2009. Related mechanisms for end processing at telomeres and DNA double-strand breaks. *Mol Cell* 35 (2):137-8.

Ivanov, E. L., and J. E. Haber. 1995. RAD1 and RAD10, but not other excision repair genes, are required for double-strand break-induced recombination in Saccharomyces cerevisiae. *Mol Cell Biol* 15 (4):2245-51.

Ivanov, E. L., N. Sugawara, J. Fishman-Lobell, and J. E. Haber. 1996. Genetic requirements for the single-strand annealing pathway of double-strand break repair in Saccharomyces cerevisiae. *Genetics* 142 (3):693-704.

Johnson, F. B., R. A. Marciniak, M. McVey, S. A. Stewart, W. C. Hahn, and L. Guarente. 2001. The Saccharomyces cerevisiae WRN homolog Sgs1p participates in telomere maintenance in cells lacking telomerase. *EMBO J* 20 (4):905-13.

Kibe, T., Y. Ono, K. Sato, and M. Ueno. 2007. Fission yeast Taz1 and RPA are synergistically required to prevent rapid telomere loss. *Mol. Biol. Cell.* 18 (6):2378-87.

Kibe, T., K. Tomita, A. Matsuura, D. Izawa, T. Kodaira, T. Ushimaru, M. Uritani, and M. Ueno. 2003. Fission yeast Rhp51 is required for the maintenance of telomere structure in the absence of the Ku heterodimer. *Nucleic Acids Res.* 31 (17):5054-63.

Kobayashi, Y., K. Sato, T. Kibe, H. Seimiya, A. Nakamura, M. Yukawa, E. Tsuchiya, and M. Ueno. 2010. Expression of mutant RPA in human cancer cells causes telomere shortening. *Biosci Biotechnol Biochem* 74 (2):382-5.

Kurokawa, Y., Y. Murayama, N. Haruta-Takahashi, I. Urabe, and H. Iwasaki. 2008. Reconstitution of DNA strand exchange mediated by Rhp51 recombinase and two mediators. *PLoS Biol* 6 (4):e88.

Larrivee, M., C. LeBel, and R. J. Wellinger. 2004. The generation of proper constitutive G-tails on yeast telomeres is dependent on the MRX complex. *Genes Dev.* 18 (12):1391-6.

Laud, P. R., A. S. Multani, S. M. Bailey, L. Wu, J. Ma, C. Kingsley, M. Lebel, S. Pathak, R. A. DePinho, and S. Chang. 2005. Elevated telomere-telomere recombination in WRN-deficient, telomere dysfunctional cells promotes escape from senescence and engagement of the ALT pathway. *Genes Dev.* 19 (21):2560-70.

Lee, J. Y., M. Kozak, J. D. Martin, E. Pennock, and F. B. Johnson. 2007. Evidence that a RecQ helicase slows senescence by resolving recombining telomeres. *PLoS Biol* 5 (6):e160.

Li, B., S. P. Jog, S. Reddy, and L. Comai. 2008. WRN controls formation of extrachromosomal telomeric circles and is required for TRF2DeltaB-mediated telomere shortening. *Mol Cell Biol* 28 (6):1892-904.

Lillard-Wetherell, K., A. Machwe, G. T. Langland, K. A. Combs, G. K. Behbehani, S. A. Schonberg, J. German, J. J. Turchi, D. K. Orren, and J. Groden. 2004. Association and regulation of the BLM helicase by the telomere proteins TRF1 and TRF2. *Hum Mol Genet* 13 (17):1919-32.

Liu, N. N., T. X. Han, L. L. Du, and J. Q. Zhou. 2010. A genome-wide screen for Schizosaccharomyces pombe deletion mutants that affect telomere length. *Cell Res* 20 (8):963-5.

Longhese, M. P., D. Bonetti, N. Manfrini, and M. Clerici. 2010. Mechanisms and regulation of DNA end resection. *EMBO J* 29 (17):2864-74.

Lydall, D. 2009. Taming the tiger by the tail: modulation of DNA damage responses by telomeres. *EMBO. J.* 28 (15):2174-2187.

Machwe, A., L. Xiao, and D. K. Orren. 2004. TRF2 recruits the Werner syndrome (WRN) exonuclease for processing of telomeric DNA. *Oncogene* 23 (1):149-56.

Makarov, V. L., Y. Hirose, and J. P. Langmore. 1997. Long G tails at both ends of human chromosomes suggest a C strand degradation mechanism for telomere shortening. *Cell* 88 (5):657-66.

Mallory, J. C., V. I. Bashkirov, K. M. Trujillo, J. A. Solinger, M. Dominska, P. Sung, W. D. Heyer, and T. D. Petes. 2003. Amino acid changes in Xrs2p, Dun1p, and Rfa2p that remove the preferred targets of the ATM family of protein kinases do not affect DNA repair or telomere length in Saccharomyces cerevisiae. *DNA Repair (Amst)* 2 (9):1041-64.

Miller, K. M., and J. P. Cooper. 2003. The telomere protein Taz1 is required to prevent and repair genomic DNA breaks. *Mol. Cell* 11 (2):303-13.

Miller, K. M., O. Rog, and J. P. Cooper. 2006. Semi-conservative DNA replication through telomeres requires Taz1. *Nature* 440 (7085):824-8.

Mimitou, E. P., and L. S. Symington. 2008. Sae2, Exo1 and Sgs1 collaborate in DNA double-strand break processing. *Nature* 455 (7214):770-4.

Mimitou, E. P., and L. S. Symington. 2009. Nucleases and helicases take center stage in homologous recombination. *Trends Biochem Sci* 34 (5):264-72.

Mimitou, E. P., and L. S. Symington. 2010. Ku prevents Exo1 and Sgs1-dependent resection of DNA ends in the absence of a functional MRX complex or Sae2. *EMBO J* 29 (19):3358-69.

Moser, B. A., L. Subramanian, Y. T. Chang, C. Noguchi, E. Noguchi, and T. M. Nakamura. 2009. Differential arrival of leading and lagging strand DNA polymerases at fission yeast telomeres. *EMBO. J.* 28 (7):810-20.

Multani, A. S., and S. Chang. 2007. WRN at telomeres: implications for aging and cancer. *J. Cell. Sci.* 120 (Pt 5):713-21.

Murayama, Y., Y. Kurokawa, K. Mayanagi, and H. Iwasaki. 2008. Formation and branch migration of Holliday junctions mediated by eukaryotic recombinases. *Nature* 451 (7181):1018-21.

Murray, J. M., H. D. Lindsay, C. A. Munday, and A. M. Carr. 1997. Role of *Schizosaccharomyces pombe* RecQ homolog, recombination, and checkpoint genes in UV damage tolerance. *Mol. Cell. Biol.* 17 (12):6868-75.

Nakamura, T. M., J. P. Cooper, and T. R. Cech. 1998. Two modes of survival of fission yeast without telomerase. *Science* 282 (5388):493-6.

Ngo, H. P., and D. Lydall. 2010. Survival and growth of yeast without telomere capping by Cdc13 in the absence of Sgs1, Exo1, and Rad9. *PLoS Genet* 6 (8):e1001072.

Nimonkar, A. V., J. Genschel, E. Kinoshita, P. Polaczek, J. L. Campbell, C. Wyman, P. Modrich, and S. C. Kowalczykowski. 2011. BLM-DNA2-RPA-MRN and EXO1-BLM-RPA-MRN constitute two DNA end resection machineries for human DNA break repair. *Genes Dev* 25 (4):350-62.

Niu, H., W. H. Chung, Z. Zhu, Y. Kwon, W. Zhao, P. Chi, R. Prakash, C. Seong, D. Liu, L. Lu, G. Ira, and P. Sung. 2010. Mechanism of the ATP-dependent DNA end-resection machinery from Saccharomyces cerevisiae. *Nature* 467 (7311):108-11.

Nora, G. J., N. A. Buncher, and P. L. Opresko. 2010. Telomeric protein TRF2 protects Holliday junctions with telomeric arms from displacement by the Werner syndrome helicase. *Nucleic Acids Res* 38 (12):3984-98.

Ono, Y., K. Tomita, A. Matsuura, T. Nakagawa, H. Masukata, M. Uritani, T. Ushimaru, and M. Ueno. 2003. A novel allele of fission yeast *rad11* that causes defects in DNA repair and telomere length regulation. *Nucleic Acids Res.* 31 (24):7141-9.

Onoda, F., M. Seki, A. Miyajima, and T. Enomoto. 2001. Involvement of SGS1 in DNA damage-induced heteroallelic recombination that requires RAD52 in Saccharomyces cerevisiae. *Mol Gen Genet* 264 (5):702-8.

Opresko, P. L. 2008. Telomere ResQue and preservation--roles for the Werner syndrome protein and other RecQ helicases. *Mech. Ageing Dev.* 129 (1-2):79-90.

Opresko, P. L., P. A. Mason, E. R. Podell, M. Lei, I. D. Hickson, T. R. Cech, and V. A. Bohr. 2005. POT1 stimulates RecQ helicases WRN and BLM to unwind telomeric DNA substrates. *J. Biol. Chem.* 280 (37):32069-80.

Opresko, P. L., M. Otterlei, J. Graakjaer, P. Bruheim, L. Dawut, S. Kolvraa, A. May, M. M. Seidman, and V. A. Bohr. 2004. The Werner syndrome helicase and exonuclease cooperate to resolve telomeric D loops in a manner regulated by TRF1 and TRF2. *Mol Cell* 14 (6):763-74.

Opresko, P. L., G. Sowd, and H. Wang. 2009. The Werner syndrome helicase/exonuclease processes mobile D-loops through branch migration and degradation. *PLoS One* 4 (3):e4825.

Opresko, P. L., C. von Kobbe, J. P. Laine, J. Harrigan, I. D. Hickson, and V. A. Bohr. 2002. Telomere-binding protein TRF2 binds to and stimulates the Werner and Bloom syndrome helicases. *J. Biol. Chem.* 277 (43):41110-9.

Ouellette, M. M., L. D. McDaniel, W. E. Wright, J. W. Shay, and R. A. Schultz. 2000. The establishment of telomerase-immortalized cell lines representing human chromosome instability syndromes. *Hum Mol Genet* 9 (3):403-11.

Palm, W., and T. de Lange. 2008. How shelterin protects mammalian telomeres. *Annu. Rev. Genet.* 42:301-34.

Paques, F., and J. E. Haber. 1997. Two pathways for removal of nonhomologous DNA ends during double-strand break repair in Saccharomyces cerevisiae. *Mol Cell Biol* 17 (11):6765-71.

Parker, A. E., R. K. Clyne, A. M. Carr, and T. J. Kelly. 1997. The *Schizosaccharomyces pombe* *rad11⁺* gene encodes the large subunit of replication protein A. *Mol. Cell. Biol.* 17 (5):2381-90.

Polotnianka, R. M., J. Li, and A. J. Lustig. 1998. The yeast Ku heterodimer is essential for protection of the telomere against nucleolytic and recombinational activities. *Curr Biol* 8 (14):831-4.

Rog, O., K. M. Miller, M. G. Ferreira, and J. P. Cooper. 2009. Sumoylation of RecQ helicase controls the fate of dysfunctional telomeres. *Mol. Cell* 33 (5):559-69.

Rossi, M. L., A. K. Ghosh, and V. A. Bohr. 2010. Roles of Werner syndrome protein in protection of genome integrity. *DNA Repair (Amst)* 9 (3):331-44.

Schramke, V., P. Luciano, V. Brevet, S. Guillot, Y. Corda, M. P. Longhese, E. Gilson, and V. Geli. 2004. RPA regulates telomerase action by providing Est1p access to chromosome ends. *Nat. Genet.* 36 (1):46-54.

Shim, E. Y., W. H. Chung, M. L. Nicolette, Y. Zhang, M. Davis, Z. Zhu, T. T. Paull, G. Ira, and S. E. Lee. 2010. Saccharomyces cerevisiae Mre11/Rad50/Xrs2 and Ku proteins regulate association of Exo1 and Dna2 with DNA breaks. *EMBO J* 29 (19):3370-80.

Smith, J., and R. Rothstein. 1995. A mutation in the gene encoding the *Saccharomyces cerevisiae* single-stranded DNA-binding protein Rfa1 stimulates a RAD52-independent pathway for direct-repeat recombination. *Mol. Cell. Biol.* 15 (3):1632-41.

Smith, J., H. Zou, and R. Rothstein. 2000. Characterization of genetic interactions with *RFA1*: the role of RPA in DNA replication and telomere maintenance. *Biochimie* 82 (1):71-8.

Sowd, G., M. Lei, and P. L. Opresko. 2008. Mechanism and substrate specificity of telomeric protein POT1 stimulation of the Werner syndrome helicase. *Nucleic Acids Res* 36 (13):4242-56.

Stewart, E., C. R. Chapman, F. Al-Khodairy, A. M. Carr, and T. Enoch. 1997. *rqh1⁺*, a fission yeast gene related to the Bloom's and Werner's syndrome genes, is required for reversible S phase arrest. *EMBO J.* 16 (10):2682-92.

Sugawara, N., G. Ira, and J. E. Haber. 2000. DNA length dependence of the single-strand annealing pathway and the role of Saccharomyces cerevisiae RAD59 in double-strand break repair. *Mol Cell Biol* 20 (14):5300-9.

Takahashi, K., R. Imano, T. Kibe, H. Seimiya, Y. Muramatsu, N. Kawabata, G. Tanaka, Y. Matsumoto, T. Hiromoto, Y. Koizumi, N. Nakazawa, M. Yanagida, M. Yukawa, E. Tsuchiya, and M. Ueno. 2011. Fission yeast Pot1 and RecQ helicase are required for efficient chromosome segregation. *Mol Cell Biol* 31 (3):495-506.

Tomita, K., T. Kibe, H. Y. Kang, Y. S. Seo, M. Uritani, T. Ushimaru, and M. Ueno. 2004. Fission yeast Dna2 is required for generation of the telomeric single-strand overhang. *Mol Cell Biol* 24 (21):9557-67.

Tomita, K., A. Matsuura, T. Caspari, A. M. Carr, Y. Akamatsu, H. Iwasaki, K. Mizuno, K. Ohta, M. Uritani, T. Ushimaru, K. Yoshinaga, and M. Ueno. 2003. Competition between the Rad50 complex and the Ku heterodimer reveals a role for Exo1 in processing double-strand breaks but not telomeres. *Mol Cell Biol* 23 (15):5186-97.

Umezu, K., N. Sugawara, C. Chen, J. E. Haber, and R. D. Kolodner. 1998. Genetic analysis of yeast RPA1 reveals its multiple functions in DNA metabolism. *Genetics* 148 (3):989-1005.

Verdun, R. E., and J. Karlseder. 2006. The DNA damage machinery and homologous recombination pathway act consecutively to protect human telomeres. *Cell* 127 (4):709-20.

Wang, R. C., A. Smogorzewska, and T. de Lange. 2004. Homologous recombination generates T-loop-sized deletions at human telomeres. *Cell* 119 (3):355-68.

Wang, X., and P. Baumann. 2008. Chromosome fusions following telomere loss are mediated by single-strand annealing. *Mol. Cell* 31 (4):463-73.

Watt, P. M., I. D. Hickson, R. H. Borts, and E. J. Louis. 1996. SGS1, a homologue of the Bloom's and Werner's syndrome genes, is required for maintenance of genome stability in Saccharomyces cerevisiae. *Genetics* 144 (3):935-45.

Wellinger, R. J., K. Ethier, P. Labrecque, and V. A. Zakian. 1996. Evidence for a new step in telomere maintenance. *Cell* 85 (3):423-33.

Wellinger, R. J., A. J. Wolf, and V. A. Zakian. 1993. *Saccharomyces* telomeres acquire single-strand $TG_{1-3}$ tails late in S phase. *Cell* 72 (1):51-60.

Wu, L., and I. D. Hickson. 2003. The Bloom's syndrome helicase suppresses crossing over during homologous recombination. *Nature* 426 (6968):870-4.

Wyllie, F. S., C. J. Jones, J. W. Skinner, M. F. Haughton, C. Wallis, D. Wynford-Thomas, R. G. Faragher, and D. Kipling. 2000. Telomerase prevents the accelerated cell ageing of Werner syndrome fibroblasts. *Nat Genet* 24 (1):16-7.

Zaug, A. J., E. R. Podell, and T. R. Cech. 2005. Human POT1 disrupts telomeric G-quadruplexes allowing telomerase extension in vitro. *Proc Natl Acad Sci U S A* 102 (31):10864-9.

Zhu, Z., W. H. Chung, E. Y. Shim, S. E. Lee, and G. Ira. 2008. Sgs1 helicase and two nucleases Dna2 and Exo1 resect DNA double-strand break ends. *Cell* 134 (6):981-94.

Zou, L., and S. J. Elledge. 2003. Sensing DNA damage through ATRIP recognition of RPA-ssDNA complexes. *Science* 300 (5625):1542-8.

# 10

# Characterization of 5'-Flanking Regions of Various Human Telomere Maintenance Factor-Encoding Genes

Fumiaki Uchiumi[1,4], Takahiro Oyama[1],
Kensuke Ozaki[1] and Sei-ichi Tanuma[2,3,4]
*[1]Department of Gene Regulation, Faculty of Pharmaceutical Sciences*
*[2]Department of Biochemistry, Faculty of Pharmaceutical Sciences*
*[3]Genome and Drug Research Center*
*[4]Research Center for RNA Science, RIST, Tokyo University of Science*
*Japan*

## 1. Introduction

Telomeres are the unique nucleoprotein complex structures located at the end of linear eukaryotic chromosomes (Blackburn, 2000; de Lange, 2006). They are composed of TTAGGG repeats that are typically 10 kb at birth and gradually shorten with cell divisions (de Lange, 2006). Telomerase is composed of the protein subunit TERT and the RNA subunit TERC (TR). It elongates the telomere by adding telomeric repeats (Greider & Blackburn, 1987). The 50 to 300 nucleotides from the terminal end of the telomeres are single stranded 3'-protluded G-overhang structures which make the t-loop configuration (de Lange, 2006; Griffith et al., 1999). Mammalian telomeres are included in heterochoromatin and attached to the nuclear matrix (Oberdoerffer & Sinclair, 2007; Gonzalez-Suarez & Gonzalo, 2008). Telomere shortening causes instability of the ends of chromosomes to lead to replicative senescence (O'Sullivan & Karlseder, 2010; Lundblad & Szostak, 1989). Therefore, the ends of telomeres should be protected from damaging or cellular activities. The t-loop structures are regulated by shelterin protein factors, TRF1, TRF2, Rap1, TIN2, TPP1, POT1 (Gilson & Geli, 2007; O'Sullivan & Karlseder, 2010), and Rec Q DNA helicases, WRN and BLM (Chu & Hickson, 2009). TRF1 and TRF2, which bind to duplex telomeric DNA and retain shelterin on the telomere repeats, were shown to interact with various functional proteins (Giannone et al., 2010). Molecular structural analysis of Rap1 revealed that its mechanism of action involves interaction with TRF2 and Taz1 proteins (Chen et al., 2011). A recent study showed that depletion of TPP1 and its partner TIN2 causes a loss of telomerase recruitment to telomeres (Abreu et al., 2010). POT1 is an important regulator of telomerase length, in stimulating the RecQ helicases WRN and BLM (Opresko et al., 2005). Tankyrase-1 (TANK1), which is classified as a poly(ADP-ribose) polymerase family protein, is also known to regulate telomere homeostasis by modifying TRF1 (Smith et al., 1998; Schreiber et al., 2006). Dyskerin, which is encoded by the *DKC1* gene, is a key auxiliary protein that is contained in a Cajal body with TERT (Cohen et al., 2007). Defects in the shelterin components and telomerase are thought to down-regulate telomere structure and length (O'Sullivan & Karlseder, 2010). The shelterin proteins also play important roles in protecting chromosomal ends from being recognized by DNA damage response (DDR) machinery (O'Sullivan & Karlseder, 2010). Although the biological significance of the shelterin complex proteins has been studied, the molecular mechanisms that regulate expression of those genes encoding telomere associated proteins is less well-characterized. We hypothesized that expressions of those telomere-associated protein-encoding genes are regulated by a similar mechanism. In order to analyze these promoter activities promptly, we isolated 200 to 300-bp of the 5'-upstream regions of these telomere regulatory protein-encoding genes and applied them to a multiple transfection assay system (Uchiumi et al., 2010a). Previously, we have observed that *WRN* and *TERT* promoter activities were up-regulated by 2-deoxy-D-glucose (2DG) and *trans*-resveratrol (Rsv) in accordance with the activation of telomerase (Zhou et al., 2009; Uchiumi et al., 2011). A potent inhibitor of glucose metabolism, 2DG is thought to mimic glucose deprivation *in vivo* such that it is mimetic of caloric restriction (CR) (Roth et al., 2001). Resveratrol (Rsv), which is a polyphenol contained in grape skins and red wine, activates sirtuin-mediated deacetylation (Stefani et al., 2007; Knutson & Leeuwenburgh, 2008). We report here that most of the promoters of the shelterin protein-encoding genes positively responded to the CR mimetic agents, 2DG and Rsv. These results suggest that telomerase and telomere maintenance factors are simultaneously regulated at the initiation of the transcription.

## 2. Materials and methods

### 2.1 Chemicals
The reagents 2-deoxy-D-glucose (2DG) and *trans*-resveratrol (Rsv) were purchased from Wako Chemicals (Tokyo, Japan) and Cayman Chemicals (Ann Arbor, MI), respectively.

### 2.2 Cells and cell culture
HeLa-S3 cells (Zhou et al., 2009) were cultured in Dulbecco's modified eagle (DME) medium supplemented with heat-inactivated 10% fetal calf serum (FCS) (Sanko-Pure Chemical, Tokyo, Japan), 2 mM L-glutamine (Invitrogen, CA, USA), penicillin (100 IU/mL) (MEIJI SEIKA, Tokyo, Japan), and streptomycin (100 μg/mL) (MEIJI SEIKA).

### 2.3 Construction of Luc-reporter plasmids
Luc reporter plasmids carrying promoter regions for the human *TERT* and *TERC* genes have been constructed and designated as pGL4-TERT, and pGL4-TERC, respectively (Zhou et al., 2009; Uchiumi et al., 2010a). Extraction of DNA from HeLa-S3 cells, and subsequent PCR for the promoter regions of interest were performed as described previously (Uchiumi et al., 2010a; Zhou et al., 2009). Primer-sets were designed against human genomic sequences from the Cross-Ref NCBI-data base (http://www.ncbi.nlm.nih.gov/sites/gquery/) for the 5'-flanking regions of the genes of interest (Table. 1). PrimeStar Taq polymerase (Takara, Kyoto, Japan) was used for all amplifications.

Amplification conditions consisted of: 30 cycles of 98°C for 10 sec, 55°C for 5 sec, and 72°C for 30 sec. PCR products were digested with *Kpn*I and *Xho*I and then separated on 0.9% agarose gels.

After electrophoresis, DNA bands of the correct length were recovered from the gel with Wizard SV Gel and PCR Clean-Up System (Promega, Madison, WI, USA) and subcloned into the *KpnI-XhoI* site of the pGL4-basic vector (pGL4[luc 2.10]) (Promega). The resultant cloned plasmids were designated pGL4-DKC1, pGL4-POT1, pGL4-RAP1, pGL4-TANK1, pGL4-TANK2, pGL4-TIN2, pGL4-TPP1, pGL4-TRF1, and pGL4-TRF2. Clone sequences were confirmed using a DNA Sequencing System (Applied Biosystems, Foster City, CA) with Rv (5'-TAGCAAAATAGGCTGTCCCC-3' and GL (5'-CTTTATGTTTTTGGCGTCTT-CC-3') primers purchased from Operon Biotechnologies (Tokyo, Japan).

| Name | Sequence |
|------|----------|
| hDysk-7065 | 5'-TCGGTACCGTGAGCCCAGGCGCAGGCGC-3' |
| AhDysk-7414 | 5'-ATCTCGAGGGAACGACCGCAGACTCCC-3' |
| hPOT-1509 | 5'-TCGGTACCTGAGAACTGAATATTGCTGTG-3' |
| AhPOT-1164 | 5'-ATCTCGAGAATATCATCTTACCAAAGAC-3' |
| hRAP1-5667 | 5'-TCGGTACCTCGCGGCGCTTCCCAGCCC-3' |
| AhRAP1-5970 | 5'-ATCTCGAGCTGTCACCGCAGACGCCTC-3' |
| hTANK1-8541 | 5'-TCGGTACCGACTGAAAGTGAGAAATGC-3' |
| AhTANK1-8860 | 5'-ATCTCGAGAGCGACGCGACGCCGCCATC-3' |
| hTANK2-4227 | 5'-TCGGTACCAGGAGAAAGGGATGTGGAAG-3' |
| AhTANK2-4519 | 5'-ATCTCGAGGCGGCGCGAAGGGTTTGTGG-3' |
| hTIN2-8835 | 5'-TCGGTACCGCAGGCTCCGCGAAGAAAGC-3' |
| AhTIN2-8508 | 5'-ATCTCGAGTGGAGAAGCTGACCGTCTC-3' |
| hTPP1-8283 | 5'-TCGGTACCTCGACGATGCTATCGGGAC-3' |
| AhTPP1-7995 | 5'-ATCTCGAGCGTGATGACGCAAGAGCGGA-3' |
| hTRF1-1070 | 5'-TCGGTACCTCCTCCTATCCTAATCTCGC-3' |
| AhTRF1-1371 | 5'-ATCTCGAGGAAACATCCTCCGCCATGTT-3' |
| hTRF2-9454 | 5'-TCGGTACCGATCCCGGCCTGTTTTTCAG-3' |
| AhTRF2-9170 | 5'-ATCTCGAGCGGGGCCCGCCGTCCCGGC-3' |

Table 1. Primers used for amplifying 5'-upstream region of various human telomere-associated genes

## 2.4 Transient transfection assay

Transient transfection of Luc-reporter plasmids was performed using multi-well culture plates that had been prepared and treated with DNA/DEAE-dextran (Uchiumi et al., 2010a). After 4 h of transfection, 2DG or Rsv was added to the culture medium (Zhou et al., 2009; Uchiumi et al., 2011). After a further incubation (19 to 24 h), cells were collected and lysed with 40 µL of 1 x Cell culture lysis reagent, mixed, and stored at -80°C. Luc assays were performed according to the manufacturer's instructions (Promega). In brief, Luc assay reagent (40 µL) was added to 10 µL of protein sample and mixed briefly. Immediately after mixing, chemiluminescence was measured for 7.5 sec with a Minilumat LB9506 luminometer (Berthold, Bad Wildbad, Germany). Protein assays were performed with the Luc sample (2.5 µL) and Protein Assay Reagent (Bio-Rad Lab., Hercules, CA, USA).

## 3. Results

### 3.1 Isolation of 5'-flanking regions of human telomere-associated protein-encoding genes

Previously, we isolated and characterized 5'-flanking regions of the human *TERT* and *TERC* genes (Zhou et al., 2009; Uchiumi et al., 2010a). In this study, those of different human telomere-associated protein-encoding genes were obtained by PCR and inserted into the MCS of the pGL4-basic (pGL4[luc 2.10]) vector. Putative transcription-factor binding elements were found by TF-SEARCH analysis. As summarized in Fig. 1, c-Ets/Elk1, Sp1/GC-box, CREB, OCT, p300, SRY, GATA, E2F, NF-κB/c-Rel, CCAAT-box, and other motifs are located within 300-bp from the 5'-upstream region of the cDNAs. Although all of these telomere-associated protein factors are commonly involved in the maintenance of telomeres, a rigid rule in the order of the *cis*-elements could not be found in their core promoter regions. However, one or more Sp1/GC-box elements are located in 5'-upstream regions of the *DKC1*, *RAP1*, *TANK1*, *TIN2*, *TPP1*, *TRF1*, *TRF2*, *TERT*, and *TERC* genes, but not in the *POT1* and *TANK2* genes. Similar to the 5'-flanking region of the *WRN* gene, all of the isolated DNA fragments have no obvious TATA-box like sequences except the 5'-flanking region of the *TERC* gene (Uchiumi et al., 2010a).

### 3.2 Effect of Rsv on the promoter activities of 5'-flanking regions of the shelterin-encoding genes

The natural compound Rsv is known to have life-span promoting properties in yeast and metazoans by affecting the insulin-signaling cascade (Fröjdö et al., 2008). In order to examine the effect of Rsv on the isolated 5'-upstream regions of the shelterin encoding genes, Luc assays were performed. Luc expression plasmids which contained 5'-flanking regions of various telomere maintenance factor-encoding genes were transfected into HeLa-S3 cells by the DEAE-dextran based multiple transfection method (Uchiumi et al., 2010a). Luc activities of reporter plasmid-transfected cells were normalized to that of the pGL4-PIF1 transfected cells, because PIF1 has been suggested to have a negative effect on telomere elongation in yeast cells (Schulz & Zakian, 1994), and it has been shown that the change in the PIF1 promoter activity is largely unaffected after treatment with Rsv (Uchiumi et al., 2011). As shown in Table 2, treatment with Rsv (10 μM) for 24 h augmented Luc activities from the cells transfected with Luc reporter plasmids. Apparent positive responses to the Rsv treatment of the 5'-flanking regions of the *TERT* and *TERC* genes were observed, consistent with the activation of telomerase by Rsv in HeLa-S3 cells (Uchiumi et al., 2011). Although no obvious GC-box like elements are found in the 300-bp 5'-upstream regions of the *POT1* and *TANK2* genes (Fig. 1), Luc activities of these plasmid-transfected cells increased 2.53- and 1.69-fold, respectively, by Rsv treatment.

### 3.3 Effect of 2DG on the promoter activities of 5'-flanking regions of the shelterin-encoding genes

2DG is known to affect life span by its CR mimetic effect on various species (Roth et al., 2001). We previously observed that treatment with 2DG induces telomerase activity along with transcriptional activation of the *TERT* and *WRN* genes in HeLa-S3 cells (Zhou et al., 2009). Therefore, we examined the effect of 2DG on the promoter activities of shelterin-encoding genes. Although most of the Luc activities of cells transfected with shelterin promoter-Luc expression constructs were diminished by 2DG, the treatment induced

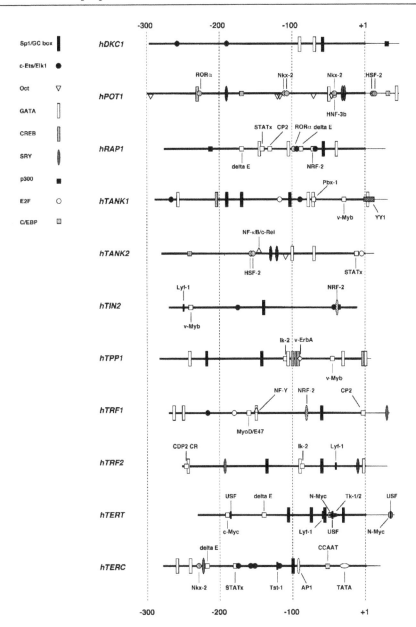

Fig. 1. Promoter regions of the human genes encoding telomere-associated proteins or shelterin protein factors. PCR-amplified 5'-flanking regions of these genes, which were inserted upstream of the *Luciferase* gene of the pGL4-basic vector (pGL4[luc 2.10]), are shown. Transcription start sites (or 5'-end of cDNAs) are designated +1. The TF-SEARCH program (http://www.cbrc.jp/research/db/TFSEARCH.html) was performed and putative transcription-factor binding-elements (score > 85) are shown schematically.

| Reporter | Rsv (10 μM) | Relative Luc activity | Fold |
|---|---|---|---|
| pGL4-PIF1 | - | 1.000 ± 0.033 | 1.00 |
| pGL4-PIF1 | + | 1.000 ± 0.088 | |
| pGL4-RTEL | - | 2.170 ± 0.119 | 1.23 |
| pGL4-RTEL | + | 2.667 ± 0.326 | |
| pGL4-DKC1 | - | 1.271 ± 0.117 | 1.88 |
| pGL4-DKC1 | + | 2.390 ± 0.325** | |
| pGL4-POT1 | - | 0.018 ± 0.003 | 2.53 |
| pGL4-POT1 | + | 0.045 ± 0.008** | |
| pGL4-RAP1 | - | 0.746 ± 0.023 | 1.84 |
| pGL4-RAP1 | + | 1.372 ± 0.164* | |
| pGL4-TANK1 | - | 0.069 ± 0.023 | 1.54 |
| pGL4-TANK1 | + | 0.107 ± 0.014 | |
| pGL4-TANK2 | - | 0.0059 ± 0.00155 | 1.69 |
| pGL4-TANK2 | + | 0.0099 ± 0.00230 | |
| pGL4-TIN2 | - | 0.128 ± 0.022 | 1.48 |
| pGL4-TIN2 | + | 0.190 ± 0.013 | |
| pGL4-TPP1 | - | 0.463 ± 0.032 | 1.54 |
| pGL4-TPP1 | + | 0.714 ± 0.115* | |
| pGL4-TRF1 | - | 0.648 ± 0.078 | 1.83 |
| pGL4-TRF1 | + | 1.189 ± 0.104*** | |
| pGL4-TRF2 | - | 0.139 ± 0.005 | 1.61 |
| pGL4-TRF2 | + | 0.224 ± 0.013*** | |
| pGL4-TERT | - | 0.794 ± 0.042 | 1.93 |
| pGL4-TERT | + | 1.532 ± 0.081*** | |
| pGL4-TERC | - | 0.557 ± 0.142 | 1.97 |
| pGL4-TERC | + | 1.096 ± 0.067* | |

Table 2. Effect of Resveratrol (Rsv) on promoter activities of telomere-associated genes in HeLa-S3 cells Various reporter plasmids were introduced into HeLa-S3 cells by multiple DEAE-dextran method transfections. After 4 h of transfection, the culture medium was discarded and changed to Rsv-containing or non-containing medium. Cells were harvested after 24 h of treatment, then Luc assays were performed. Relative values represent Luc activities compared with that of the pGL4-PIF1 transfected cells. Results show means ± S.D. from three independent samples (N=3). Significance of differences between control and Rsv treated cells were analyzed by Student's $t$-test (*p<0.05, **p<0.01, ***p<0.005).

relatively positive values compared to that of the pGL4-PIF1-transfected cells (Table 3). Similar to the response to Rsv (Table 2), the *TERT* and *TERC* promoters were activated by the 2DG treatment. The increase in relative promoter activity (compared with that of the pGL4-PIF1-transfected cells) after 2DG (8 mM) treatment was significant for the *RTEL*, *DKC1*, *POT1*, *RAP1*, *TANK1*, *TIN2*, *TPP1*, and *TRF1* promoters (Table 3). These results suggest that the CR mimetic compound 2DG affects the balance of gene expression to protect telomeres.

| Reporter | | Relative Luc activity | |
|---|---|---|---|
| | 2DG (mM) | 4 | 8 |
| pGL4-PIF1 | - | 1.000 ± 0.206 | 1.000 ± 0.141 |
| pGL4-PIF1 | + | 1.000 ± 0.230 | 1.000 ± 0.148 |
| pGL4-RTEL | - | 1.800 ± 0.802 | 2.550 ± 0.648 |
| pGL4-RTEL | + | 4.011 ± 0.917 | 6.651 ± 1.958* |
| pGL4-DKC1 | - | 1.136 ± 0.111 | 2.560 ± 0.265 |
| pGL4-DKC1 | + | 8.767 ± 4.556 | 6.698 ± 0.921*** |
| pGL4-POT1 | - | 0.030 ± 0.008 | 0.027 ± 0.010 |
| pGL4-POT1 | + | 0.139 ± 0.065 | 0.102 ± 0.015*** |
| pGL4-RAP1 | - | 0.993 ± 0.247 | 2.201 ± 0.236 |
| pGL4-RAP1 | + | 5.456 ± 1.411* | 4.977 ± 0.749*** |
| pGL4-TANK1 | - | 0.047 ± 0.012 | 0.106 ± 0.012 |
| pGL4-TANK1 | + | 0.272 ± 0.110 | 0.567 ± 0.150* |
| pGL4-TANK2 | - | 0.012 ± 0.003 | 0.006 ± 0.002 |
| pGL4-TANK2 | + | 0.066 ± 0.011*** | 0.033 ± 0.032 |
| pGL4-TIN2 | - | 0.130 ± 0.038 | 0.213 ± 0.023 |
| pGL4-TIN2 | + | 0.686 ± 0.273 | 0.474 ± 0.093** |
| pGL4-TPP1 | - | 0.604 ± 0.151 | 0.751 ± 0.099 |
| pGL4-TPP1 | + | 3.211 ± 0.237*** | 5.721 ± 1.302* |
| pGL4-TRF1 | - | 0.853 ± 0.131 | 1.355 ± 0.279 |
| pGL4-TRF1 | + | 2.178 ± 0.408** | 3.442 ± 0.567** |
| pGL4-TRF2 | - | 0.173 ± 0.073 | 0.232 ± 0.022 |
| pGL4-TRF2 | + | 0.378 ± 0.036* | 0.693 ± 0.244 |
| pGL4-TERT | - | 0.586 ± 0.094 | 1.707 ± 0.316 |
| pGL4-TERT | + | 1.844 ± 0.498* | 3.456 ± 0.963* |
| pGL4-TERC | - | 0.651 ± 0.120 | 0.897 ± 0.119 |
| pGL4-TERC | + | 1.878 ± 0.426** | 2.516 ± 0.507** |

Table 3. Effect of 2-deoxy-D-glucose (2DG) on promoter activities of telomere-associated genes in HeLa-S3 cells Various reporter plasmids were introduced into HeLa-S3 cells by multiple DEAE-dextran method transfections. After 4 h of transfection, the culture medium was discarded and changed to 2DG-containing (4 and 8 mM) or non-containing medium. Cells were harvested after 24 h (4 mM) or 19 h (8 mM) of the 2DG treatment, then Luc assays were performed. Relative values represent Luc activities compared with that of the pGL4-PIF1 transfected cells. Results show means ± S.D. from three independent samples (N=3). Significance of differences between control and 2DG treated cells were analyzed by Student's $t$-test (*$p<0.05$, **$p<0.01$, ***$p<0.005$).

## 4. Discussion

### 4.1 The promoter regions of the shelterin-encoding genes coordinately respond to CR mimetic drugs

In the present study, 5'-flanking regions of different human telomere-associated protein factor-encoding genes were isolated, and these Luc reporter plasmids were used for transient transfection assays. The shelterin- or telomere-associated protein-encoding genes, including *TERT*, *TERC*, *DKC1*, and double-stranded break repair protein-encoding genes, such as *ATM* and *ATR*, are conserved among human, mouse and yeast (Stern & Bryan, 2008). Given that these telomere-associated proteins are localized to the telomere t-loop to protect the specific structure, and appear to act in co-operation with each other (O'Sullivan & Karlseder, 2010), their gene expression should be regulated synchronously when the telomeric region needs to be protected. Aging or cellular senescence are thought to be controlled by a genomic maintenance regulatory system (Vieg, 2007). Our hypothesis is that aging or longevity affecting reagents might have an effect on the expression of the telomere-associating protein-encoding genes. The results (Tables 2 and 3) indicate that promoter activities of the shelterin-encoding genes are simultaneously up-regulated by Rsv and 2DG in HeLa-S3 cells when they are compared with *PIF1* promoter activity. Previously, we have reported that multiple GC-boxes are commonly located in the human *TERT* and *WRN* promoter regions and that might play a role in the positive response to Rsv and 2DG in HeLa-S3 cells (Uchiumi et al., 2010c). Although there are no canonical roles of transcription factor binding elements or their order in these promoter regions, Sp1 binding elements or GC-boxes are found in all of them except 5'-upstream of the *POT1* and *TANK2* genes (Fig. 1). Therefore, GC-box binding factors may up-regulate this telomere-associated gene expression. However, there are no GC-box like motifs in the 300-bp up-stream regions of the *POT1* and *TANK2* genes, which are relatively AT-rich and contain Oct-1 binding sites. This observation suggests that POU family proteins might also be involved in the positive regulation of these genes. Apparent up-regulation of promoter activities by Rsv and 2DG treatment was observed in the cells transfected with the Luc reporter plasmids containing 200-bp 5'-upstream regions of the *TERT* and *TERC* genes (Tables 2 and 3). It is noteworthy that the duplicated GGAA-motifs are found in both promoter regions (Uchiumi et al., 2010a, Uchiumi et al., 2011b), suggesting that the GGAA-motif binding factors, including Ets family proteins, might be involved in the positive response to the aging or longevity affecting signals.

Previously, we observed elevation of the human *WRN* promoter activity in accordance with activation of telomerase after Rsv and 2DG treatment of HeLa-S3 cells (Uchiumi et al., 2011; Zhou et al., 2009). 2DG suppresses glucose metabolism to establish a limit for the usage or uptake of glucose into cells (Roth et al., 2001). On the other hand, Rsv is known to activate sirtuin family protein deacetylases (Kaeberlein, 2010). It is thought that both Rsv and 2DG are CR mimetic drugs (Stefani et al., 2007; Roth et al., 2001), and that CR can extend the mean and maximum life spans of numerous organisms (Carvallini et al., 2008; Roth et al., 2001). The present study suggests that induction of telomerase activity in concert with up-regulation of the telomere-associated protein- or shelterin-encoding gene expression may play a role in regulating the aging process through the telomere maintainance system.

### 4.2 A possible role for telomere maintenance system in aging/senescence regulation

Aging or senescence is a complicated biological process involving various regulatory factors (Campisi & d'Adda di Fagagna, 2007; Kuningas et al., 2008; Sanz & Stefanatos, 2008). Aging could be explained by a mitochondrial free radical theory (Benz & Yau, 2008). On the other hand, cellular senescence could be caused by DNA damage or the associated signals on chromosomes (Vieg, 2007). It is well known that cellular senescence is correlated with the cell growth arrest (Campisi & d'Adda di Fagagna, 2007). DNA damage signals activate ATM or ATR, and then phosphrylate p53 to induce transcription of the *CDKN1A* gene that encodes cyclin-dependent kinase inhibitor 1A (p21). These sequentially occurring events arrest the cell cycle at G1-phase (Meek, 2009). Repair of DNA damage will occur at this stage, unless the cell has initiated apoptosis. Thus, aging is thought to be controlled through both reactive oxygen species (ROS) generated by mitochondria and damages to DNA including telomeric regions of the chromosomes (Sahin & DePinho, 2010). Recently, it was shown that telomere dysfunction causes activation of p53 which directly represses PGC-1α and PGC-1β, leading to mitochondrial compromise (Sahin et al., 2011). Moreover, an experiment to reactivate telomerase in telomerase-deficient mice ameliorated DNA damage signaling and reversed neurodegeneration (Jaskelioff et al., 2011). These lines of evidence suggest that telomeres exert signals to affect mitochondria along with DNA repair systems. The concept that telomere length-associated signaling stimulates mitochondrial function might have combined the mitochondrial free radical theory with the molecular mechanism of chromosomal maintenance system against DNA damaging stresses. Rsv has been shown to have effect activation of PGC-1α to improve mitochondrial function in mouse brown adipose tissue and muscle (Lagouge et al., 2006). The present study indicates that shelterin protein-encoding gene promoters are simultaneously activated by Rsv treatment. Thereby, accumulation of shelterin proteins might lead to stabilization of telomeric regions of the chromosome and activation of PGC-1α.

### 4.3 Hormesis, the beneficial effects from low doses of toxic stresses, might be a determinant of longevity

The deficiencies in RecQ DNA helicases, including WRN and BLM, are known to cause premature aging (Chu & Hickson, 2009). In the present study, we have observed that CR mimetic drug treatment activates promoters of the shelterin protein-encoding genes, suggesting that CR evokes functions of the telomere maintenance machinery. Hormesis is a phenomenon that generally refers to the beneficial effects from low level toxic or other harmful damage, such as irradiation, heat shock, or food restriction (Schumacher, 2009). High doses of 2DG and Rsv have harmful or toxic effects on cells, leading to cell death or apoptosis (Lin et al., 2003; Cosan, et al., 2010). In contrast, relatively low doses of these CR mimetic reagents, as used in the present study, have effects similar to hormesis. Therefore, resistance to stresses eventually provoked by prolonged low doses of CR mimetic reagents could promote the longevity of organisms. Thus, the results obtained in the present study are consistent with the concept of hormesis.

### 4.4 Molecules that are involved in the regulation of the aging process

From studies of life spans of the *C. elegans*, it has been suggested that the insulin/IGF-1 pathway influences aging (Kenyon, 2010). In this signaling system, DAF-16 (FoxO transcription factor) plays a role in activating genes that act to extend life span (van der Horst & Burgering, 2007). AMP-activated protein kinase (AMPK), which is known to extend the life span of nematodes (Apfeld et al., 2004), phosphorylates FoxO, PGC-1α, and CREB to induce various genes encoding mitochondrial and oxidative metabolism regulating factors (Cantó & Auwerx, 2010). The other biologically important function of AMPK is that it blocks the mTOR (mammalian target of rapamycin) pathway (Cantó & Auwerx, 2010). A recent study suggested that mTOR is a prime target in the genetic control of aging to determine life span and aging in yeast, worms, flies, and mice (Zoncu et al., 2011). The mTOR pathway accelerates growth by regulating signals downstream of insulin/IGF-1 receptors (Zoncu et al., 2011). Activation of mTOR is thought to speed up aging in adulthood, and reduced mTOR signaling would have the opposite effect, acting downstream of dietary restriction. Thus the anti-aging effect could be expected by mTOR inhibition, such as dietary restriction, rapamycin, introduction of the AMPK expression vector, and genetic inactivation of mTOR by techniques such as RNA interference.

## 5. Conclusions and future perspectives

It would be advantageous for cells to estimate the state of chromosomes just by monitoring telomeric regions. Monitoring the somatic genes, including promoter, exon, intron or other regions that harbors genetic information, would not work for that purpose, because single or multiple mutations might be lethal to the cell. Thus, microsatellite regions, including telomeres, would be suitable for a DNA damage monitoring system. Recently, it was shown that telomere length regulates mitochondrial function by activating PGC-1α (Sahin et al., 2011). This effect is the same as Rsv treatment (Lagouge et al., 2006). The CR mimetic drugs may have a common role in strengthening telomere maintenance. In the present study, we performed a multiple transfection experiment, which showed that shelterin protein-encoding gene promoters simultaneously respond to CR mimetic drugs in HeLa-S3 cells. Given that anti-aging drugs induce or activate the DNA repair system, especially by maintenance of telomeres, this multiple transfection system has demonstrable potential to contribute to the evaluation and development of such drugs.

## 6. Acknowledgments

The authors are grateful to Ryosuke Akiyama for outstanding technical assistance. This work was supported in part by a Research Fellowship from the Research Center for RNA Science, RIST, Tokyo University of Science, Tokyo, Japan.

## 7. References

Abreu, E., Aritonovska, E., Reichenbach, P., Cristofari, G., Culp, B., Terns, R.M., Lingner, J. & Terns, M.P. (2010). TIN2-tethered TPP1 recruits human telomerase to telomeres *in vivo*, *Molecular and Cellular. Biology* 30.(12): 2971–2982.

Blackburn, E.H. (2000). The end of the (DNA) line, *Nature Structural Biology* 7.(10): 847–850.

Campisi, J. & d'Adda di Fagagna, F. (2007). Cellular senescence: when bad things happen to good cells, *Nature Reviews. Molecular Cell Biology* 8.(9): 729–740.

Cantó, C., Auwerx, J. (2010). AMP-activated protein kinase and its downstream transcriptional pathways, *Cellular and Molecular Life Sciences* 67.(20): 3407–3423.

Cavallini, G., Donati, A., Gori, Z. & Bergamini, E. (2008). Towards an understanding of the anti-aging mechanism of caloric restriction, *Current Aging Science* 1.(1): 4–9.

Chen, Y., Rai, R., Zhou, Z.R., Kanoh, J., Ribeyre, C., Yang, Y., Zheng, H., Damay, P., Wang, F., Tsuji, H., Hiraoka, Y., Shore, D., Hu, H.Y., Chang, S. & Lei, M. (2011). A conserved motif within RAP1 has diversified roles in telomere protection and regulation in different organisms, *Nature Structural & Molecular Biology* 18.(2): 213–221.

Chu, W.K. & Hickson, I.D. (2009). RecQ helicases: multifunctional genome caretakers, *Nature Reviews. Cancer* 9.(9): 644–654.

Cohen, S.B., Graham, M.E., Lovrecz, G.O., Bache, N., Robinson, P.J. & Reddel, R.R. (2007). Protein composition of catalytically active human telomerase from immortal cells, *Science* 315.(5820): 1850–1853.

de Lange, T. (2006). Mammalian telomeres, *in* de Lange, T., Lundblad, V. & Blackburn, E. (ed.), *Telomeres (second ed.)*, Cold Spring Harbor Laboratory Press, New York, pp. 387–431.

Fröjdö, S., Durand, C. & Pirola L. (2008). Metabolic effects of resveratrol in mammals--a link between improved insulin action and aging, *Current Aging Science* 1.(3): 145–151.

Giannone, R.J., McDonald, H.W., Hurst, G.B., Shen, R.F., Wang, Y. & Liu, Y. (2010). The protein network surrounding the human telomere repeat binding factors TRF1, TRF2, and POT1, *PLoS One* 5.(8): e12407.

Gilson, E. & Géli, V. (2007). How telomeres are replicated, *Nature Reviews. Molecular Cell Biology* 8.(10): 825–838.

Gonzalez-Suarez, I. & Gonzalo, S. (2008). Crosstalk between chromatin structure, nuclear compartmentalization, and telomere biology, *Cytogenetic and Genome Research* 122.: 202–210.

Greider, C.W. & Blackburn, E.H. (1987). The telomere terminal transferase of *Tetrahymena* is a ribonucleoprotein enzyme with two kinds of primer specificity, *Cell* 51.(6): 887–898.

Griffith, J.D., Comeau, L., Rosenfield, S., Stansel, R.M., Bianchi, A., Moss, H. & de Lange, T. (1999). Mammalian telomeres end in a large duplex loop, *Cell* 97.(4): 503–514.

Jaskelioff, M., Muller, F.L., Paik, J.H., Thomas, E., Jiang, S., Adams, A.C., Sahin, E., Kost-Alimova, M., Protopopov, A., Cadiñanos, J., Horner, J.W., Maratos-Flier, E. & Depinho, R.A. (2011). Telomerase reactivation reverses tissue degeneration in aged telomerase-deficient mice, *Nature* 469.(7328): 102–106.

Kaeberlein, M. (2010). Resveratrol and rapamycin: are they anti-aging drugs? *Bioessays* 32.(2): 96–99.

Kenyon CJ. (2010). The genetics of aging, *Nature* 464. (7288): 504–512.

Knutson, M.D., & Leeuwenburgh, C. (2008). Resveratrol and novel potent activators of SIRT1: effects on aging and age-related diseases, *Nutrition Reviews* 66.(10): 591–596.

Kuningas, M., Mooijaart, S.P., van Heemst, D., Zwaan, B.J., Slagboom, P.E. & Westendorp, R.G. (2008). Genes encoding longevity: from model organisms to humans, *Aging Cell* 7.(2): 270–280.

Lagouge, M., Argmann, C., Gehart-Hines, Z., Meziane, H., Lerin, C., Daussin, F., Messadeq, N., Milne, J., Lambert, P., Elliott, P., Geny, B., Laakso, M., Puigserver, P. & Auwerx, J. (2006). Resveratrol improves mitochondrial function and protects against metabolic disease by activating SIRT1 and PGC-1alpha, *Cell* 127.(6): 1109–1122.

Lin, X., Zhang, F., Bradbury, C.M., Kaushal, A., Li, L., Spitz, D.R., Aft, R.L. & Gius, D. (2003). 2-Deoxy-D-glucose-induced cytotoxicity and radiosensitization in tumor cells is mediated via disruptions in thiol metabolism, *Cancer Research* 63.(12): 3413–3417.

Lundblad, V. & Szostak, J.W. (1989). A mutant with a defect in telomere elongation leads to senescence in yeast, *Cell* 57.(4): 633–643.

Meek, D.W. (2009). Tumor suppression by p53: a role for the DNA damage response? *Nature Reviews. Cancer* 9.(10): 714–723.

Oberdoerffer, P. & Sinclair, D.A. (2007). The role of nuclear architecture in genomic instability and aging, *Nature Reviews. Molecular Cell Biology* 8.(9): 692–702.

Opresko, P.L., Mason, P.A., Podell, E.R., Lei, M., Hickson, I.D., Cech, T.R. & Bohr, V.A. (2005). POT1 stimulates RecQ helicases WRN and BLM to unwind telomeric DNA substrates, *Journal of Biological Chemistry* 280.(37): 32069–32080.

O'Sullivan, R.J. & Karlseder, J. (2010). Telomeres: protecting chromosomes against genome instability, *Nature Reviews. Molecular Cell Biology* 11.(3): 171–181.

Roth, G.S., Ingram, D.K. & Lane, M.A. (2001). Caloric restriction in primates and relevance to humans, *Annals of the New York Academy of Sciences* 928.: 305–315.

Sahin, E. & Depinho, R.A. (2010). Linking functional decline of telomeres, mitochondria and stem cells during aging, *Nature* 464.(7288): 520–528.

Sahin, E., Colla, S., Liesa, M., Moslehi, J., Müller, F.L., Guo, M., Cooper, M., Kotton, D., Fabian, A.J., Walkey, C., Maser, R.S., Tonon, G., Foerster, F., Xiong, R., Wang, Y.A., Shukla, S.A., Jaskelioff, M., Martin, E.S., Heffernan, T.P., Protopopov, A., Ivanova, E., Mahoney, J.E., Kost-Alimova, M., Perry, S.R., Bronson, R., Liao, R., Mulligan, R., Shirihai, O.S., Chin, L. & DePinho, R.A. (2011). Telomere dysfunction induces metabolic and mitochondrial compromise, *Nature* 470.(7334): 359–365.

Sanz, A. & Stefanatos, R.K. (2008). The mitochondrial free radical theory of aging: A critical view, *Current Aging Science* 1.(1): 10–21.

Schumacher, B. (2009). Transcription-blocking DNA damage in aging: a mechanism for hormesis, *Bioessays* 31.(12): 1347–1356.

Schreiber, V., Dantzer, F., Ame, J.-C. & de Murcia, G. (2006). Poly(ADP-ribose): novel functions for an old molecule, *Nature Reviews. Molecular Cell Biology* 7.(7): 517–528.

Schulz, V.P. & Zakian, V.A. (1994). The saccharomyces PIF1 DNA helicase inhibits telomere elongation and de novo telomere formation, *Cell* 76.(1): 145–155.

Smith, S., Giriat, I., Schmitt, A. & de Lange, T. (1998). Tankyrase, a poly(ADP-ribose) polymerase at human telomeres, *Science* 282.(5393): 1484–1487.

Stefani, M., Markus, M.A., Lin, R.C., Pinese, M., Dawes, I.W. & Morris, B.J. (2007). The effect of resveratrol on a cell model of human aging, *Annals of the New York Academy of Sciences* 1114.: 407–418.

Stern, J.L. & Bryan, T.M. (2008). Telomerase recruitment to telomeres, *Cytogenetic and Genome Research* 122.: 243–254.

Cosan, D.T., Soyocak, A., Basaran, A., Degirmenci, I., Gunes, H.V. & Sahin, F.M. (2011). Effects of various agents on DNA fragmentation and telomerase enzyme activities in adenocarcinoma cell lines, *Molecular Biology Reports* 38.(4): 2463-2469.

Uchiumi, F., Sakakibara, G., Sato, J. & Tanuma, S. (2008). Characterization of the promoter region of the human *PARG* gene and its response to PU.1 during differentiation of HL-60 cells, *Genes to Cells* 13.(12): 1229–1247.

Uchiumi, F., Watanabe, T. & Tanuma, S. (2010a). Characterization of various promoter regions of the human DNA helicase-encoding genes and identification of duplicated *ets* (GGAA) motifs as an essential transcription regulatory element, *Experimental Cell Research* 316.(9): 1523–1534.

Uchiumi, F., Enokida, K., Shiraishi, T., Masumi, A. & Tanuma, S. (2010b). Characterization of the promoter region of the human *IGHMBP2* (*Sμbp-2*) gene and its response to TPA in HL-60 cells, *Gene* 436.(1-2): 8–17.

Uchiumi, F., Higami, Y. & Tanuma, S. (2010c). Regulations of telomerase activity and *WRN* gene expression, *in* Gagnon, A.N. (ed.), *Telomerase: Composition, Functions and Clinical Implications*, Nova Science Publishers, Inc., Hauppauge, NY, pp. 95–103.

Uchiumi, F., Watanabe, T., Hasegawa, S., Hoshi, T., Higami, Y. & Tanuma, S. (2011a). The effect of resveratrol on the werner syndrome RecQ helicase gene and telomerase activity, *Current Aging Science* 4.(1): 1–7.

Uchiumi, F., Miyazaki, S. & Tanuma, S. (2011b). The possible functions of duplicated ets (GGAA) motifs located near transcription start sites of various human genes, *Cellular and Molecular Life Sciences* 68.(12): 2039-2051.

van der Horst, A. & Burgering, B.M. (2007). Stressing the role of FoxO proteins in lifespan and disease, *Nature Reviews. Molecular Cell Biology* 8.(6): 440–450.

Vijg, J. (2007). Genome instability and accelerated aging, *in* Vijg, J. (ed.), *Aging of the Genome*, Oxford University Press, Oxford, pp. 151–180.

Zhou, B., Ikejima, T., Watanabe, T., Iwakoshi, K., Idei, Y., Tanuma, S. & Uchiumi, F. (2009). The effect of 2-deoxy-D-glucose on Werner syndrome RecQ helicase gene, *FEBS Letters* 583.(8): 1331–1336.

Zoncu, R., Efeyan A., & Sabatini, D.M. (2011). mTOR: from growth signal integration to cancer, diabetes and aging, *Nature Reviews. Molecular Cell Biology* 12.(1): 21–35.

# Part 4

# Measuring DNA Repair Capacity

# 11

# DNA Repair Measured by the Comet Assay

Amaya Azqueta[1], Sergey Shaposhnikov[2] and Andrew R. Collins[2]
*[1]University of Navarra*
*[2]University of Oslo*
*[1]Spain*
*[2]Norway*

## 1. Introduction

The stability of the genome is of crucial importance, and yet the DNA molecule is prone to spontaneous loss of bases, and damage from exogenous and endogenous sources – with potentially mutagenic consequences. Damage can take the form of small alterations to bases (alkylation or oxidation); breaks in the sugar-phosphate backbone involving one or both strands (single or double strand breaks – SSBs or DSBs); bulky adducts combined with bases; and covalent bonds between adjacent bases (intra-strand cross-links), across the double helix (inter-strand cross-links), or between DNA and protein. These lesions can disrupt replication, or cause incorporation of the wrong base.

Cells possess repair enzymes that correct almost all the damage before it can result in permanent change to the genome. Different pathways deal with the various kinds of damage. Repair of SSBs is in most cells a rapid process, consisting of little more than ligation. DSBs are more complicated (and potentially more serious) since the continuity of the double helix is disrupted. Homologous recombination ensures restoration of the correct DNA sequence by using the DNA of the sister chromatid or homologous chromosome as a template, while non-homologous end-rejoining is less precise and can entail loss of sequence. Base excision repair (BER) is concerned with small base alterations and starts with removal of the damaged base by a more or less specific glycosylase, leaving a base-less sugar or AP-site (apurinic/apyrimidinic site). An AP endonuclease or lyase cleaves the DNA at this site, and – after trimming of the broken ends of DNA – the one-nucleotide gap is filled by DNA polymerase β. Ligation is the final stage. Nucleotide excision repair (NER) is a more complex affair, involving recognition of a bulky adduct or helix distortion (such as is caused by the dimerisation of adjacent pyrimidines by UV(C) radiation), endonucleolytic incision on each side of the lesion, and removal of an oligonucleotide containing the damage. This is then filled in by DNA polymerase δ, κ or ε and the new patch of nucleotides is ligated into the DNA, completing the repair. NER enzymes are also involved in repair of inter-strand cross-links, removing the linking molecule from one strand, leaving it attached to the other strand as a mono-adduct to be removed in a second NER reaction (according to the simplest, and possibly simplistic, model).

Individual DNA repair capacity is regarded as a biomarker of susceptibility to mutation and cancer. A person with high repair rate is assumed to be at lower risk than one with low repair rate. DNA repair is partially determined genetically, and polymorphisms in repair genes will affect overall repair activity. However, this variation cannot account for the wide range of individual repair rates as measured in human populations. The intrinsic repair rate is likely to be affected by environmental conditions such as the presence of DNA-damaging agents that induce repair activity, and there is accumulating evidence that nutritional and lifestyle factors – for instance, micronutrients – can also modulate DNA repair.

Levels of mRNA corresponding to DNA repair pathways are frequently assessed by DNA microarray techniques, or by RT-PCR for selected genes. However, gene expression does not necessarily correlate with enzyme activity, and there is no substitute for measurement of repair capacity, i.e. phenotype. This is where the comet assay can be most usefully applied.

The comet assay, with modifications, can measure various kinds of damage, and the corresponding repair pathways. The basic comet assay detects strand breaks (see section 2.1. "The comet assay"), and so is readily applied to SSB repair by monitoring the rejoining of breaks. With a modification to detect particular classes of damage by incorporating a digestion with lesion-specific endonuclease, repair of oxidised and alkylated bases, as well as dimerised pyrimidines, can be followed. There are other specialised modifications of the assay to study cross-link repair. In addition to these assays based on following the removal of damage, there is a method for measuring NER in cells in culture by blocking repair synthesis and accumulating incision events as DNA breaks. Another approach to measuring BER or NER involves an 'in vitro' assay in which a cell extract is incubated with a DNA substrate containing specific lesions, and again the occurrence of breaks is monitored.

A quite distinct application of the comet assay is to the study of repair rates in different genes, taking advantage of the ability to identify – by the use of specific hybridisation probes – particular regions of the genome.

Here we will describe the different methods, and give examples of their application to cell culture, animal and human studies, where appropriate, without providing an exhaustive review of the literature.

## 2. Methods

### 2.1 The comet assay

The comet assay (single cell gel electrophoresis) is a simple, sensitive, economical method for measuring DNA SBs. Cells are embedded in agarose on a microscope slide, lysed, and electrophoresed. Broken DNA is drawn towards the anode, forming a 'comet tail'; it is stained with a DNA-binding dye and observed with fluorescence microscopy (Figure 1a). The assay depends on the fact that DNA in the mammalian nucleus is organised as a series of DNA loops, attached to the nuclear framework, or matrix, at intervals. The DNA is (negatively) supercoiled, by virtue of its arrangement as nucleosomes, and each supercoiled loop should be regarded as a structural unit. Lysis of cells with detergent and high salt (removing membranes, soluble cell components and most histones), leaves the DNA still attached to the matrix, and known as a nucleoid; the supercoiling is still present, and when this supercoiling is relaxed by a DNA SB, only the loop containing the break is affected. The assay can be carried out at 'neutral' pH (around 10 - not high enough to denature DNA [Ostling & Johanson, 1984] or at high pH above pH 13 [Singh et al., 1988]). Both neutral and alkaline versions detect SSBs, since a single SB is sufficient to relax supercoiling. The assay does not depend on alkaline denaturation to reveal SSBs (unlike other assays such as neutral/alkaline elution, and alkaline unwinding), but the apparent analogy has led to much confusion, and it is often stated that the neutral assay only detects DSBs. The neutral and alkaline comet assays do, however, differ in one important respect; at a high pH, AP-sites are converted to breaks.

The more breaks are present, the more loops are relaxed, and the more intense is the fluorescence of the comet tail relative to the nucleoid core when the nucleoids are stained with an appropriate DNA-binding dye (Figure 2). Comets (normally 30 to 100 per gel) are scored, most commonly, by computer-based image analysis, with '% tail DNA' as the preferred parameter, although an alternative 'visual scoring' technique is still widely used (Collins, 2004). For statistical analysis, the unit of analysis is the mean or median % tail DNA from the comets representing one independent sample of cells. % Tail DNA can be converted to 'real' units such as breaks per $10^9$ Da by use of a calibration curve, based on $\gamma$- or X-irradiation of cells, since the breakage rate per Gy is known.

The comet assay can be applied to virtually any eukaryotic cell type that can be obtained as a single cell or nuclear suspension. Cell cultures and white blood cells are widely used, but also methods have been developed for disaggregating many kinds of tissue without causing damage to the cells' DNA. Sperm, with highly compacted DNA, can be subjected to comet analysis after treating with protease or dithiothreitol. The most commonly adopted strategy with plant cells is to release the nuclei by simply chopping the plant tissue with a sharp blade. The presence of chloroplasts in leaf tissue can lead to release of free radicals and oxidative damage to DNA unless the isolation is carried out under safelight conditions.

The basic comet assay is limited in its usefulness because only strand breaks (and alkali-labile sites) are detected. An additional step - digestion of the nucleoid DNA, after lysis, with a lesion-specific enzyme - converts various other kinds of DNA damage to DNA breaks (Figure 1b). Thus formamidopyrimidine DNA glycosylase (FPG) recognises oxidised purines, principally 8-oxoguanine (8-oxoG), but also ring-opened purines or formamidopyrimidines (and in addition some alkylated bases). Endonuclease III (EndoIII) converts oxidised pyrimidines to breaks, while 3-methyladenine DNA glycosylase II (AlkA) acts on alkylated bases (principally 3-methyladenine). UV-induced cyclobutane pyrimidine dimers are detected by the UV endonuclease, T4 endonucleaseV (T4endoV).

## 2.2 Measuring DNA repair with the "challenge assay"

The simplest assay for DNA repair is the so-called 'challenge assay' (Au et al., 2010), whereby cells are treated with a damaging agent and the removal of the damage is monitored over time to study the kinetics of repair. Different assays can be used to assess the level of damage remaining at different time points; the comet assay is one of them. It is commonly used to monitor rejoining of SBs by cells, but by incorporating the digestion of DNA (nucleoids) with a lesion-specific endonuclease the removal of different DNA lesions can also be assessed. With this aim FPG is used to convert oxidised purines into SBs, Alk A to convert the alkylated bases and T4endoV to convert the cyclobutane pyrimidine dimers induced by UV. Using all the possibilities, this assay allows us to measure SSB rejoining, BER (removal of oxidised and alkylated bases) and NER (removal of UV-induced cyclobutane dimers).

Different agents are used to induce the desired type of lesion in the DNA depending on the repair pathway to be studied. SSBs are easily induced by a brief treatment with $H_2O_2$ or by irradiation with X- or $\gamma$-rays. Oxidized purines, mainly 8-oxoG, are induced by treating the cells with the photosensitiser Ro 19-8022 plus visible light. Methyl methanesulfonate (MMS) can be used to produce alkylated bases and UV(C) radiation induces cyclobutane dimers.

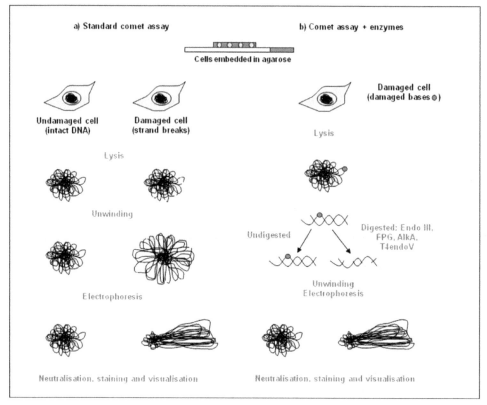

Fig. 1. Scheme of the standard comet assay (a), and the modified assay including digestion with lesion-specific enzymes (b).

The conditions of the treatment can vary depending on the cell type, and it is recommended first to establish optimal conditions; a high level of induced lesions, but not enough to saturate the assay or the capacity of the cells to repair the damage without entering apoptosis.

After the treatment cells are incubated in the appropriate cell culture medium and conditions (normally in an incubator at 37°C with 5% of $CO_2$) for different times. Just after the treatment (time 0) an aliquot of the cells is taken to check the level of induced damage. Further aliquots are taken at different times of incubation, including times soon after the start of incubation in order to estimate the initial rate of repair accurately.

In the case of adherent cells it is necessary to set up as many cell cultures as there are time-points (in multi-well plates or petri dishes) because at each time-point cells should be trypsinized. If cells are growing in suspension, an aliquot can be removed from the whole cell culture at each time-point. Setting the right times is a very important issue, influenced by the cell type and the repair pathway to be studied and so a prior investigation should be done on this topic also.

To avoid continuing repair of DNA damage while processing the cells after sampling at the different time-points, cells should be kept on ice during their manipulation. This is particularly important when very short intervals of time are tested.

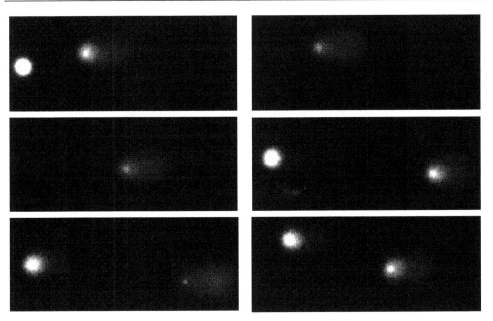

Fig. 2. Comet images with different levels of DNA damage.

The comet assay is done as described above; either the basic version (to assess SSB rejoining) or with an enzyme digestion (to assess BER or NER). The lysis step of the comet assay can last between 1 h and 24 h (or even longer) so gel-embedded cells/nucleoids can be kept in the lysis solution until all of the samples have been processed. Then samples from all time-points can be run in the same experiment, which as well as being practically convenient, avoids experimental variability.

To be able to compare different kinetics of repair, the half time of damage removal ($t_{1/2}$) should be calculated. To obtain an accurate estimation of this parameter the choice of the different time points is crucial. Generally the repair of SSBs is rapid, with a $t_{1/2}$ of 10 minutes or so while the repair of oxidized and alkylated bases and UV-induced cyclobutane dimers takes a few hours (Lorenzo et al., 2009). Another useful parameter is the initial repair rate, but this is difficult to estimate accurately if repair is rapid.

As in all of the assays, proper controls should be included to interpret the results correctly. A non-damaged cell culture should be included at all time points (including time 0) to check for any variation in or problem with experimental conditions.

### 2.2.1 Applications of the challenge assay

The challenge assay is used in cell culture experiments to check the influence of different compounds on the cellular repair rate. It is also used in animal studies and in human biomonitoring, normally studying lymphocytes.

The residual damage should always be measured at several time points after the incubation, so that the kinetics of the repair can be quantified and compared between different cell types or experimental conditions. Ideally, residual damage should be measured at shorter intervals immediately after treatment, since the initial rate of removal of damage is considered the defining step of the process. Another option, as explained before, is to calculate the $t_{1/2}$ for lesion removal. Measuring residual damage at a unique late point when most of the damage has been repaired, as is often reported, gives limited and ambiguous information.

For a valid comparison of different cell types or lymphocyte samples, the level of induced damage to be removed should ideally be the same in all cells/samples in the study, a state that in many cases is not easily achieved. It is a good assay to use with cell lines for examining the effect of an agent on repair when the compound to be tested does not affect the level of induced DNA damage. But sometimes cell cultures can be protected from DNA damage by the compound being studied; thus, for example, when an antioxidant micronutrient is tested for an effect on repair, it will obviously decrease the level of induced oxidative damage.

Compared with cell lines, animals and humans have more variability that can affect the level of damage achieved with the challenge compound. In biomonitoring, one subject group can have a higher antioxidant status that protects them against the damaging agent. This problem may well arise and is very difficult to solve. One possibility is to arrange for different doses of damage to each group to ensure the same initial level of lesions but this is in general impracticable.

Another disadvantage of this assay is that it involves a lot of cell culturing, specially when adherent cells are used and trypsinization is needed at all time points; the scheduled times to carry out the assay of residual damage can be inconvenient, and overall the experiment is complicated to perform. This is especially the case in biomonitoring studies, since the large number of samples to be tested precludes such complicated procedures – and there is inevitably day-to-day variation in culture conditions and results. On the other hand its endpoint is the removal of lesions and restoration of normal DNA structure, i.e. overall repair, whereas other methods tend to look only at one step in the repair process.

## 2.2.2 The challenge assay in cell culture studies

The "challenge assay" is the most suitable comet assay-based approach to measure DNA repair in cell culture and it has been used with different purposes. In 2003, Blasiak et al. demonstrated the temperature-dependence of the DNA repair process with the aim of using hyperthermia in the modulation of cancer therapy. They treated human peripheral lymphocytes and two variants of a human myelogenous leukemia cell line (K562 and its doxorubicin-resistant variant) with doxorubicin and studied the removal of the damage at 37°C and 41°C. They found an increase in the repair rate of the cells incubated at 41°C compared with 37°C. Tsai-Hsiu et al. (2003) studied the effect of S-adenosylhomocysteine (SAH), an inhibitor of most methyltransferases, on the repair rate of a mouse endothelial cell line and a human intestinal cell line. Cells were treated with $H_2O_2$ before incubating them with different concentrations of SAH or homocysteine as control and the removal of the damage was monitored. They showed that SAH decreased the DNA repair rate in a dose-dependent manner.

Ramos et al. (2008) studied the chemoprotective effects of the flavonoids quercitin and rutin, and the phytochemical ursolic acid, on the DNA damage induced by tert-butyl hydroperoxide (t-BHP) in a human hepatoma cell line. They checked the removal of the DNA damage induced by t-BHP after incubating the cells with different concentrations of quercitin, rutin or ursolic acid for 24h. There was an increase in the DNA repair rate of cells incubated with quercitin and ursolic acid, when the remaining lesions were measured 2 h after the treatment. The same group showed an enhancement in the repair rate when the human colon carcinoma cell line Caco-2 was preincubated with Salvia extracts or luteonil-7-glucoside before $H_2O_2$ treatment; but there was no effect when the compounds were just present during the repair time (Ramos et al., 2010a).

The repair of both SBs and oxidized bases (induced by treatment with $H_2O_2$ or with a photosensitiser plus visible light, respectively) were assessed in the human cervical cancer cell line HeLa and in Caco-2 cells incubated with different concentrations of the carotenoid $\beta$-cryptoxanthin (Lorenzo et al., 2009). This carotenoid induced a faster removal of both kinds of lesions in both cell lines at very low concentrations. The effect of $\beta$-cryptoxanthin on the removal of the 8-oxoG in Caco-2 cells is shown in Figure 3a.

Rejoining of X-ray-induced SBs by mouse leukocytes was studied by Gudkov et al. (2009) The natural ribonucleosides guanosine and inosine were present during the repair period, and SBs were measured after irradiation. Both ribonucleosides increased the repair rate. Moreover, in the presence of the repair inhibitor nicotinamide (prior to the irradiation and during the repair process), repair was slower and ribonucleosides did not induce any effect.

### 2.2.3 The challenge assay in animal studies

Although this approach is not ideal for application in *in vivo* animal studies, the lack of a good alternative makes it very common. Gover et al. (2001) studied the repair of DNA lesions induced by different concentrations of the fungicide mercuric chloride in leucocytes of rats. The comet assay was performed in whole blood at different times after an oral administration of a single dose. The level of the DNA damage decreased from 48 h and reached the control level at 2 weeks after the treatment. Very similar studies have been done to check the effects of the insecticide JS-118 (Zhang et al., 2010) and of copper sulfate (Saleha Banu et al., 2004) in mice.

This approach has also been applied to DNA repair in organs. Cells from the liver, kidney and bone marrow of mice were used to check the effect of the intraperitoneal administration of (MMS), a known genotoxic compound, and acetaminophen, an analgesic drug (Oshida et al., 2008). The level of DNA damage found at 4 h was less than at 24 h in all the organs and with both compounds. According to the authors this decrease can be due to detoxification, repair of the lesions induced by the treatment, or cell turnover.

The effect of intraperitoneal administration of the phytochemical feluric acid on repair of the DNA damage induced in lymphocytes by whole body $\gamma$-irradiation of mice was studied by Maurya et al. (2005). The disappearance of the induced SBs was faster in animals which received feluric acid compared to controls.

In these studies the challenging agent is given to the animals and repair occurs in physiological conditions (inside the animal). The assay has also been applied in animal studies where the challenge occurs *ex vivo*. Miranda et al. (2008) studied the protective effect of intragastric administration of aqueous extracts form Yerba mate tea in mice over a period of 60 days. After this period cells were isolated from liver, kidney and bladder and embedded in agarose before treating them with $H_2O_2$. There was an enhancement of DNA repair in liver cells.

### 2.2.4 The challenge assay in human studies

As explained before, the challenge assay presents many inconveniences when used in humans, but there are several studies that use this approach to measure the DNA repair capacity of individuals.

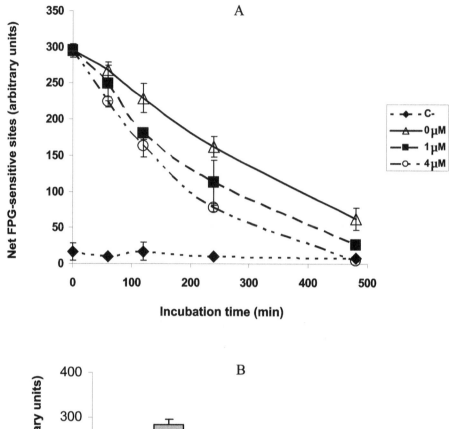

Fig. 3. Effect of β-cryptoxanthin on BER in Caco-2 cells: "challenge assay" (a) and *in vitro* repair assay (b). From Lorenzo Y, Azqueta A, Luna L, Bonilla F, Dominguez G, Collins AR (2009) The carotenoid β-cryptoxanthin stimulates the repair of DNA oxidation damage in addition to acting as an antioxidant in human cells. Carcinogenesis 30 (2):308-314, by permission of Oxford University Press.

In a case-control study the DNA repair capacity of lymphocytes from 44 healthy donors and 38 patients with squamous cell carcinoma of head and neck (before treatment) was measured (Palyvoda et al., 2003). Repair of $\gamma$-ray induced lesions showed a high variability between individuals ($t_{1/2}$ from about 10 min to more than 1 h). Lymphocytes from patients showed lower repair rates and a higher amount of non-repaired damage after the incubation period.

This approach has also been used to monitor DNA repair in relation to occupation, environment or lifestyle. The repair rates of stimulated lymphocytes from 10 nuclear power plant workers chronically exposed to low doses of ionizing radiation and 10 controls were assessed (Touil et al., 2002). The interindividual variation in the rates of repair of $\gamma$-irradiation-induced DNA damage was high but there were no significant differences between groups. In a similar study, the repair rate of lymphocytes from 104 asbestos-exposed workers and 101 control workers was studied (Zhao et al., 2006). Lymphocytes from asbestos-exposed workers showed slower repair of $H_2O_2$ induced damage.

The challenge assay has also been applied in nutritional studies to check the influence of phytochemicals or whole foods on DNA repair. To study the effect of lutein, lycopene and $\beta$-carotene, 8 healthy volunteers were given supplements daily during 1 week in a cross over study with a wash-out period of 3 weeks (Torbergsen & Collins, 2000). Lutein did not have any effect on the DNA repair rate, but lycopene and $\beta$-carotene apparently accelerated the rejoining of the SBs. However, an increase in the level of SBs in non-irradiated cells during approximately the first 4 h of the incubation period was seen. This could be due to the oxidative stress that lymphocytes suffer from sudden exposure to atmospheric oxygen. This transient increase was less pronounced in lymphocytes taken after lycopene or $\beta$-carotene supplementation so the apparent acceleration of repair could be explained by an antioxidant protection exerted by the presence of carotenoids. In another study (Astley et al., 2004) healthy volunteers followed a dietary intervention with a mixed carotene capsule, a daily portion of cooked minced carrots, a portion of mandarin oranges, a vitamin C tablet or a matched placebo (about 10 volunteers   per group).   Only the lymphocytes from individuals taking mixed carotene capsules showed an improvement in rate of repair of $H_2O_2$ induced DNA damage.

As explained above, crucial information is lost when just one time of recovery is used in the challenge assay, and – especially  if starting levels of damage are not the same – it can be misleading to compare repair rates on the basis of residual damage.

### 2.3 Measuring NER by inhibiting DNA synthesis

Many years ago, inhibitors of the DNA polymerase species that participate in NER were employed to block repair synthesis after UV(C) irradiation of cells in culture: the earlier steps of repair continue, leading to an accumulation of DNA breaks which normally occur as very transient repair intermediates. Aphidicolin, or cytosine arabinoside in combination with hydroxyurea, are equally effective as inhibitors. As methods for measuring DNA breaks, alkaline unwinding and alkaline elution were used. The principle was then combined with the comet assay, and used as early as 1992 (Gedik et al., 1992), detecting the accumulation of breaks in HeLa cells irradiated with 0.5 Jm$^{-2}$ of UV(C) and incubated for just 5 min.

This approach was adapted by Speit et al. (2004) as a way of enhancing the detection of damage done to DNA by a range of different agents (benzo[a]pyrene diolepoxide [BPDE], bischloroethylnitrosourea, and MMS). It is particularly useful in the detection of 'bulky adducts', which are repaired by NER, but are not recognised by T4endoV, and so are not amenable to the enzyme-modified comet assay. (The bacterial enzyme complex, uvrABC, detects bulky DNA adducts as well as UV-induced pyrimidine dimers. Many efforts have been made to incorporate this enzyme complex to the comet assay but, until now, it seems to detect only a small fraction of the available lesions [Dusinska & Collins, 1996]).

This inhibitor-based incision assay can be used as a simple and sensitive method to measure repair capacity, reflected in the rate of accumulation of breaks. Incision is generally considered to be the rate-limiting step of NER. Before the introduction of the comet assay, the accumulation of incision events was used to investigate the molecular defects in the disease xeroderma pigmentosum (Squires et al., 1982) and to characterise DNA repair-defective mutant cell lines (Stefanini et al., 1991).

## 2.3.1 Applications of the inhibitor assay for NER

This assay has not been widely used but it has considerable potential, particularly in human studies. Actually it seems that in the case of freshly isolated lymphocytes, the DNA breaks present as NER intermediates persist long enough to be detected with the comet assay without using aphidicolin or cytosine arabinoside (Collins et al., 1995; Green et al., 1994). Repair synthesis is unable to proceed due to the lack of enough DNA precursors (dNTPs). If deoxyribonucleosides are added to the medium, breaks are no longer detected. However, it seems wise to include aphidicolin or cytosine arabinoside to ensure that DNA resynthesis is completely blocked and so to be sure of detecting all the breaks.

Cipollini *et al.* (2006) treated lymphocytes with 1.5 Jm$^{-2}$ of UV(C) and observed breaks accumulating to a maximum at about 60 min with then a decrease. This decline can be due to eventual completion of repair even with the low concentrations of precursors, or to a synthesis of DNA precursors induced as a response to the DNA damage. Experimental variation and inter-individual differences in kinetics were seen in 4 subjects. This could be explained by individual differences in the precursor pool size rather than differences in the repair capacity. The characterisation of several UV-sensitive rodent mutant cell lines included the measurement of their ability to carry out incision after irradiation with 0.1 Jm$^{-2}$ of UV(C) (Collins et al., 1997).

The assay has been used to look for effects of *in vivo* exposure to different genotoxic agents by looking for an enhanced level of breaks when lymphocytes are incubated with DNA sythesis inhibitors. Crebelli et al. (2002) found a higher level of breaks (with cytosine arabinoside) in aluminium workers compared with controls; while Speit et al. (2003) did not detect such a difference between smokers and non-smokers.

The best use of the assay in human biomonitoring is probably as an *ex vivo* assay, i.e. treating the subjects' lymphocytes with UV(C) (or some other agent whose damage is repaired by NER) and incubating them with inhibitor *in vitro*. The Kirsch-Volders group recently carried out a pilot study with 22 subjects, treating peripheral blood mononucleated cells with BPDE for 2 h with and without preincubation with aphidicolin (Vande Loock *et al.*, 2010). They quantified repair capacity as the amount of SBs induced by BPDE with aphidicolin, minus the SBs induced by aphidicolin (a very small amount) and by BPDE alone – reckoning that this equates to the incision activity of the NER enzymes. (UV(C) is a cleaner agent to use, since it does not directly induce significant levels of SBs; all the SBs detected are NER intermediates.)

As a biomonitoring assay for human studies, the inhibitor assay for NER is still in the development phase. A comparison of results from this assay and from a UV challenge assay and an *in vitro* NER assay would be very informative.

## 2.4 Measuring BER and NER with an *in vitro* assay
## 2.4.1 Practical details

The comet assay has been modified to measure the excision repair activity in an extract of cells (or a nuclear extract). In this *in vitro* approach a substrate, in the form of agarose-embedded nucleoids derived by lysis of cells containing a specific lesion, is incubated with the extract whose excision repair activity is to be measured by the comet assay (Collins et al., 2001; Gaivão et al., 2009; Langie et al., 2006) (Figure 4). The nature of the DNA lesion in the substrate defines the repair pathway that is measured. Substrate containing 8-oxoG is used to measure the BER activity of 8-oxoG DNA glycosylase (OGG) in the extracts tested (Collins et al., 2001); if substrate contains bulky adducts or cyclobutane pyrimidine dimers NER is measured (Gaivão et al., 2009; Langie et al., 2006). More recently an assay for cross-link repair has been developed (Herrera et al., 2009). In all cases the enzymes contained in the extract will carry out the initial steps of repair by recognizing the lesion and introducing a break at or near its site.   The rate of accumulation of breaks, assessed by alkaline electrophoresis, is a measure of the repair capacity of the cells.

The substrate nucleoids should contain a high level of specific base damage so that the enzymes in the extract have an excess of lesions to work on. This level should be more than enough to saturate the comet assay but the background level of breaks as well as other unwanted lesions should be very low. To reach this equilibrium is not always an easy task and as a result it is not always possible to produce substrate with the desired lesion.

Furthermore breaks do not continue to increase indefinitely but reach a saturation. This means that the longer the incubation, the less difference will be detected in activity between different extracts - so the time of incubation is crucial.

The description of the assay is divided into 5 steps: preparation of cells for the substrate, preparation of cells for the extract, preparation of the substrate nucleoids, preparation of the extract and incubation of the substrate with the extract.

*Preparation of cells for the substrate for BER and NER:* A substrate to measure BER is prepared by treating the cells with the photosensitiser Ro 19-8022 plus visible light to induce oxidized purines, mainly 8-oxoG. For NER, cells are irradiated with UV(C) to induce cyclobutane pyrimidine dimers (they can also be treated with BPDE or oxaliplatin to produce bulky adducts or cross-links respectively but we do not have experience with such treatments). Non-treated cells should be used to prepare a control substrate.   The cell type used for substrate is not important.

After treatment cells are slowly frozen in aliquots and kept at -80°C (for months or even years).

*Preparation of cells for the extract:* Extract is normally prepared from lymphocytes or cultured cells (recently animal tissue has been successfully used [Langie et al., 2011] in this assay but this will not be covered in this article). In the order of 5-10 million cells are needed to perform about 6 determinations of repair activity.

Cells are washed, spun, suspended at $10^7$ per 100 µl in extraction buffer and aliquots flash-frozen in liquid nitrogen before storage (for at least months) at -80°C. In fact there are three alternative methods to prepare cells for making extract: (1) direct preparation in extraction buffer, as above; (2) preparation of a dry pellet of the cells, snap-frozen and kept at 80°C (the rest of the extraction being done on the day of the assay); (3) cells frozen slowly in freezing medium to maintain viability, with complete extraction procedure carried out on day of experiment. Each method gives comparable results but experience shows that with method (2) the presence of residual supernatant when the pellet is thawed presents problems.

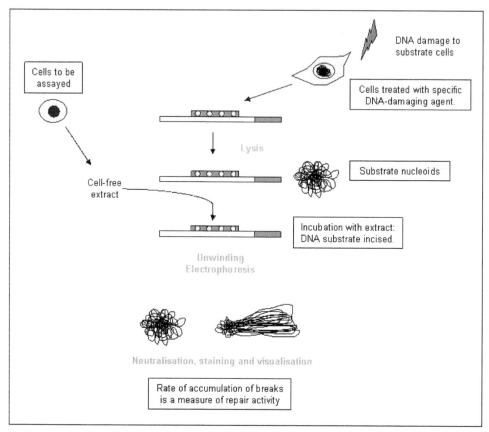

Fig. 4. Scheme of the *in vitro* repair assay

*Preparation of the substrate nucleoids:* Cells are thawed and embedded in agarose on a microscope slide as explained before (see section 2.1. "The comet assay"). Then slides are placed in the lysis solution for at least one hour to produce nucleoids. Control cells, without any induced damage, should always be included to check for unspecific nuclease activity in the extracts.

*Preparation of the extracts:* On the day of the experiment, after preparing substrate gels, an aliquot of the cells frozen for preparing the extract is thawed and kept on ice. Triton X-100 is added (final concentration 0.2%) to destabilize the membranes and complete lysis. After centrifugation at high speed to remove nuclei and cell debris, the supernatant is then diluted 4-fold in reaction buffer. All the procedures should be done on ice to avoid loss of activity.

*Incubation of the substrate with the extracts:* Substrate gels should be washed with the reaction buffer before the incubation with the extracts. Extracts are incubated with the substrate gels for between 10 and 30 min at 37°C in a humidified atmosphere. The optimal time should be established in each laboratory. If the 2 gels per slide format is used about 45 µl of extract is added on the top of the gel and a cover slip (or a Parafilm square) is placed on top. After that the standard comet assay protocol is followed.

Substrate nucleoids should be also treated with a buffer control (extraction buffer + Triton + reaction buffer) and with enzymes to test the level of DNA damage contained in the nucleoids. FPG is used for BER substrates and T4endoV for NER. Therefore the standard experiment should include substrate for BER or NER and substrate without damage, incubated with extracts, the buffer control and FPG or T4endoV.

To express the results the values of the activity of the extracts and controls obtained from the substrate without damage, representing the non-specific activity, should be subtracted from the values obtained from the BER or NER substrate.

It is crucial that the number of cells to prepare the extract is the same in all the samples. Counting is always time consuming and not accurate; it is useful to measure the protein concentrations in the extract residues left after an experiment, and to express repair activity relative to the protein concentration, thus allowing for variation in cell numbers. However, this correction is valid only over a narrow range, since repair activity deviates from linearity at high concentrations. Therefore, care must be taken to work with similar and appropriate protein concentrations or cell densities in all extracts.

### 2.4.2 Applications of the *in vitro* DNA repair assay

The *in vitro* DNA repair assay has been used in some cell culture and animal studies but it is mostly used in human biomonitoring. It is a very useful tool in this type of study since it can be used on extracts prepared from lymphocyte samples and stored frozen until a batch of extracts are ready to be measured at the same time.

This approach does not measure the whole process, but only the initial step of repair.

### 2.4.3 Cell culture studies with the *in vitro* assay

Very few examples of the *in vitro* repair assay applied to cell culture studies can be found in the literature. Ramos et al. (2010a) evaluated the incision activity of extract from Caco-2 cells treated for 24 h with water extracts of *Salvia* species, rosmarinic acid and luteonil-7-glucoside on nucleoid substrate containing 8-oxoG. All extracts from treated cells showed an increase. It was significant in the case of one of the *Salvia* species and with luteonil-7-glucoside. In the same way, Ramos and colleagues showed an increase in the BER activity of extract from Caco-2 cells incubated with ursolic acid (a triterpenoid) while the incubation with luteolin (a flavonoid) had no effect (Ramos et al., 2010b). The effect of β-cryptoxanthin in BER was also assessed in extracts from HeLa and Caco-2 cells incubated for 2 h with different concentrations of β-cryptoxanthin, on HeLa nucleoids. A significant increase in BER was shown by extracts of cells treated with β-cryptoxanthin, even at very low concentration (Lorenzo et al., 2009). They also incubated β-cryptoxanthin with nucleoids to check whether β-cryptoxanthin present in the extract could directly induce breaks in the nucleoids, but did not find any effect. The increase in the incision activity of Caco-2 cells incubated with β-cryptoxanthin is shown in Figure 3b.

The effect of STI571, the most used drug in the treatment of chronic myeloid leukemia, on NER was assessed using the *in vitro* DNA repair assay (Sliwinski et al., 2008). STI571 inhibits the activity of the BCR/ABL oncogenic kinase, and so 3 different cell lines were used: human myeloid leukemic cells expressing BCR/ABL, human lymphoid leukemia cells also expressing BCR/ABL, and human lymphoid leukemic cells which do not express BCR/ABL. Extracts were prepared after treating the cells with STI571 for 2 h, and incubated with UV-treated nucleoid DNA. The NER activity of extract from BCR/ABL cells showed a drastic and highly significant decrease after treatment with the drug – in contrast with the control cells, not expressing BCR/ABL, in which the drug did not induce any change in NER activity.

### 2.4.4 Animal studies with the *in vitro* assay

The *in vitro* repair assay has been mostly used on humans; applications in animal studies are extremely limited. Obtaining sufficient lymphocytes from rodents to prepare extract is not easy.

Recently Langie et al. (2011) have optimised the assay to measure the BER in extracts from rodent tissues. Various attempts have been made before, but they were hampered by the high non-specific nuclease activity present in the extracts. Langie et al. successfully measured the incision activity of extracts from liver and brain from C57/BL mice. Optimisation of the protein concentration in the tissue extract as well as the use of aphidicolin were the key steps to get rid of the non-specific activity. The assay was validated by using tissues from BER deficient OGG1 knockout mice where a low activity was found, significantly lower than with wild-type mice. In the same paper the assay was used to determine the effect of aging on the incision activity of extracts from mouse brain and the effect of diet on the incision activity of extracts from mouse liver. The BER activity of brain decreases with age, and in liver it is induced under dietary restriction.

### 2.4.5 Human studies

The *in vitro* repair assay has been widely used to measure BER and NER activities in human lymphocytes. The BER capacity of lymphocytes from 86 workers in a plastics factory, exposed to styrene, and 52 controls was studied (Vodicka et al., 2004a). The incision activity on HeLa nucleoids containing 8-oxoG was higher in styrene-exposed workers. The same group carried out a similar study to determine the effect of the occupational exposure to different xenobiotics from a tire plant on the DNA repair capacity of the workers (Vodicka et al., 2004b). No differences in repair activity were reported in lymphocytes of 15 workers with a high risk of exposure, 11 with a low risk and 12 employed in checking and quality control. In the same way the BER repair capacity was determined in 61 exposed workers from an asbestos cement plant and 21 controls (Dusinska et al., 2004). Females exposed to asbestos showed a decrease in their BER capacity compared with non-exposed ones but there were no differences between exposed and non-exposed males.

This approach has also been widely used in nutritional intervention studies. BER activity was measured in lymphocytes from 6 subjects before and after the intake of coenzyme $Q_{10}$ during 1 week (Tomasetti et al., 2001). A significantly higher OGG1 activity was seen after the supplementation. Very recently a nutritional intervention study has been published, where not only BER but also NER in UV-treated nucleoids was measured (Brevik et al., 2011). A randomized parallel study with 3 groups (a high phytochemical group with a high intake of a variety of antioxidant-rich plant products, a kiwifruit group supplemented with three kiwifruits per day and a control group without supplementation) and 8 weeks of intervention was carried out. BER activity was measured in lymphocytes from 23, 25, and 21 subjects from each group respectively, while NER was measured in lymphocytes from 13, 11 and 12 subjects from each group. BER showed an increase in both supplemented groups, being significant in the high phytochemical group, while NER showed significant decreases in both these groups. The control group did not show any changes.

It is important to include in *in vitro* repair experiments an incubation of extract with an undamaged substrate, to check for possible non-specific nuclease activities in the extract. It is not always clear in publications whether this has been done.

## 2.5 Following DNA repair at the level of the gene (FISH-comet assay)

The special feature of the comet assay is the ability to study DNA damage in individual cells. By combining fluorescent *in situ* hybridization (FISH) and applying labelled probes to particular DNA sequences, an even finer level of resolution can be achieved. Fig 5 illustrates general principles of the comet assay combined with FISH.

Depending on the target sequence, different probes are applied to comets. The most widely used FISH probes are centromere, telomere and ribosomal DNA repeats, short interspersed repetitive elements (SINEs) and long interspersed repetitive elements (LINEs). Those repetitive probes produce strong signals and are often commercially available. Other popular commercially available non-gene-specific probes are chromosome arm- or band-specific painting probes (DNA from microdissected chromosomes) and whole-chromosome painting probes (DNA from flow-sorted chromosomes).

Probes for specific DNA sequences can consist of PCR products, cDNAs or genomic DNA cloned in cosmids, P1 artificial chromosomes, bacterial artificial chromosomes (BACs) or yeast artificial chromosomes. Large unique probes that are not commercially available can be prepared for FISH using standard molecular biology techniques. Another useful design of probe is the 'padlock probe'—a linear oligonucleotide designed so that the two end segments, connected by a linker region, are complementary—in opposite orientations—to adjacent target sequences (Larsson et al., 2004). On hybridization, the two juxtaposed probe ends can be joined by a DNA ligase, circularizing the padlock probe and leaving it physically catenated to the target sequence. The reaction requires a perfect match between the probe ends and the target sequence and therefore, it is stable and extremely specific. The crucial feature of these probes when applied to comets is that the reaction steps are performed at 37°C so that there is no tendency for the agarose to melt or become unstable.

When analysing FISH-comets results, visualization and scoring depend entirely on direct observation. In most cases it is not possible to score the signals automatically because of the complexity of the preparations (for instance, the occurrence of signals in the same cell in different optical planes). Figure 5 illustrates different appearances of the signals: a linear array or separate spots.

An important question related to FISH signal visualization, is how many signals to expect. Based on our hypothesis that DNA organisation in comets reflects the DNA loop organization in living cells, it is reasonable to expect that the number of signals detected in comets will be related to the number of signals observed on chromosome spread preparations. Our results with chromosome 16 probes confirmed this hypothesis: twice as many signals were observed in the alkaline version of the assay relative to the number seen under neutral conditions (Shaposhnikov et al., 2008). This can be explained by DNA denaturation in alkaline comets since each strand of DNA will act as a target for the FISH probe. Furthermore, the average numbers of signals seen per cell corresponded closely with the numbers expected according to the gene copy number in a random interphase cell population.

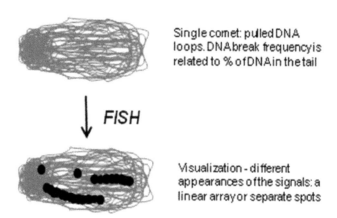

Single comet: pulled DNA loops. DNA break frequency is related to % of DNA in the tail

↓ *FISH*

Visualization - different appearances of the signals: a linear array or separate spots

Fig. 5. General principles of the FISH-comet assay

Santos et al. (1997) published the first successful results of combining FISH with the (neutral) comet assay. Their aim was to investigate how centromeric and telomeric DNA behaves under electrophoresis. Probes to all centromeres, all telomeres, as well as chromosome-specific centromere and telomere DNA, and 3 segments of the gene *MGMT* (coding for the repair enzyme O6-methylguanine DNA methyltransferase) were used. Telomere probes were seen mostly over the comet head, consistent with their attachment to the nuclear membrane. The signals from the much larger centromere DNA (1000 kb in size) appeared as long strings of dots, extending well into the comet tail. *MGMT* gave signals that were found in the head as well as the tail, the 3 segments generally forming a linear array.

### 2.5.1 Gene-specific DNA repair

The FISH-comet assay can be used to monitor gene-specific DNA repair by following the 'retreat' of the gene-specific signals from tail to head during the incubation period. Thus it is possible to compare the kinetics of overall genomic and gene-specific repair.

McKenna et al. (2003) examined the repair of γ-ray-induced SBs in human cells. Using a probe for the *TP53* gene, they found that the number of signals increased immediately after irradiation (most being in the comet tails), and decreased over the first 15 min at which point most were in comet heads. By 60 min, the normal, lower number of signals was restored, while in contrast the % tail DNA (representing total DNA) was still elevated. Thus *TP53* repair was faster than total genomic repair.

We studied the repair of the *DHFR* gene (coding for dihydrofolate reductase), *MGMT*, and the *TP53* gene using a different approach (Horvathova et al., 2004). Probes were designed for each end of the gene and detected using antibodies giving different coloured signals so that the gene had red and green ends after hybridization. After $H_2O_2$-treatment, Chinese hamster ovary (CHO) cells gave comets with about 50% of the DNA in the tail. Almost all *DHFR* probe signals were in comet heads, whereas we had expected them to have a similar distribution to total DNA. The probable explanation is that a 'matrix associated region' or MAR is present in this gene, and this prevents the DNA from escaping from the head. For the *MGMT* gene, CHO cells were treated either with $H_2O_2$ or with Ro 19-8022 and light to create 8-oxoG residues (FPG-sensitive sites). In contrast to the *DHFR* result, signals appeared over tail DNA - though they were predominantly green dots, while almost all red dots were located over the head. Thus one end of the gene appeared to be attached to the matrix. Green signals were restored to the head region of the comets with similar kinetics to the total DNA, indicating similar time courses for total DNA repair and repair of *MGMT*. $H_2O_2$-treated human lymphocytes, hybridized with *TP53* probes, gave signals of both colours in the tail; after 20 min incubation, virtually all *TP53* signals were in the head, while the % of total DNA in the tail had decreased by only about one-third. Thus, the region of DNA containing *TP53* was apparently repaired significantly faster than genomic DNA overall. Kumaravel et al. (2005) also reported preferential repair of *TP53*, after ionising radiation or $H_2O_2$ treatment.

We recently used padlock probes and rolling circle amplification (RCA) to investigate the repair of two DNA repair genes, 8-oxoguanine-DNA glycosylase-1 (*OGG1*) and xeroderma pigmentosum group D (*XPD*), and the housekeeping gene for hypoxanthine-guanine phosphoribosyltransferase (*HPRT*) (Henriksson et al., 2011). The repair rates of these genes after $H_2O_2$ damage were compared with the repair rates of *Alu* repeats and of total genomic DNA. The signals were mainly detected in comet tails. The *HPRT* gene showed rapid repair compared to total DNA, and after approximately 10 min the *HPRT* gene signals were almost completely absent, whereas the mean % tail DNA, indicating total DNA damage, decreased from 67 to 43% over 2 h (consistent with the slow repair of SBs by lymphocytes, as described by Torbergsen and Collins, 2000). *HPRT* and *XPD* were repaired more rapidly and *OGG1* more slowly than *Alu* repeats.

## 3. Conclusion

The comet assay has proved to be remarkably versatile. Far from being just another way of measuring DNA breaks, it can give quantitative information about base damage if lesion-specific endonucleases are included in the protocol, and by extension it can be used to monitor the cellular repair of such damage (the challenge assay). The NER pathway for helix distortions and bulky adducts can be blocked at the repair synthesis stage by DNA polymerase inhibitors, and this leads to an accumulation of SBs – readily measured with the comet assay. A more biochemical approach to DNA repair is exemplified by the *in vitro* repair assay, in which a cell extract is incubated with a specifically damaged DNA substrate – again leading to an accumulation of DNA breaks – repair intermediates, for which the comet assay is ideally suited as a detector. These different approaches have found application in cell culture studies (e.g. investigating inhibitors and enhancers of repair, and repair mutant phenotypes), in animal experiments, and in human biomonitoring (particularly in relation to occupational exposure, and nutrition). Finally, DNA repair has been examined at the level of specific genome regions, using fluorescent *in situ* hybridisation with probes recognising different genes; it is clear that the rate of repair varies greatly between genes – as it does between people.

## 4. References

Astley, S., Elliott, R.M., Archer, D.B., & Southon, S. (2004). Evidence that dietary supplementation with carotenoids and carotenoid-rich foods modulates the DNA damage: repair balance in human lymphocytes. *British Journal of Nutrition*, Vol. 91, No. 1, (January 2004), pp. (63-72), 0007-1145

Au, W.W., Giri, A.A., & Ruchirawat, M. (2010). Challenge assay: A functional biomarker for exposure-induced DNA repair deficiency and for risk of cancer. *International Journal of Hygiene and Environmental Health*, Vol. 213, No. 1, (January 2010), pp. (32-39), 1438-4639

Blasiak, J., Widera, K., & Pertyński, T. (2003). Hyperthermia can differentially modulate the repair of doxorubicin-damaged DNA in normal and cancer cells. *Acta Biochimica Polonica*, Vol. 50, No. 1, (month and year of the edition), pp. (191-195), 0001-527X

Brevik, A., Karlsen, A., Azqueta, A., Estaban, A.T., Blomhoff, R., & Collins, A.R. (2011). Both base excision repair and nucleotide excision repair in humans are influenced by nutritional factors. *Cell Biochemistry and Function*, Vol. 29, No. 1, (January 2011), pp. (36-42), 0263-6484

Cipollini, M., He, J., Rossi, P., Baronti, F., Micheli, A., Rossi, A.M., & Barale, R. (2006). Can individual repair kinetics of UVC-induced DNA damage in human lymphocytes be assessed through the comet assay? *Mutation Research*, Vol. 601, No. 1-2, (October 2006), pp. (150-161), 0027-5107

Collins, A.R. (2004). The comet assay for DNA damage and repair: principles, applications, and limitations. *Molecular Biotechnology*, Vol. 26, No. 3, (March 2004), pp. (249-261), 1073-6085

Collins, A.R., Dusinska, M., Horvathova, E., Munro, E., Savio, M., & Stetina, R. (2001). Inter-individual differences in DNA base excision repair activity measured *in vitro* with the comet assay. *Mutagenesis*, Vol. 16, No. 4, (July 2001), pp. (297-301), 0267-8357

Collins, A.R., Ma, A., & Duthie, S.J. (1995). The kinetics of repair of oxidative DNA damage (strand breaks and oxidised pyrimidines) in human cells. *Mutation Research*, Vol. 336, No. 1, (January 1995), pp. (69-77), 0027-5107

Collins, A.R., Mitchell, D.L., Zunino, A., de Wit, J., & Busch, D. (1997). UV-sensitive rodent mutant cell lines of complementation groups 6 and 8 differ phenotypically from their human counterparts. *Environmental and Molecular Mutagenesis*, Vol. 29, No. 2, (1997), pp. (152-160), 0893-6692

Crebelli, R., Carta, P., Andreoli, C., Aru, G., Dobrowolny, G., Rossi, S., & Zijno, A. (2002). Biomonitoring of primary aluminium industry workers: detection of micronuclei and repairable DNA lesions by alkaline SCGE. *Mutation Research*, Vol. 516, No. 1-2, (April 2002), pp. (63–70), 0027-5107

Dusinska, M., & Collins, A.R. (1996). Detection of oxidised purines and UV-induced photoproducts in DNA of single cells, by inclusion of lesion-specific enzymes in the comet assay. *Alternatives to Laboratory Animals*, Vol. 24, (1996), pp. (405-411), 0261-1929.

Dusinska, M., Collins, A.R., Kazimirova, A., Barancokova, M., Harrington, V., Volkovova, K., Staruchova, M., Horska, A., Wsolova, L., & Kocan, A. (2004). Genotoxic effects of asbestos in humans, *Mutation Research*, Vol. 553, No. 1-2, (September 2004), pp. (91-102), 0027-5107

Gaivão, I., Piasek, A., Brevik, A., Shaposhnikov, S., & Collins, A.R. (2009). Comet assay-based methods for measuring DNA repair *in vitro*; estimates of inter- and intra-individual variation. *Cell Biology and Toxicology*, Vol. 25, No. 1, (February 2009), pp. (45-52), 0742-2091

Gedik, C.M., Ewen, S.W., & Collins, A.R. (1992). Single-cell gel electrophoresis applied to the analysis of UV-C damage and its repair in human cells. *International Journal of Radiation Biology*, Vol. 62, No. 3, (September 1992), pp. (313-320), 0955-3002

Green, M.H.L., Waugh, A.P.W., Lowe, J.E., Harcourt, S.A., Cole, J., & Arlett, C.F. (1994). Effect of deoxyribonucleosides on the hypersensitivity of human peripheral blood lymphocytes to UV-B and UV-C irradiation. *Mutation Research*, Vol. 315, No. 1, (July 1994), pp. (25-32), 0027-5107

Grover, P., Banus, B.S., Devi, K.D., & Begum, S. (2001). *In vivo* genotoxic effects of mercuric chloride in rat peripheral blood leucocytes using comet assay. *Toxicology*, Vol. 167, No. 3, (October 2001), pp. (191–197), 0300-483X

Gudkov, S.V., Gudkova, O.Y., Chernikov, A.V., & Bruskov, V.I. (2009). Protection of mice against X-ray injuries by the post-irradiation administration of guanosine and inosine. *International Journal of Radiation Biology*, Vol. 85, No. 2, (February 2009), pp. (116-125), 0955-3002

Henriksson, S., Shaposhnikov, S., Nilsson, M., & Collins, A.R. (2011). Study of gene-specific DNA repair in the comet assay with padlock probes and rolling circle amplification. *Toxicology Letters*, Vol. 202, No. 2, (April 2011), pp. (142-147), 0378-4274

Herrera, M., Dominguez, G., Garcia, J.M., Peña, C., Jimenez, C., Silva, J., Garcia, V., Gomez, I., Diaz, R., Martin, P., & Bonilla, F. (2009). Differences in repair of DNA cross-links between lymphocytes and epithelial tumor cells from colon cancer patients measured *in vitro* with the comet assay. *Clinical Cancer Research*, Vol. 15, No. 17, (September 2009), pp. (5466-5472), 1078-0432

Horvathova, E., Dusinska, M., Shaposhnikov, S., & Collins, A.R. (2004). DNA damage and repair measured in different genomic regions using the comet assay with fluorescent *in situ* hybridization. *Mutagenesis*, Vol. 19, No. 4, (July 2004), pp. (269-276), 0267-8357

Kumaravel, T.S., & Bristow, R.G. (2005). Detection of genetic instability at HER-2/neu and p53 loci in breast cancer cells using comet-FISH. *Breast Cancer Research and Treatment*, Vol. 91, No. 1, (May 2005), pp. (89-93), 0167-6806

Langie, S.A., Cameron, K.M., Waldron, K.J., Fletcher, K.P., von Zglinicki, T., & Mathers, J.C. (2011). Measuring DNA repair incision activity of mouse tissue extracts towards singlet oxygen-induced DNA damage: a comet-based *in vitro* repair assay. *Mutagenesis*, Vol. 26, No. 3, (May 2011), pp. (461-471), 0267-8357

Langie, S.A., Knaapen, A.M., Brauers, K.J.J., van Berlo, D., van Schooten, F.J., & Godschalk, R.W.L. (2006). Development and validation of a modified comet assay to phenotypically assess nucleotide excision repair. *Mutagenesis*, Vol. 21, No. 2, (March 2006), pp. (153-158), 0267-8357

Larsson, C., Koch, J., Nygren, A., Janssen, G., Raap, A. K., Landegren, U., & Nilsson, M. (2004). *In situ* genotyping individual DNA molecules by target-primed rolling-circle amplification of padlock probes. *Nature Methods*, Vol. 1, No. 3, (December 2004), pp. (227-232), 1548-7091

Lorenzo, Y., Azqueta, A., Luna, L., Bonilla, F., Dominguez, G., & Collins, A.R. (2009). The carotenoid β-cryptoxanthin stimulates the repair of DNA oxidation damage in addition to acting as an antioxidant in human cells. *Carcinogenesis*, Vol. 30, No. 2, (February 2009), pp. (308-314), 0143-3334

Maurya, D.K., Salvi, V.P., & Nair, C.K. (2005). Radiation protection of DNA by ferulic acid under *in vitro* and *in vivo* conditions. *Molecular and Cellular Biochemistry*, Vol. 280, No. 1-2, (December 2005), pp. (209-217), 0300-8177

McKenna, D.J., Rajab, N.F., McKeown, S.R., McKerr, G., & McKelvey-Martin, V.J. (2003). Use of the comet-FISH assay to demonstrate repair of the *TP53* gene region in two human bladder carcinoma cell lines. *Radiation Research*, Vol. 159, No. 1, (January 2003), pp. (49-56), 0033-7587

Miranda, D.D., Arçari, D.P., Pedrazzoli, J.Jr., Carvalho, O., Cerutti, S.M., Bastos, D.H., & Ribeiro, M.L. (2008). Protective effects of mate tea (*Ilex paraguariensis*) on $H_2O_2$-induced DNA damage and DNA repair in mice. *Mutagenesis*, Vol. 23, No. 4, (July 2008), pp. (261-265), 0267-8357

Oshida, K., Iwanaga, E., Miyamoto-Kuramitsu, K., & Miyamoto, Y. (2008). An *in vivo* comet assay of multiple organs (liver, kidney and bone marrow) in mice treated with methyl methanesulfonate and acetaminophen accompanied by hematology and/or blood chemistry. *The Journal of Toxicological Sciences*, Vol. 33, No. 5, (December 2008), pp. (515-524), 0388-1350

Ostling, O., & Johanson, K.J., (1984). Microelectrophoretic study of radiation-induced DNA damages in individual mammalian cells. *Biochemical and Biophysical Research Communications*, Vol. 123, No. 1, (August 1984), pp. (291-298), 0006-291X

Palyvoda, O., Polanska, J., Wygoda, A., & Rzeszowska-Wolny, J. (2003). DNA damage and repair in lymphocytes of normal individuals and cancer patients: studies by the comet assay and micronucleus tests. *Acta Biochimica Polonica*, Vol. 50, No. 1, (2003), pp. (181-190), 0001-527X

Ramos, A.A., Azqueta, A., Pereira-Wilson, C., & Collins A.R. (1010a). Polyphenolic compounds from Salvia species protect cellular DNA from oxidation and stimulate DNA repair in cultured human cells. *Journal of Agricultural and Food Chemistry*, Vol. 58, No. 12, (June 2010), pp. (7465-7471), 0021-8561

Ramos, A.A., Lima, C.F., Pereira, M.L., Fernandes-Ferreira, M., & Pereira-Wilson, C. (2008). Antigenotoxic effects of quercetin, rutin and ursolic acid on HepG2 cells: evaluation by the comet assay. *Toxicology Letters*, Vol. 177, No. 1, (February 2008), pp. (66–73), 0378-4274

Ramos, A.A., Pereira-Wilson, C., & Collins, A.R. (2010b). Protective effects of ursolic acid and luteolin against oxidative DNA damage include enhancement of DNA repair in Caco-2 cells. *Mutation Research*, Vol. 692, No. 1-2, (October 1010), pp. (6-11), 0027-5107

Saleha Banu, B., Ishaq, M., Danadevi, K., Padmavathi, P., & Ahuja, Y.R. (2004). DNA damage in leukocytes of mice treated with copper sulfate. *Food and Chemical Toxicology*, Vol. 42, No. 12, (December 2004), pp. (1931-1936), 0278-6915

Santos, S.J., Singh, N.P. & Natarajan, A.T. (1997). Fluorescence in situ hybridization with comets. *Experimental Cell Research*, Vol. 232, No. 2, (May 1997), pp. (407-411), 0014-4827

Shaposhnikov, S.A., Salenko, V.B., Brunborg, G., Nygren, J., & Collins, A.R. (2008). Single-cell gel electrophoresis (the comet assay): loops or fragments?. *Electrophoresis*, Vol. 29, No. 14, (July 2008), pp. (3005-3012), 0173-0835

Singh, N.P., McCoy, M.T., Tice, R.R., & Schneider, E.L. (1988). A simple technique for quantitation of low levels of DNA damage in individual cells. *Experimental Cell Research*, Vol. 175, No. 1, (March 1988), pp. (184-191), 0014-4827

Sliwinski, T., Czechowska, A., Szemraj, J., Morawiec, Z., Skorski, T., & Blasiak, J. (2008). STI571 reduces NER activity in BCR/ABL-expressing cells. *Mutation Research*, Vol. 654, No. 2, (July 2008), pp. (162-167), 0027-5107

Speit, G., Schütz, P., & Hoffmann, H. (2004). Enhancement of genotoxic effects in the comet assay with human blood samples by aphidicolin. *Toxicology Letters*, Vol. 153, No. 3, (November 2004), pp. (303-310), 0378-4274

Speit, G., Witton-Davies, T., Heepchantree, W., Trenz, K., & Hoffmann, H. (2003). Investigations on the effect of cigarette smoking in the comet assay. *Mutation Research*, Vol. 542, No. 1-2, (December 2003), pp. (33–42), 0027-5107

Squires, S., Johnson, R.T., & Collins, A.R. (1982). Initial rates of DNA incision in UV-irradiated human cells; differences between normal, xeroderma pigmentosum and tumour cells. *Mutation Research*, Vol. 95, No. 2-3, (August 1982), pp. (389-404), 0027-5107

Stefanini, M., Collins, A.R., Riboni, R., Klaude, M., Botta, E., Mitchell, D.L., & Nuzzo, F. (1991). Novel Chinese hamster ultraviolet-sensitive mutants for excision repair form complementation groups 9 and 10. *Cancer Research*, Vol. 51, No. 15, (August 1991), pp. (3965-3971), 008-5472

Tomasetti, M., Alleva, R., Borghi, B., & Collins, A.R. (2001). *In vivo* supplementation with coenzyme $Q_{10}$ enhances the recovery of human lymphocytes from oxidative DNA damage. *The FASEB Journal*, Vol. 15, No. 8, (June 2001), pp. (1425-1427), 0892-6638

Torbergsen, A.C., & Collins, A.R. (2000). Recovery of human lymphocytes from oxidative DNA damage; the apparent enhancement of DNA repair by carotenoids is probably simply an antioxidant effect. *European Journal of Nutrition*, Vol 39, No. 2, (April 2000), pp. (80-85), 1436-6207

Touil, N., Aka, P.V., Buchet, J.P., Thierens, H., & Kirsch-Volders, M. (2002). Assessment of genotoxic effects related to chronic low level exposure to ionizing radiation using biomarkers for DNA damage and repair. *Mutagenesis*, Vol. 17, No. 3, (May 2002), pp. (223-232), 0267-8357

Tsai-Hsiu, Y., Nae-Cherng, Y., & Miao-Lin, H. (2003). S-adenosylhomocysteine enhances hydrogen peroxide-induced DNA damage by inhibition of DNA repair in two cell lines. *Nutrition and Cancer*, Vol. 47, No. 1, (2003), pp. (70-75), 0163-5581

Vande Loock, K., Decordier, I., Ciardelli, R., Haumont, D., & Kirsch-Volders, M. (2010). An aphidicolin-block nucleotide excision repair assay measuring DNA incision and repair capacity. *Mutagenesis*, Vol. 25, No. 1, (January 2010), pp. (25-32), 0267-8357

Vodicka, P., Kumar, R., Stetina, R., Musak, L., Soucek, P., Haufroid, V., Sasiadek, M., Vodickova, L., Naccarati, A., Sedikova, J., Sanyal, S., Kuricova, M., Brsiak, V., Norppa, H., Buchanova, J., & Hemminki, K. (2004b). Markers of individual susceptibility and DNA repair rate in workers exposed to xenobiotics in a tire plant. *Environmental and Molecular Mutagenesis*, Vol. 44, No. 4, (2004), pp. (283-292), 0893-6692

Vodicka, P., Tuimala, J., Stetina, R., Kumar, R., Manini, P., Naccarati, A., Maestri, L., Vodickova, L., Kuricova, M., Jarventaus, H., Majvaldova, Z., Hirvonen, A., Imbriani, M., Mutti, A., Migliore, L., Norppa, H., & Hemminki, K. (2004a).

Cytogenetic markers, DNA single-strand breaks, urinary metabolites, and DNA repair rates in styrene-exposed lamination workers. *Environmental Health Perspectives,* Vol. 112, No. 8, (June 2004), pp. (867-871), 0091-6765

Zhang, T., Hu, J., Zhang, Y., Zhao, Q., & Ning, J. Leucocytes DNA damage in mice exposed to JS-118 by the comet assay. *Human & Experimental Toxicology.* DOI:10.1177/0960327110388960

Zhao, X.H., Jia, G., Liu, Y.Q., Liu, S.W., Yan, L., Jin, Y., & Liu, N. (2006). Association between polymorphisms of DNA repair gene XRCC1 and DNA damage in asbestos-exposed workers. *Biomedical and Environmental Sciences,* Vol. 19, No. 3, (June 2006), pp. (232-238), 0895-3988

# Permissions

The contributors of this book come from diverse backgrounds, making this book a truly international effort. This book will bring forth new frontiers with its revolutionizing research information and detailed analysis of the nascent developments around the world.

We would like to thank Inna Kruman, for lending her expertise to make the book truly unique. She has played a crucial role in the development of this book. Without her invaluable contribution this book wouldn't have been possible. She has made vital efforts to compile up to date information on the varied aspects of this subject to make this book a valuable addition to the collection of many professionals and students.

This book was conceptualized with the vision of imparting up-to-date information and advanced data in this field. To ensure the same, a matchless editorial board was set up. Every individual on the board went through rigorous rounds of assessment to prove their worth. After which they invested a large part of their time researching and compiling the most relevant data for our readers. Conferences and sessions were held from time to time between the editorial board and the contributing authors to present the data in the most comprehensible form. The editorial team has worked tirelessly to provide valuable and valid information to help people across the globe.

Every chapter published in this book has been scrutinized by our experts. Their significance has been extensively debated. The topics covered herein carry significant findings which will fuel the growth of the discipline. They may even be implemented as practical applications or may be referred to as a beginning point for another development. Chapters in this book were first published by InTech; hereby published with permission under the Creative Commons Attribution License or equivalent.

The editorial board has been involved in producing this book since its inception. They have spent rigorous hours researching and exploring the diverse topics which have resulted in the successful publishing of this book. They have passed on their knowledge of decades through this book. To expedite this challenging task, the publisher supported the team at every step. A small team of assistant editors was also appointed to further simplify the editing procedure and attain best results for the readers.

Our editorial team has been hand-picked from every corner of the world. Their multi-ethnicity adds dynamic inputs to the discussions which result in innovative outcomes. These outcomes are then further discussed with the researchers and contributors who give their valuable feedback and opinion regarding the same. The feedback is then collaborated with the researches and they are edited in a comprehensive manner to aid the understanding of the subject.

Apart from the editorial board, the designing team has also invested a significant amount of their time in understanding the subject and creating the most relevant covers. They scrutinized every image to scout for the most suitable representation of the subject and create an appropriate cover for the book.

The publishing team has been involved in this book since its early stages. They were actively engaged in every process, be it collecting the data, connecting with the contributors or procuring relevant information. The team has been an ardent support to the editorial, designing and production team. Their endless efforts to recruit the best for this project, has resulted in the accomplishment of this book. They are a veteran in the field of academics and their pool of knowledge is as vast as their experience in printing. Their expertise and guidance has proved useful at every step. Their uncompromising quality standards have made this book an exceptional effort. Their encouragement from time to time has been an inspiration for everyone.

The publisher and the editorial board hope that this book will prove to be a valuable piece of knowledge for researchers, students, practitioners and scholars across the globe.

# List of Contributors

**Xuefeng Pan, Peng Xiao and Dongxu Zhao**
School of Life Science, Beijing Institute of Technology, Beijing, China

**Xuefeng Pan, Hongqun Li and Fei Duan**
Health Science Center, Hebei University, Baoding, China

**Sergey Korolev**
Saint Louis University School of Medicine, USA

**Radhika Pankaj Kamdar and Yoshihisa Matsumoto**
Research Laboratory for Nuclear Reactors, Tokyo Institute of Technology, Tokyo, Japan

**Radhika Pankaj Kamdar**
Department of Human Genetics, Emory University, Atlanta, Georgia, USA

**Kaoru Sugasawa**
Biosignal Research Center, Organization of Advanced Science and Technology Kobe University, Japan

**Xi-Dai Long, Yong-Zhi Huang and Xiao-Yin Huang**
Department of Pathology, Youjiang Medical College for Nationalities, China

**Xi-Dai Long, Zhi Zeng and Guo-Hui Fu**
Department of Pathology, Shanghai Jiao Tong University School of Medicine, China

**Jin-Guang Yao, Cen-Han Huang, Pinhu Liao and Zan-Song Huang**
Department of Medicine, Youjiang Medical College for Nationalities, China

**Fu-Zhi Ban**
Department of Medicine, The Southwestern Affiliated Hospital of Youjiang Medical College for Nationalities, China

**Li-Min Yao**
Department of Imaging Medicine (Grade 2008), Youjiang Medical College for Nationalities, China

**Lu-Dan Fan**
Department of Clinic Medicine (Grade 2009), Youjiang Medical College for Nationalities, China

**Rikke Dalgaard Hansen and Ulla Vogel**
Danish Cancer Society & National Research Centre for the Working Environment, Denmark

**Ravindran Ankathil**
Human Genome Center, School of Medical Sciences, University Sains Malaysia, Health Campus, Kubang Kerian, Kelantan, Malaysia

**Fabio Coppedè**
Contract Professor at Faculty of Medicine, University of Pisa, Italy

**Masaru Ueno**
Hiroshima University, Japan

**Fumiaki Uchiumi, Takahiro Oyama, Kensuke Ozaki and Sei-ichi Tanuma**
Department of Gene Regulation, Faculty of Pharmaceutical Sciences, Japan

**Sei-ichi Tanuma**
Department of Biochemistry, Faculty of Pharmaceutical Sciences, Japan
Genome and Drug Research Center, Japan

**Sei-ichi Tanuma and Fumiaki Uchiumi**
Research Center for RNA Science, RIST, Tokyo University of Science, Japan

**Amaya Azqueta**
University of Navarra, Spain

**Sergey Shaposhnikov and Andrew R. Collins**
University of Oslo, Norway

Printed in the USA
CPSIA information can be obtained
at www.ICGtesting.com
JSHW011431221024
72173JS00004B/762